Studies in Classification, Data Analysis, and Knowledge Organization

For further volumes:
http://www.springer.com/series/1564

Antonio Giusti • Gunter Ritter
Maurizio Vichi
Editors

Classification and Data Mining

Springer

Editors
Prof. Antonio Giusti
Department of Statistics
University of Florence
Florence, Italy

Prof. Dr. Gunter Ritter
Faculty for Informatics and Mathematics
University of Passau
Passau, Germany

Prof. Maurizio Vichi
Department of Statistics,
Probability and Applied Statistics
University of Rome "La Sapienza"
Rome, Italy

ISSN 1431-8814
ISBN 978-3-642-28893-7 ISBN 978-3-642-28894-4 (eBook)
DOI 10.1007/978-3-642-28894-4
Springer Heidelberg New York Dordrecht London

Library of Congress Control Number: 2012952267

Printed on acid-free paper

Springer is part of Springer Science+Business Media (www.springer.com)

Preface

Following a biannual tradition of organizing joint meetings between classification societies, the Classification and Data Analysis Group of the Italian Statistical Society, CLADAG, has organized its international meeting together with the German Classification Society, GfKl, at Firenze, Italy, September 8–10, 2010. The Conference was originally conceived as a German-Italian event, but it counted the participation of researchers from several nations and especially from Austria, France, Germany, Great Britain, Italy, Korea, the Netherlands, Portugal, Slovenia, and Spain. The meeting has shown once more the vitality of data analysis and classification and served as a forum for presentation, discussion, and exchange of ideas between the most active scientists in the field. It has also shown the strong bonds between the two classification societies and has greatly helped to deepen relationships.

The conference program included 4 Plenary, 12 Invited, and 31 Contributed Sessions. This book contains selected and peer-reviewed papers presented at the meeting in the area of "Classification and Data Mining." Browsing through the volume, the reader will see both methodological articles showing new original methods and articles on applications illustrating how new domain-specific knowledge can be made available from data by clever use of data analysis methods. According to the title, the book is divided into three parts:

1. Classification and Data Analysis
2. Data Mining
3. Applications

The methodologically oriented papers on classification and data analysis deal, among other things, with robustness, analysis of spatial data, and application of Monte Carlo Markov Chain methods. Variable selection and clustering of variables play an increasing role in applications where there are substantially more variables than observations. Support vector machines offer models and methods for the analysis of complex data structures that go beyond classical ones. Special discussed topics are association patterns and correspondence analysis.

Automated methods in data mining, producing knowledge discovery in huge data structures such as those associated with new media (e.g., Internet), digital images, or genomes in Genetics, continue to represent, in the near future, a big challenge for data analysis. Information is readily retrieved in these fields; however, interpreting it and identifying relevant results is not a straightforward task at all. Especially data produced by the Internet, genetics studies on genomes, and proteomes have a particular appeal as objects of analysis and are studied in this book. Furthermore, there are applications of the Markov chains model, to a new brand of problems such as the knowledge discovery in the Internet, the analysis of large biomedical data sets, and in more general sensor data. Moreover, the automatic online processing of data streams is becoming increasingly important. In sociology and market research, opinion mining on a large number of expressed preferences plays an important role. All these data typologies require algorithmic methods in the interface between statistics and computer science. Other contributions in the book focus on the application of the singular value decomposition to structural learning in Bayesian networks and on molecular simulation for drug design.

The last part of the book contains interesting applications to various fields of research such as sociology, market research, environment, geography, and music: estimation in demographic data, description of professional profiles, metropolitan studies such as income in municipalities, labor market research, environmental energy consumption, geographical data such as seismic time series, auditory models in speech and music, application of mixture models to multi-state data, and visualization techniques.

We hope that this short description stimulates the reader to take a closer look at some of the articles. Our thanks go to Andrea Giommi and his local organizing team who have done a great job (Bruno Bertaccini, Matilde Bini, Anna Gottard, Leonardo Grilli, Alessandra Mattei, Alessandra Petrucci, Carla Rampichini, Emilia Rocco). We gratefully acknowledge the Faculty of Economics and the "Ente Cassa di Risparmio di Firenze" for financial support, and desire to express our special thanks to Chiara Bocci for her valuable contribution to the organization of the meeting and for her assistance in producing this book. Also on behalf of our colleagues we may say that we have very much enjoyed having been their guests in Firenze. The dinner with a view to the Dome was excellent and we appreciate it very much.

We wish to express our gratitude to the other members of the Scientific Programme Committee: Daniel Baier, Reinhold Decker, Filippo Domma, Luigi Fabbris, Christian Hennig, Carlo Lauro, Berthold Lausen, Hermann Locarek-Junge, Isabella Morlini, Lars Schmidt-Thieme, Gabriele Soffritti, Alfred Ultsch, Rosanna Verde, Donatella Vicari, and Claus Weihs.

We also thank the section organizers for having put together such strong sections. The Italian tradition of discussants and rejoinders has been a new experience for GfKl. Thanks go to the referees for their important job. Last but not least, we thank all speakers and all who came to listen and to discuss with them.

Florence, Italy — Antonio Giusti
Passau, Germany — Gunter Ritter
Rome, Italy — Maurizio Vichi

Contents

Contributors

Antonio Balzanella Seconda Università degli Studi di Napoli, Caserta, Italy

Hans-Georg Bartel Department of Chemistry, Humboldt University Berlin, Berlin, Germany

Federico Benassi Department of Statistics "G. Parenti", University of Florence, Florence, Italy

Bruno Bertaccini Department of Statistics "G. Parenti", University of Florence, Florence, Italy

Pietro Bertuccelli Department of Economics, Statistics, Mathematics e Sociology, University of Messina, Messina, Italy

Chiara Bocci Department of Statistics "G. Parenti", University of Florence, Florence, Italy

Hamparsum Bozdogan Department of Statistics, Operations, and Management Science, University of Tennessee, Knoxville, TN, USA

Sergio Camiz Sapienza Università di Roma, Roma, Italy

Paolo Chirico Department of Applied Statistics e Mathematics, University of Turin, Italy

Hamdi Chouikha TU Dortmund University, Dortmund, Germany

Andreas Christmann Department of Mathematics, University of Bayreuth, Bayreuth, Germany

Fabrizio Cipollini Department of Statistics "G. Parenti", University of Florence, Florence, Italy

Gaetano De Luca INGV (Istituto Nazionale di Geofisica e Vulcanologia), Catania, Italy

Laura Deldossi Dipartimento di Scienze Statistiche, Università Cattolica del Sacro Cuore, Milano, Italy

José G. Dias UNIDE, ISCTE – University Institute of Lisbon, Lisbon, Portugal

Carlo Drago University of Naples Federico II, Naples, Italy

Emanuela Dreassi Department of Statistics "G. Parenti", University of Florence, Florence, Italy

Markus Eichhoff Chair of Computational Statistics, TU Dortmund, Germany

Camilla Ferretti Department of Economics and Social Sciences, Università Cattolica del Sacro Cuore, Piacenza, Italy

Klaus Friedrichs Chair of Computational Statistics, TU Dortmund, Germany

Zeno Gantner University of Hildesheim, Hildesheim, Germany

Piero Ganugi Department of Economics and Social Sciences, Università Cattolica del Sacro Cuore, Piacenza, Italy

Gastão Coelho Gomes Universidade Federal do Rio de Janeiro, Rio de Janeiro, Brazil

Mario Rosario Guarracino High Performance Computing and Networking, National Research Council of Italy, Naples, Italy

Center for Applied Optimization, University of Florida, Gainesville, FL, USA

Robert Hable Department of Mathematics, University of Bayreuth, Bayreuth, Germany

Jukka Heikkonen Department of Information Technology, University of Turku, Turku, Finland

J. Andrew Howe Department of Statistics, Operations, and Management Science, University of Tennessee, Knoxville, TN, USA

Maria Iannario Department of Statistical Sciences, University of Naples Federico II, Naples, Italy

Alfonso Iodice D'Enza Università di Cassino, Cassino, Italy

Antonio Irpino Dipartimento di Studi Europei e Mediterranei, Seconda Università degli Studi di Napoli, Caserta, Italy

Suman Katragadda Department of Statistics, Operations, and Management Science, University of Tennessee, Knoxville, TN, USA

Sonja Kuhnt TU Dortmund University, Dortmund, Germany

Anna Langovaya TU Dortmund University, Dortmund, Germany

Yves Lechevallier INRIA, Le Chesnay Cedex, France

Caterina Liberati Dipartimento di Scienze Statistiche 'P. Fortunati', Università di Bologna, Rimini, Italy

Cristiana Martini University of Modena and Reggio Emilia, Modena, Italy

Jorge Mateu Departamento de Matematicas, Universitat Jaume I, Castellon de la Plana, Spain

Angelo Mazza Dipartimento di Impresa, Culture e Società, Università di Catania, Catania, Italy

Mario Mezzanzanica Department of Statistics, Milano Bicocca University, Milano, Italy

Carlos Morales-Merino Curt-Engelhorn-Zentrum Archäometrie, Mannheim, Germany

Isabella Morlini Department of Economics, University of Modena and Reggio Emilia, Modena, Italy

Massimo Mucciardi Department of Economics, Statistics, Mathematics e Sociology, University of Messina, Messina, Italy

Hans-Joachim Mucha Weierstrass Institute for Applied Analysis and Stochastics (WIAS), Berlin, Germany

Eugenia Nissi Department of quantitative methods and economic theory, "G. d'Annunzio" University of Chieti-Pescara, Pescara, Italy

Sergio Palermi ARTA (Agenzia Regionale per la Tutela dell'Ambiente dell'Abruzzo), Palermi, Italy

Francesco Palumbo Università degli Studi di Napoli Federico II, Naples, Italy

Roberta Paroli Dipartimento di Scienze Statistiche, Università Cattolica del Sacro Cuore, Milano, Italy

Domenico Perrotta EC Joint Research Centre, Ispra site, Ispra, Italy

Alessandra Petrucci Department of Statistics "G. Parenti", University of Florence, Florence, Italy

Domenico Piccolo Department of Statistical Sciences, University of Naples Federico II, Naples, Italy

Antonio Punzo Dipartimento di Impresa, Culture e Società, Università di Catania, Catania, Italy

Marco Riani University of Parma, Parma, Italy

Antonio Angelo Romano Department of Statistics and Mathematics for Economic Research, University of Naples "Parthenope", Naples, Italy

Elvira Romano Dipartimento di Studi Europei e Mediterranei, Seconda Università degli Studi di Napoli, Napels, Italy

Karin Sahmer Groupe ISA, Lille Cedex, France

AnnaLina Sarra Department of quantitative methods and economic theory, "Gabriele d'Annunzio" University of Chieti-Pescara, Pescara, Italy

Giuseppe Scandurra Department of Statistics and Mathematics for Economic Research, University of Naples "Parthenope", Naples, Italy

Germana Scepi University of Naples Federico II, Naples, Italy

Lars Schmidt-Thieme University of Hildesheim, Hildesheim, Germany

Ivan Arcangelo Sciascia Dipartimento di Statistica e Matematica applicata, Università di Torino, Turin, Italy

Federico M. Stefanini Department of Statistics "G. Parenti", University of Florence, Florence, Italy

Nguyen Thai-Nghe University of Hildesheim, Hildesheim, Germany

Francesca Torti University of Milano Bicocca, Milan, Italy

Roberta Varriale Department of Statistics "G. Parenti", University of Florence, Florence, Italy

Igor Vatolkin Chair of Algorithm Engineering, TU Dortmund, Germany

Rosanna Verde Dipartimento di Studi Europei e Mediterranei, Seconda Università degli Studi di Napoli, Caserta, Italy

Marcus Weber Zuse Institute Berlin (ZIB), Berlin, Germany

Claus Weihs Chair of Computational Statistics, TU Dortmund, Germany

Sergio Zani Department of Economics, University of Parma, Parma, Italy

Part I
Classification and Data Analysis

Robust Random Effects Models: A Diagnostic Approach Based on the Forward Search

Bruno Bertaccini and Roberta Varriale

Abstract This paper presents a robust procedure for the detection of atypical observations and for the analysis of their effect on model inference in random effects models. Given that the observations can be outlying at different levels of the analysis, we focus on the evaluation of the effect of both first and second level outliers and, in particular, on their effect on the higher level variance which is statistically evaluated with the Likelihood-Ratio Test. A cut-off point separating the outliers from the other observations is identified through a graphical analysis of the information collected at each step of the Forward Search procedure; the Robust Forward LRT is the value of the classical LRT statistic at the cut-off point.

1 Introduction

Outliers in a dataset are observations which appear to be inconsistent with the rest of the data (Hampel et al., 1986; Staudte and Sheather, 1990; Barnett and Lewis, 1993) and can influence the statistical analysis of such a dataset leading to invalid conclusions. Starting from the work of Bertaccini and Varriale (2007), the purpose of this work is to implement the Forward Search method proposed by Atkinson and Riani (2000) in the random effects models, in order to detect and investigate the effect of outliers on model inference.

While there is an extensive literature on the detection and treatment of single and multiple outliers for ordinary regression, these topics have been little explored in the area of random effects models. In this work, we propose a new diagnostic method based on the Forward Search approach, in order to detect both first and second level outliers. We focus our attention on the effect of outliers on the Likelihood-Ratio

B. Bertaccini (✉) · R. Varriale
Department of Statistics "G. Parenti", University of Florence, Florence, Italy
e-mail: bertaccini@ds.unifi.it; roberta.varriale@ds.unifi.it

A. Giusti et al. (eds.), *Classification and Data Mining*, Studies in Classification, Data Analysis, and Knowledge Organization, DOI 10.1007/978-3-642-28894-4_1,

Test, that is used in the multilevel framework in order to detect the significance of the second level variance.

The basic idea of this approach is to fit the hypothesis model to an increasing subset of units, where the order of entrance of observations into the subset is based on their closeness to the previously fitted model. During the search, parameter estimates, residuals and other informative statistics are collected and these information are analysed in order to identify a cut-off point separating the outliers from the other observations. Differently from the other robust approaches, the robustness of the method does not derive from the choice of a particular estimator with a high breakdown point, but from the progressive inclusion of units into a subset which, in the first steps, is intended to be outlier free (Atkinson and Riani, 2000). Our procedure can detect the presence of more than one outlier; the membership of almost all the outliers to the same group (factor level) suggests the presence of an outlying unit at the higher level of the analysis.

2 The Random Effects Model

The simplest random effects model is a two level linear model without covariates, also known as a random effects ANOVA. Forward Search for fixed effects ANOVA has already been proposed by the authors Bertaccini and Varriale (2007); in the following, we will extend this work to the random effects framework.

Let y_{ij} be the observed outcome variable of individual i ($i = 1, 2, \ldots, n_j$) within group, or factor level, j ($j = 1, 2, \ldots, J$) where J is the total number of groups and $N = \sum_{j=1}^{J} n_j$ is the total number of individuals. The simplest linear model in this framework is expressed by:

$$y_{ij} = \mu + u_j + e_{ij} = \mu + \xi_{ij} \tag{1}$$

where μ is the grand mean outcome in the population, u_j is the group effect associated with unit j and e_{ij} is the residual error at the lower level of the analysis. When u_j are the effects attributable to a infinite set of levels of a factor of which only a random sample are deemed to occur in the data, we have a random effects model (Searle et al., 1992). In this approach, each observed response y_{ij} differs from the grand mean μ by a total residual ξ_{ij} given by the sum of two random error components, u_j and e_{ij}, representing, respectively, the residual error at the higher and lower level of the analysis. Under the usual assumptions for the random effects model (Searle et al., 1992), it is possible to show that var $(y_{ij}) = var(u_j) + var(e_{ij}) = \tau^2 + \sigma^2$. Thus, τ^2 and σ^2, representing respectively the variability between and within groups, are components of the total variance of y_{ij}.

In many applications of hierarchical analysis, one common research question is whether the variability of the random effects at group level is significatively equal to 0, namely:

$$H_0 : \tau^2 = 0. \tag{2}$$

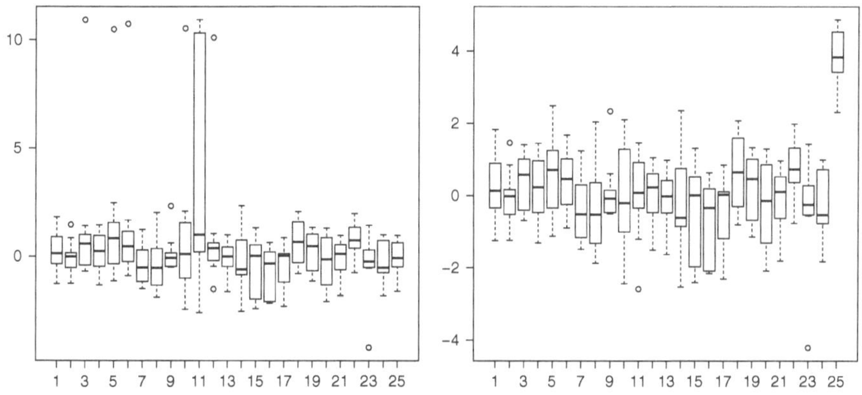

Fig. 1 Boxplot showing the compositions of the described datasets

In the maximum likelihood estimation framework, comparison of nested models is typically performed by means of the LR Test which under certain regularity conditions follows a chi-squared distribution. In random effects models, when there is only one variance being set to zero in the reduced model, the asymptotic distribution of the LRT statistic is a 50 : 50 mixture of a χ^2_k and χ^2_{k+1} distributions, where k is the number of the other restricted parameters in the reduced model that are unaffected by boundary conditions (Self and Liang, 1987). A rule of thumb in applied research is to divide by 2 the asymptotic p-value of the Chi-squared LRT statistic distribution (Skrondal and Rabe-Hesketh, 2004). As the only alternative strategy to our knowledge, Heritier et al. (2009) suggest to perform the LRT by computing bootstrapping techniques, but this method can fail when applied to "classical" robust estimators. In the following, we will use the former strategy to test the null hypothesis as in Eq. (2). Due to the presence of outliers in the data, the value of the LRT statistic can erroneously suggest to reject the null hypothesis H_0 even when there is no second level residual variability. As an example, consider the two balanced datasets represented in Fig. 1 with $n_{ij} = 10$ first level units in each group j and the total number of groups, J, equal to 25. While the bulk of the data has been generated by the model $y_{ij} = \mu + e_{ij}$ in both cases, with $\mu = 0$ and $e_{ij} \sim N(0, 1)$, the outliers have very different features: in the first case there are more than one first level outliers, while in the second case there is only one outlier at the second level of the analysis. In particular, the eight outliers in the first case have been generated from a Uniform $U(10, 11)$ distribution, while in the second case, the first level units belonging to the outlier group have been generated by the $N(0 + \gamma, 1)$ distribution where γ is an observation from the $U(4, 5)$ distribution. In both cases, the LRT statistic for testing H_0 has one degree of freedom and its value – respectively 4.8132 and 94.4937 with halved p-value of 0.0141 and $<$0.0001 falls in the rejection region due to the contamination. Obviously, in these datasets outliers are so different from the bulk of the data that they are easily identifiable by any approach; these were done only to introduce the problem more clearly.

Table 1 Classical LR test: approximation of the true type I error probability with contaminated data

		ϵ					
J	n_{ij}	0.05	0.06	0.07	0.08	0.09	0.1
15	10	0.0578	0.0655	0.0747	0.0960	0.1153	0.1513
	15	0.0607	0.0767	0.1088	0.1385	0.1754	0.2183
	20	0.0781	0.1053	0.1436	0.1931	0.2546	0.3355
20	10	0.0662	0.0814	0.1005	0.1287	0.1680	0.2104
	15	0.0795	0.1108	0.1441	0.1889	0.2601	0.3381
	20	0.1038	0.1477	0.2067	0.2837	0.3816	0.4809
25	10	0.0678	0.0938	0.1334	0.1668	0.2077	0.2772
	15	0.0959	0.1261	0.1895	0.2620	0.3523	0.4627
	20	0.1277	0.1872	0.2657	0.3708	0.4896	0.6119
30	10	0.0781	0.1019	0.1313	0.1767	0.2265	0.2902
	15	0.0966	0.1433	0.2051	0.2704	0.3557	0.4656
	20	0.1278	0.1947	0.2815	0.3823	0.5098	0.6380

The LR Test can often lead to erroneous conclusions also in the presence of "lighter" contamination. Let us consider some datasets with an increasing number of balanced groups ($J = 15, 20, 25, 30$) and an increasing number of observations for each group ($n_{ij} = 10, 15, 20$). While $(1 - \epsilon)N$ observations are generated by a Standard Normal distribution and are randomly assigned to all groups, the ϵN outliers are generated by a Normal $N(2, 1)$ distribution and are randomly assigned to the first half of the total number of groups. Table 1 shows the relative frequencies $rf_{(J,n_{ij},\epsilon)}$ over 10,000 simulations in which the LRT statistic falls in the rejection area at the nominal significance level of $\alpha = 0.05$. For example, for $J = 20, n_{ij} = 15$ and $\epsilon = 0.08$ the classical LRT rejects the null hypothesis (2) 1,889 times giving a "real" α value of 0.1889. Obviously, the larger the ϵ is and the stronger the effect of the contamination on the LRT is.

In the following, we will focus on the effect of outliers on the LRT performed with halved p-value used to test (2).

3 Forward Search

The Forward Search is a statistical methodology initially proposed by Atkinson and Riani (2000) useful both to detect and investigate observations that differ from the bulk of the data and to analyse their effect on the estimation of parameters and on model inference. The basic idea of this "forward" procedure is to fit the hypothesized model to an increasing subset of units until all data are fitted. In particular, the entrance order of the observations into the subset is based on their closeness to the fitted model that is expressed by the residuals.

The Forward Search algorithm consists of three steps: the first concerns the choice of an initial subset, the second refers to the way in which the Forward Search progresses and the third relates to the monitoring of the statistics during the search. In this work, the methodology is adapted to the peculiarity of the random effects model taking into account the presence of groups in the data structure. In particular, focusing on the inferential issue expressed in Eq. (2), we propose a procedure to obtain a Robust Forward LR Test (LRT_F), by individuating a cut-off point of all the classical LRT values computed during the search, cut-off point that divides the group of outliers from the other observations.

3.1 Step 1: Choice of the Initial Subset

The first step of the forward procedure consists in the choice of an initial subset of observations supposed to be outliers free, S^*. Many robust methods were proposed to sort the data into a clean and a potentially contaminated part and the Forward Search is not sensitive to the method used to select the initial subset, provided unmasked outliers are not included at the start (Atkinson and Riani, 2000). In the random effects framework, our proposal is to include in the initial subset of observations the y_{ij} which satisfy:

$$min|y_{ij} - med_j| \;\; \forall j = 1, 2, \ldots, J \qquad (3)$$

where med_j is the group j sample median. We impose that every group has to be represented by at least two observations; in this way, every group contributes to the estimation of the within random effects and the initial subset dimension, $m^* = \sum_{j=1}^{J} m_j^*$, is at least $2 * J$, where m_j^* is the number of observations in group j at the first step of the search.

3.2 Step 2: Adding Observations During the Search

At each step, the Forward Search algorithm adds to the subset the observations closer to the previously fitted model. Formally, given the subset $S^{(m)}$ of dimension $m = \sum_{j=1}^{J} m_j$, where m_j is the number of observations in group j at step $(m - m^* + 1)$, the Forward Search moves to $S^{(m+1)}$ in the following way: after the random effects model is fitted to the $S^{(m)}$ subset, all the n_{ij} observations are ordered inside each group according to their squared total residuals $\hat{\xi}_{ij}^2 = (y_{ij} - \hat{y}_{ij,S^{(m)}})^2$. Since $\hat{y}_{ij,S^{(m)}} = \hat{\mu}_{S^{(m)}}$, the total residuals express the closeness of each unit to the grand mean estimate, making possible the detection of both first and second level outliers. For each group j, we choose the first m_j ordered observations and we add the observation with the smallest squared residual among the remaining. The random effects model is now fitted to $S^{(m+1)}$ and the procedure ends when all the N observations are entered into the model. In moving from $S^{(m)}$ to $S^{(m+1)}$, while most of the time just one new unit joins the previous subset, it may also happen that two or more new units enter $S^{(m+1)}$ as one or more leave, given that all the groups have to be always represented in the subset with at least two observations.

The procedure allows the choice between different parameters' estimators; available estimators are ANOVA, ML and REML (default is ML).

3.3 Step 3: Monitoring the Search

At each stage of the search, it is possible to collect information on parameter estimates, residuals and other relevant statistics, to guide the researcher in the outliers detection. In order to illustrate the application and the advantages of the Forward Search approach we show the methodology using the two datasets described in Fig. 1. In both cases, the LRT computed with the classical approach "erroneously" falls in the rejection area of the null hypothesis expressed in Eq. (2).

Figure 2 shows how the observations join the subset $S^{(m)}$ during the search. The last observations joining $S^{(m)}$ belong to different second level units (right panel of Fig. 2), precisely to the groups 3, 6, 10, 11 and 12, and are represented by the bold lines that lie under the other lines at the end of the search; this suggests the possible presence of outliers in these groups.

Figure 3a shows the N absolute total residuals $\hat{\xi}_{ij}$ computed at each step of the Forward Search. Throughout the search, all the residuals are very small except those related to the last eight entered observations. These units can be considered outliers in any fitted subset and even when they are included in the algorithm in the last steps of the search their residuals decrease only slightly. Furthermore, Fig. 3a clearly highlights the sensitivity of the Forward Search that also recognises the presence of an additional anomalous observation generated randomly from the Standard Normal distribution; this observation belongs to the group 23 and join the search at step 242 just before the other eight outlier observations.

Finally, Fig. 3b represents the halved p-value obtained, at each step of the search, from the LRT for the null hypothesis: $H_0 : \tau^2 = 0$. During almost all the search the halved p-value is very high, clearly suggesting that the second level variance is equal to 0. In the last steps of the search it erroneously moves to the rejection area and it reaches the value 0.0141 at step 250, as indicated in Sect. 2.

The second example is characterized by the presence of one second level outlier. In this case, the observations joining $S^{(m)}$ during the last steps of the search belong to the same second level unit, 25, suggesting the presence of an anomalous group of observations.

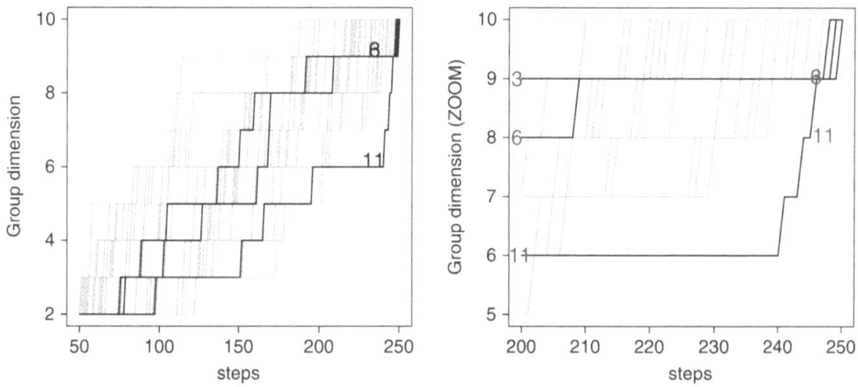

Fig. 2 Forward plots of groups dimensions: during the search (**a**) and zoom of the last 50 steps (**b**)

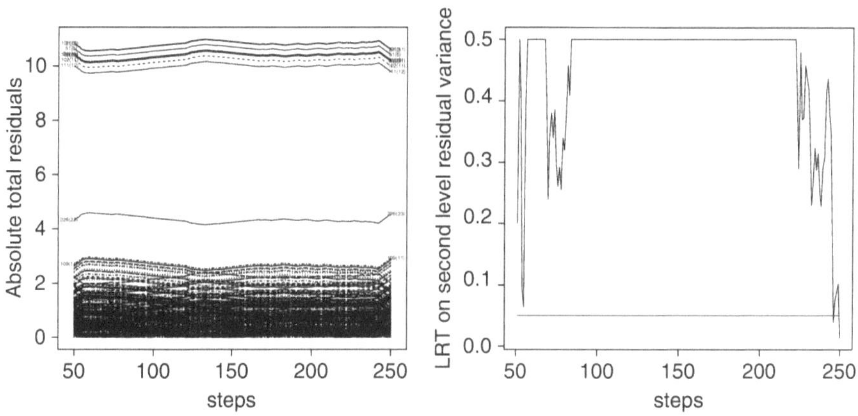

Fig. 3 First dataset: Forward plot of the estimated absolute residuals (**a**); Forward plot of the Likelihood-Ratio Test. The *horizontal line* represents the chosen halved α value (**b**)

Figure 4a shows the total residuals computed during the search, highlighting the presence of two opposite patterns of lines. This feature is due to the fact that at least two observations belonging to the outlier group are in the initial subset S^*. For this reason, the estimated grand mean is relatively high in the first steps of the search; then it starts to decrease as the number of clean observations joining $S^{(m)}$ increases and it increases again at the end of the search when all the other outliers join $S^{(m)}$.

Finally, Fig. 4b represents a very interesting behaviour of the halved p-value obtained with the LRT. Contrary to the first example, during the search the p-value is always very low since the units belonging to the outlier group that are in $S^{(m)}$ lead to the wrong conclusion of the presence of second level variability. Then, the LRT correctly increases as the number of non outlying units entering the subset $S^{(m)}$ increases, and it obviously sharply decreases when the units of the outlier group finally enter the search.

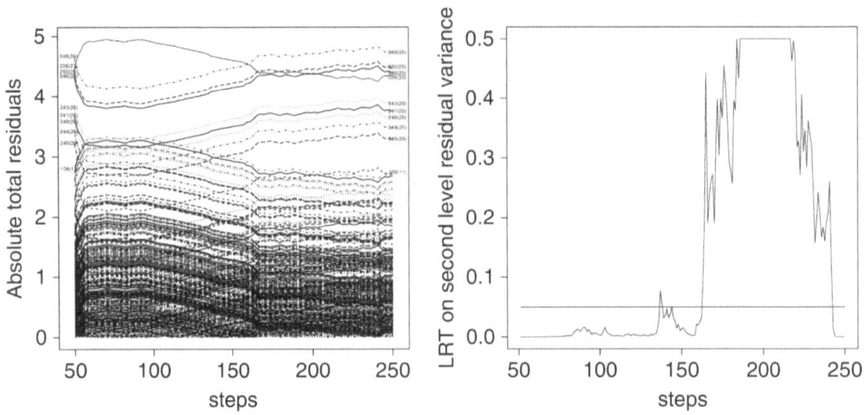

Fig. 4 Second dataset: Forward plot of the estimated absolute residuals (**a**); Forward plot of the Likelihood-Ratio Test. The *horizontal line* represents the chosen halved α value (**b**)

References

Atkinson, A. C., & Riani, M. (2000). *Robust diagnostic regression analysis*. New York: Springer.

Barnett, V., & Lewis, T. (1993). *Outliers in statistical data* (3rd ed.). New York: Wiley.

Bertaccini, B., & Varriale, R. (2007). Robust ANalysis Of VAriance: An approach based on the Forward Search. *Computational Statistics and Data Analysis, 51*, 5172–5183.

Hampel, F. R., Ronchetti, E. M., Rousseeuw, P. J., & Stahel, W. A. (1986). *Robust statistics: The approach based on influence functions*. New York: Wiley.

Heritier, S., Cantoni, E., Copt, S., & Victoria-Feser, M. (2009). *Robust methods in biostatistics*. Chichester: Wiley.

Searle, S. R., Casella, G., & McCulloch, C. E. (1992). *Variance components*. New York: Wiley.

Self, S. G., & Liang, K. Y. (1987). Asymptotic properties of maximum likelihood estimators and likelihood ratio tests under nonstandard conditions. *Journal of the Acoustical Society of America, 82*, 605–610.

Skrondal, A., & Rabe-Hesketh, S. (2004). *Generalized latent variable modeling: Multilevel, longitudinal and structural equation models*. Boca Raton: Chapman & Hall/CRC.

Staudte, R. G., & Sheather, S. J. (1990). *Robust estimation and testing*. New York: Wiley.

Joint Correspondence Analysis Versus Multiple Correspondence Analysis: A Solution to an Undetected Problem

Sergio Camiz and Gastão Coelho Gomes

Abstract The problem of the proper dimension of the solution of a Multiple Correspondence Analysis (*MCA*) is discussed, based on both the re-evaluation of the explained inertia *sensu* Benzécri (Les Cahiers de l'Analyse des Données 4:377–379, 1979) and Greenacre (Multiple correspondence analysis and related methods, Chapman and Hall (Kluwer), Dordrecht, 2006) and a test proposed by Ben Ammou and Saporta (Revue de Statistique Appliquée 46:21–35, 1998). This leads to the consideration of a better reconstruction of the off-diagonal sub-tables of the Burt's table crossing the nominal characters taken into account. Thus, Greenacre (Biometrika 75:457–467, 1988) Joint Correspondence Analysis (*JCA*) is introduced, the results obtained on an application are shown, and the quality of reconstruction of both *MCA* and *JCA* solutions are compared to that of a series of Simple Correspondence Analyses run on the whole set of two-way tables. It results that *JCA*'s reduced-dimensional reconstruction is much better than the *MCA*'s one, that reveals highly biased and non-monotone, but also than the *MCA*'s re-evaluation, as suggested by Greenacre (Multiple correspondence analysis and related methods, Chapman and Hall (Kluwer), Dordrecht, 2006).

1 Introduction

The identification of the dimension of a data table under study is a crucial issue in most multidimensional scaling techniques, in particular in the linear methods, since most of the analyses that follow the scaling depend on this choice. To quote

S. Camiz (✉)
Sapienza Università di Roma, Rome, Italy
e-mail: sergio.camiz@uniroma1.it

G.C. Gomes
Universidade Federal do Rio de Janeiro, Rio de Janeiro, Brazil
e-mail: gastao@im.ufrj.br

A. Giusti et al. (eds.), *Classification and Data Mining*, Studies in Classification, Data Analysis, and Knowledge Organization, DOI 10.1007/978-3-642-28894-4_2,

only some, the number of factors to be interpreted, those on which to attempt a classification, the dimension in which to search for a non-linear solution or for a factor analysis, etc.

In this paper, we focus on this problem in Multiple Correspondence Analysis (*MCA*, Benzécri et al., 1973–1982; Greenacre, 1984), in particular considering its alternative, the Joint Correspondence Analysis (*JCA*, Greenacre, 1988), whose solution depends on an a priori selected dimensionality, and on the partial reconstruction of the original data that results by the application of reconstruction formulas.

The application of these methods to a small example taken from a recent study (Camiz and Gomes, 2009) will show unexpected results when comparing the reconstruction: even if *JCA* was supposed to perform better, the results of *MCA*, in comparison with those of *JCA*, would seriously get questionable its use, unless without some adjustments. Indeed, the application to the Burt's table of the chi-square metrics, and the following correspondence analysis, biases the results, by improving the reconstruction of the diagonal blocks while raising the bias of the off-diagonal ones that contain the most interesting information.

2 Theoretical Framework

In exploratory multidimensional scaling the identification of the proper dimension of the solution is strictly tied to the crucial distinction between relevant and non-relevant information, something similar to the identification of errors in classical statistics, but not the same. For metric scaling, the percentage of explained inertia is usually taken as a measure of information, also tied to its interpretability. Thus, taking into account a large share of inertia is the most often used rule of thumb, but without a good theoretical grounding. Indeed, in literature stopping rules may be found: for Principal Component Analysis, Jackson (1993) compared some of the existing ones. For Simple Correspondence Analysis (*SCA*, Benzécri et al., 1973–1982; Greenacre, 1984) a classical test for goodness of fit (Kendall and Stuart, 1961) may be applied as approximated by the Malinvaud (1987) test (see also Saporta and Tambrea, 1993):

$$\widetilde{Q}_\alpha = \sum_{ij} \frac{\left(n_{ij} - \widetilde{n}_{\alpha ij}\right)^2}{n r_i c_j} = \chi^2 - \sum_{\beta=1}^{\alpha} \chi^2_\beta = \sum_{\gamma=\alpha+1}^{\min(r,c)-1} \lambda_\gamma,$$

where $\widetilde{n}_{\alpha ij}$ is the cell value estimated by the α-dimensional solution. $\widetilde{Q}_\alpha$, asymptotically chi-square distributed with $(r-\alpha-1)\times(c-\alpha-1)$ degrees of freedom, tests the independence of the residuals in respect to the α-dimensional representation. This is possible because the eigenvalues of *SCA* sum, up to the grand total, to the table chi-square, namely

$$\chi^2 = n \sum_{\alpha=1}^{\min(r,c)-1} \lambda_\alpha = \sum_{\alpha=1}^{\min(r,c)-1} \chi^2_\alpha.$$

2.1 Multiple Correspondence Analysis

It is well known that *MCA* is but a generalization of *SCA* and it is based on *SCA* of either the indicator matrix Z, gathering all characters involved, or the Burt's table $B = Z'Z$, that includes the diagonal tables with the marginals. The eigenvectors of both Z and B are the same, whereas the B's eigenvalues are the squares of Z's (also called B's singular values): $\mu_\alpha^2 = \nu_\alpha$. As *SCA*, it may be shown that, given a Burt matrix B, *MCA* may be defined as the weighted least-squares approximation of B by another matrix H of lower rank, that minimizes

$$n^{-1}Q^{-2}\text{trace}\left(D_r^{-1}(B-H)D_r^{-1}(B-H)'\right).$$

that is, considering the subtables of B, that minimizes

$$n^{-1}\sum_{i=1}^{Q}\sum_{j=1}^{Q}\|N_{ij}-H_{ij}\|_{ij}^2. \tag{1}$$

where the norm $\|A_{ij}\|_{ij}^2 = \text{trace}\left(D_i^{-1}A_{ij}\ D_j^{-1}\ A'_{ij}\right)$ is the usual chi-square. Indeed, in *SCA* this is limited to only one table.

In *MCA* the identification of the dimensionality is particularly difficult: indeed, for B, crossing Q characters with $J = \sum_{i=1}^{Q} l_i$ pooled levels (with l_i the number of levels of the i-th character) a statistic may again be calculated as if it were a contingency table

$$\chi_B^2 = 2\sum_{i=2}^{Q}\sum_{j=1}^{i-1}\chi_{ij}^2 + n(J-Q), \tag{2}$$

where χ_{ij}^2 is the chi-squared statistic for the off-diagonal subtable $N_{ij} = Z_i'Z_j$, and $n(J-Q)$ is that of the diagonal subtables. As χ_B^2 is not chi-square distributed, no test is possible. Thus, the current users refer to the total inertia of Z: $I_z = \frac{J-Q}{Q}$, and consider its share explained by the highest level eigenvectors, although it is very low, due to their high number of pooled levels. In practice, they are satisfied when the first factors are enough larger than the following, regardless of the figures involved, as it is generally admitted that the explained inertia is "highly underestimated". This underestimation was raised by Benzécri (1979) argumented by the arbitrary number of levels and by the relation between the eigenvalues issued by either *SCA* or *MCA* of Z applied on a two characters table: the relation $\mu_\alpha = \frac{1\pm\sqrt{\lambda_\alpha}}{2}$ is thus interpreted to limit attention to the eigenvalues larger than the trivial average $\frac{1}{2}$, the smaller considered as "artifacts". This argument is generalized to consider in *MCA* only the eigenvalues larger than their mean, that is $\mu \geq \overline{\mu}_\alpha = \frac{1}{Q}$. As a consequence, each factor inertia is re-evaluated as the average deviation from the mean eigenvalue, according to the formula

$$\rho\left(\mu_\alpha\right)=\left(\frac{Q}{Q-1}\right)^2\left(\mu_\alpha-\overline{\mu}\right)^2, \quad \mu_\alpha \geq \overline{\mu}=\frac{1}{Q}. \tag{3}$$

and its share of total inertia is based on the inertias sum, thus taking the ratio $\frac{\rho(\mu_\alpha)}{\sum_{\alpha>\frac{1}{q}} \rho(\mu_\alpha)}$. Greenacre (1988, 2006) too suggests to re-evaluate the inertia according to (3), but compares each one to the total off-diagonal inertia of the table, that is

$$\frac{Q}{Q-1}\left(\sum_{\mu_\alpha} \mu_\alpha^2-\frac{J-Q}{Q^2}\right),$$

a share that results always lower than Benzécri's one.

Regardless of the re-evaluation, to decide the number of factors to take into account, the only test currently available is proposed by Ben Ammou and Saporta (1998), based on the distribution of the average eigenvalue under the null hypothesis of independence: its expected variance is

$$\sigma^2=E[S_\lambda^2]=\frac{1}{n_{..}Q^2(J-Q)} \sum_{i \neq j}\left(l_i-1\right)\left(l_j-1\right),$$

so that one may assume for $\frac{1}{Q}$ the confidence interval at 95% level $\frac{1}{Q} \pm 2\sigma$. Indeed, since the kurtosis is lower than for a normal distribution, the actual proportion is larger than 95%.

2.2 Joint Correspondence Analysis

In order to remove the bias due to the diagonal submatrices, Greenacre (1988) proposes the *Joint Correspondence Analysis* (*JCA*) as a better generalization of *SCA*. *JCA* fits only the off-diagonal contingency tables by minimizing, instead of (1),

$$n^{-1} \sum_{i=1}^{Q} \sum_{j=1}^{i-1}\left\|N_{ij}-H_{ij}\right\|_{ij}^2, \tag{4}$$

and considers as measure of inertia, instead of (2), the sum of the chi-squares of all off-diagonal tables

$$\chi_J^2=\sum_{i=1}^{Q} \sum_{j=1}^{i-1} \chi_{ij}^2,$$

that unfortunately may not be tested for significance. *JCA* is an alternating weighed least-squares algorithm that reminds the *MINRES* method for least-squares

Table 1 Burt's table of the three-characters data set of 2,000 words

	L2	L3	L4	WN	WV	WA	TC	TR	TD	TS
L2	1,512	0	0	788	483	241	433	385	399	295
L3	0	375	0	203	23	149	64	82	86	143
L4	0	0	113	62	9	42	3	29	21	60
WN	788	203	62	1,053	0	0	229	284	273	267
WV	483	23	9	0	515	0	174	133	125	83
WA	241	149	42	0	0	432	97	79	108	148
TC	433	64	3	229	174	97	500	0	0	0
TR	385	82	29	284	133	79	0	496	0	0
TD	399	86	21	273	125	108	0	0	506	0
TS	295	143	60	267	83	148	0	0	0	498
	L2	L3	L4	WN	WV	WA	TC	TR	TD	TS

Table 2 First one-dimensional layer of the layers by kind of words table, one-dimensional reconstruction, and corresponding residuals of *SCA*

	Layer				Reconstruction				Residual			
	TC	TR	TD	TS	TC	TR	TD	TS	TC	TR	TD	TS
L2	57	7	17	−80	435	382	400	296	−2	3	−1	−1
L3	−33	−4	−10	47	60	89	85	141	4	−7	1	2
L4	−23	−3	−7	33	5	25	22	61	−2	4	−1	−1

factor analysis, where the off-diagonal elements of a correlation matrix are fitted (Thomson, 1934). In the special case $Q = 2$, the solution is exactly the *SCA* of the off-diagonal table $N = N_{12}$.

3 An Application

To show the different behavior of the different correspondence analyses, we refer to a data set taken from Camiz and Gomes (2009), consisting in 2,000 words taken from four different kind of periodic reviews (*Childish (TC), Review (TR), Divulgation (TD),* and *Scientific Summary (TS)*), classified according to their grammatical kind (*Verb (WV), Noun (WN),* and *Adjective (WA)*) and the number of internal layers (*Two- (L2), Three- (L3),* and *Four and more layers (L4)*), as a measure of the word complexity (Table 1). All the computations have been performed with the *ca* package (Nenadic and Greenacre, 2006, 2007) contained in the *R* environment (R-project, 2009).

We first limit attention to the table crossing Layers by Kind of words, with a chi-square = 125.262 with six degrees of freedom, thus highly significant (test value = 10.177). According to Malinvaud (1987) its *SCA* gives only one significant eigenvalue (0.061891, test-value = 10.439) summarizing 98.82 of total inertia. The one-dimensional reconstruction is reported in Table 2, with a reduction of absolute

Table 3 Results of *MCA* on the Burt's table crossing two characters: singular values and eigenvalues, percentages of inertia, total and off-diagonal residuals of the corresponding reconstruction, re-evaluated inertia and percentages, total and off-diagonal residuals of the corresponding reconstruction

	Singular	Eigen	Perc.	Cumul.	Reconstruction		Re-evaluation		Reconstruction	
N.	value	value	Inertia	Perc.	Total	Off-diag	Inertia	Perc.	Total	Off-diag
0					5,215	328			5,215	328
1	0.389863	0.624390	24.98	24.98	4,357	483	0.061891	98.82	4,125	29
2	0.263783	0.513598	20.54	45.52	3,978	730	0.000740	1.18	4,026	0
3	0.250000	0.500000	20.00	65.52	3,102	730				
4	0.236587	0.486402	19.46	84.98	1,946	487				
5	0.141083	0.375610	15.02	100.00	0	0				

residuals from 328, in respect to independence, to only 29. Indeed, the two-dimensional solution has no residuals and identical results are found with *JCA*, as expected.

The *MCA*, applied to the corresponding 2×2 Burt's table, gives the results shown in Table 3. In the table, both singular values and eigenvalues are reported with their percentage to the trace (=2.5), the absolute residuals of the total and off-diagonal reconstructions, then the re-evaluated inertias with the corresponding reconstructions, limited to the two main eigenvalues larger than the mean (0.5). According to Ben Ammou and Saporta (1998) only the first factor should be taken into account, since the confidence interval for the mean eigenvalue is $0.47658 < \lambda < 0.52342$.

In the last two columns of Table 3 the absolute residuals for the re-evaluated *MCA*, both total and off-diagonal, are reported according to the dimension, the 0 corresponding to the deviation from independence: the results are identical to those of *SCA*. Instead, looking at the columns 6 and 7, we have a surprise: whereas the total residuals of the reconstruction decrease monotonically to zero, the off-diagonal ones immediately increase, until the mean eigenvalue, then monotonically decrease, with a better approximation only at the last step. That is, only the total reconstruction is better that the independent table in estimating the table itself.

If we apply both *MCA* and *JCA* to the three-characters data table from which the previous table was extracted, we find a similar but worst pattern. Here, only 3 out of 7 *MCA* eigenvalues are above the mean, with only one significant, as the confidence interval at 95% level is now ($0.30146 < \lambda < 0.36521$), and a second one non-significant but very close to its upper bound. This is in agreement with the Malinvaud (1987) test applied to the three two-way tables, only one of which has a significant second factor. In Table 4 total and off-diagonal absolute residuals for normal *MCA*, *JCA*, and re-evaluated *MCA* inertias are reported according to the dimension (the 0 corresponds to the independence).

Observing the table one may note the same pattern of the residuals of *MCA* as before: a monotone reduction of the total residuals and an increase of the off-diagonal ones until the average eigenvalue, then a reduction of the latter, so that only a six-dimensional solution shows off-diagonal residuals lower than the

Table 4 Total and off-diagonal absolute residuals of normal *MCA*, *JCA*, and re-evaluated *MCA* on the Burt's table crossing three characters

	MCA		JCA		Re-evaluated MCA	
Dim	Total	Off-diag.	Total	Off-diag.	Total	Off-diag.
0	8,905	954	8,905	954	8,905	954
1	7,557	1,044	6,629	240	6,885	311
2	7,378	1,537	6,206	145	6,581	232
3	7,089	1,813	5,836	18	6,509	214
4	5,949	1,572				
5	3,675	977				
6	2,335	729				
7	0	0				

independence. On the opposite, the re-evaluated inertias get a monotone pattern but far from the quality of adjustment of *JCA*, that performs quite well. Indeed, the re-evaluated *MCA* needs two dimensions to approach the one-dimensional solution of *JCA*, but never reaching the two-dimensional one.

4 Conclusion

The results of this experimentation show that the Ben Ammou and Saporta (1998) test reveals useful for estimating the suitable dimension of an *MCA* solution. Instead, the reconstruction of the Burt's table performed by normal *MCA* is so biased that it is not the case to keep on using *MCA* as it is normally performed. The re-evaluated inertias avoid the dramatic bias introduced by the diagonal blocks, but its quality of reconstruction, limited to the factors whose eigenvalue is larger than the mean, is far from being acceptable. In particular, it is so poor in respect to *JCA* that one may wonder why not eventually shift to this method. Indeed, some questions may arise whether the chi-square metrics would be really suitable for a Burt's table, but this is a question that deserves a broader discussion.

Acknowledgements This work was mostly carried out during the reciprocal visits of both authors in the framework of the bilateral agreement between Sapienza Università di Roma and Universidade Federal do Rio de Janeiro, of which both authors are scientific responsible. The first author was also granted by his Faculty, the Facoltà d'Architettura ValleGiulia of La Sapienza. All grants are gratefully acknowledged.

References

Ben Ammou, S., & Saporta, G. (1998). Sur la normalité asymptotique des valeurs propres en ACM sous l'hypothèse d'indépendance des variables. *Revue de Statistique Appliquée, 46*(3), 21–35.

Benzécri, J. P. (1979). Sur les calcul des taux d'inertie dans l'analyse d'un questionnaire. *Les Cahiers de l'Analyse des Données, 4*(3), 377–379.

Benzécri, J. P., et al. (1973–1982). *L'Analyse des données*, Tome 1. Paris: Dunod.
Camiz, S., & Gomes, G. C. (2009). Correspondence analyses for studying the language complexity of texts. In *VIII Congreso Chileno de Investigación Operativa, OPTIMA, Concepción (Chile)*, on CD-ROM.
Greenacre, M. J. (1984). *Theory and application of correspondence analysis*. London: Academic.
Greenacre, M. J. (1988). Correspondence analysis of multivariate categorical data by weighted least squares. *Biometrika, 75*, 457–467.
Greenacre, M. J. (2006). From simple to multiple correspondence analysis. In M. J. Greenacre, J. Blasius (Eds.), *Multiple correspondence analysis and related methods* (pp. 41–76). Dordrecht: Chapman and Hall (Kluwer).
Greenacre, M. J., & Blasius, J. (Eds.). (2006). *Multiple correspondence analysis and related methods*. Dordrecht: Chapman and Hall (Kluwer).
Jackson, D. A. (1993). Stopping rules in principal component analysis: A comparison of heuristical and statistical approaches. *Ecology, 74*(8), 2204–2214.
Kendall, M. G., & Stuart, A. (1961). *The advanced theory of statistics* (Vol. 2). London: Griffin.
Malinvaud, E. (1987). Data analysis in applied socio-economic statistics with special consideration of correspondence analysis. In *Marketing science conference*. Joy en Josas: HEC-ISA.
Nenadic, O., & Greenacre, M. (2006). Computation of multiple correspondence analysis, with code in R. In M. J. Greenacre & J. Blasius (Eds.), *Multiple correspondence analysis and related methods* (pp. 523–551). Dordrecht: Chapman and Hall (Kluwer).
Nenadic, O., & Greenacre, M. (2007). Correspondence analysis in R, with two- and three-dimensional graphics: The *ca* package. *Journal of Statistical Software, 20*(3), 1–13.
R-project (2009). http://www.r-project.org/
Saporta, G., & Tambrea, N. (1993). About the selection of the number of components in correspondence analysis. In J. Janssen & C.H. Skiadas (Eds.), *Applied stochastic models and data analysis* (pp. 846–856). Singapore: World Scientific.
Thomson, G. H. (1934). Hotelling's method modified to give Spearman's *g*. *Journal of Educational Psychology, 25*, 366–374.

Inference on the CUB Model: An MCMC Approach

Laura Deldossi and Roberta Paroli

Abstract We consider a special finite mixture model for ordinal data expressing the preferences of raters with regards to items or services, named CUB (Covariate Uniform Binomial), recently introduced in statistical literature. The mixture is made up of two components that belong to different families of distributions: a shifted Binomial and a discrete Uniform. Bayesian analysis of the CUB model naturally comes from the elicitation of some priors on its parameters. In this case the parameters estimation must be performed through the analysis of the posterior distribution. In the theory of finite mixture models complex posterior distributions are usually evaluated through computational methods of simulation such as the Markov Chain Monte Carlo (MCMC) algorithms. Since the mixture type of the CUB model is non-standard, a suitable MCMC algorithm has been developed and its performance has been evaluated via a simulation study and an application on real data.

1 Introduction

Statistical models for ordinal data are an active research area in recent years from many alternative points of view (see for example Bini et al., 2009, for a general review). Ordinal data can be obtained by surveys on consumers or users who express preferences or evaluations on a list of known items or on objects or services. Applications on the perception of the value or of the quality of objects are common in various fields: teaching evaluation, health system or public services, risk analysis, university services performances, measurement system analysis and many others. One of the innovative tools in the evaluation analysis area assumes that the ordinal

L. Deldossi (✉) · R. Paroli
Dipartimento di Scienze Statistiche, Università Cattolica del Sacro Cuore, Milano, Italy
e-mail: laura.deldossi@unicatt.it; roberta.paroli@unicatt.it

A. Giusti et al. (eds.), *Classification and Data Mining*, Studies in Classification, Data Analysis, and Knowledge Organization, DOI 10.1007/978-3-642-28894-4_3,

results can be thought as the final outcome of an unobserved choice mechanism with two latent components: the *feeling* with the items or the objects, which is a latent continuous random variable discretized by the judgment or the rate, and the *uncertainty* in the choice of rates, which is related to several individual factors such as knowledge or ignorance of the problem, personal interests, opinions, time spent in the decision and so on. From these assumptions, the CUB model has been recently derived by D'Elia and Piccolo (2005) and Piccolo (2006). Ordinal data are modeled as a two components mixture distribution whose parameters are connected with the two latent components of *feeling* and *uncertainty*. Classical inference is performed by the maximum likelihood method. In D'Elia and Piccolo (2005) the maximum likelihood estimates of parameters are obtained by means of the E-M algorithm.

The innovative contribution of this paper is that inference is performed in a Bayesian framework and a suitable and new ad hoc MCMC scheme is developed. Bayesian approach to mixture models has obtained strong interest since the end of the last century (McLachlan and Peel, 2000), due to the believe that the Bayesian paradigm is particularly suited to solve the computational difficulties and the non-standard problems in their inference. This paper is organized as follows: in Sect. 2 we introduce the notations of the model with or without covariates; in Sect. 3 Bayesian inference is performed and our suitable MCMC algorithm is illustrated. Finally, in Sects. 4 and 5, some simulation results will be used to check the statistical performances of the algorithm and an application to a real data set will be illustrated. Some concluding remarks and topics for future work end the paper.

2 The CUB Model

Let R be the ordinal random variable that describes the rate assigned by a respondent to a given item of a preferences' test, with $r \in \{1, \ldots, m\}$. R may be modeled as a mixture of a shifted Binomial$(m-1, 1-\xi)$ and a discrete Uniform(m), whose probability distribution is therefore defined as:

$$P(R=r) = \pi \binom{m-1}{r-1} (1-\xi)^{r-1} \xi^{m-r} + (1-\pi)\frac{1}{m}. \tag{1}$$

Note that the only unknown parameters are the mixture proportion $\pi \in (0, 1]$ and the shifted Binomial parameter $\xi \in [0, 1]$ since the maximum rate m, that has to be greater than 3 due to the identifiability conditions (Iannario, 2010), is fixed. This mixture is non-standard because its components belong to two different families of distributions and the second component is fully known having assigned a value to m.

In the context of the preferences analysis the Uniform component may express the degree of uncertainty in judging an object on the categorical scale, while the shifted Binomial component may represent the behavior of the rater with respect to the liking/disliking feeling for the object under evaluation. For any items we are interested in estimating the parameters ξ, that is a proxy of the rating measure, and π, that is inversely related to the uncertainties in the rating process.

Fitting to observed ordinal data may be improved adding individual information (covariates) on each respondents i, for $i = 1, \dots, n$, to relate both the feeling ξ_i and the uncertainty π_i to the respondent's features. The general formulation of a CUB(p, q) model is then expressed by a stochastic component:

$$P(R_i = r; Y_i, W_i) = \pi_i \binom{m-1}{r-1} (1-\xi_i)^{r-1} \xi_i^{m-r} + (1-\pi_i)\frac{1}{m} \tag{2}$$

where $r = 1, 2, \dots, m$, Y_i and W_i are the subject's covariates vectors of dimension $p+1$ and $q+1$ explaining π_i and ξ_i respectively , and a systematic component that links the covariates to π_i and ξ_i. This component is modeled as a logistic function:

$$\pi_i = \frac{\exp(Y'\beta)}{1+\exp(Y'\beta)}, \qquad \xi_i = \frac{\exp(W'\gamma)}{1+\exp(W'\gamma)} \tag{3}$$

where the vectors $\boldsymbol{\beta} = (\beta_0, \beta_1, \dots, \beta_p)\prime$ and $\boldsymbol{\gamma} = (\gamma_0, \gamma_1, \dots \gamma_q)\prime$ are the parameters to be estimated. Due to the choice of the logistic function, the parametric space of π_i and ξ_i are restricted to $\xi_i \in (0, 1)$ and $\pi_i \in (0, 1)$. In an objective Bayesian perspective we place non-informative independent priors on the parameters: we assume that each entry of vector $\boldsymbol{\beta}$ is Normal with known hyperparameters μ_B and σ_B^2 ($\beta_j \sim \mathcal{N}(\mu_B, \sigma_B^2)$, for any $j = 0, \dots, p$); each entry of vector $\boldsymbol{\gamma}$ is Normal with known μ_G and σ_G^2 ($\gamma_j \sim \mathcal{N}(\mu_G, \sigma_G^2)$, for any $j = 0, \dots, q$).

3 Bayesian Inference

Bayesian approach to inference of complex statistical models uses probability to quantify the beliefs of the observer about the model parameters, given the observed data. Inference of mixture models is now feasible using posterior simulation via recently developed MCMC methods (see McLachlan and Peel, 2000, for a fully comprehensive review). For the CUB model, given a sample of n subjects, the posterior distribution is

$$\pi(\boldsymbol{\beta}, \boldsymbol{\gamma} | R; Y, W) \propto P(R | \boldsymbol{\beta}, \boldsymbol{\gamma}, Y, W) p(\boldsymbol{\beta}) p(\boldsymbol{\gamma}), \tag{4}$$

where $P(R|\boldsymbol{\beta}, \boldsymbol{\gamma}, Y, W)$ is the likelihood function and $p(\boldsymbol{\beta})$ and $p(\boldsymbol{\gamma})$ are the prior distributions. The likelihood function is defined as

$$P(R|\boldsymbol{\beta}, \boldsymbol{\gamma}, Y, W) = \prod_{i=1}^{n} \left\{ \pi_i \binom{m-1}{r_i-1} (1-\xi_i)^{r_i-1} \xi_i^{m-r_i} + (1-\pi_i)\frac{1}{m} \right\} = \tag{5}$$

$$= \prod_{i=1}^{n} \left\{ \frac{\exp(Y_i \boldsymbol{\beta})}{1+\exp(Y_i \boldsymbol{\beta})} \binom{m-1}{r_i - 1} \frac{\exp(W_i \boldsymbol{\gamma})^{m-r_i}}{(1+\exp(W_i \boldsymbol{\gamma}))^{m-1}} + \frac{1}{1+\exp(Y_i \boldsymbol{\beta})} \cdot \frac{1}{m} \right\}. \tag{6}$$

Bayesian inference will be executed by sampling from the posterior density through a suitable MCMC algorithm. Such methods allow the construction of an ergodic Markov chain with stationarity distribution equals to the posterior distribution of the parameters of interest. The simplest method is the Gibbs sampling that simulates and updates each parameters in turn by sampling from its corresponding full conditional distribution. However, since the conditional distributions for the CUB parameters are not generally of standard form (being here in a logit model), it is more convenient to use the Metropolis-Hastings algorithm.

We now introduce our MCMC algorithm which consists of two Metropolis steps. Its scheme is briefly the following: given vectors $\boldsymbol{\beta}^{(k-1)}$ and $\boldsymbol{\gamma}^{(k-1)}$ generated at the $(k-1)$-th iteration, the steps of the generic k-th iteration are:

1. The parameters $\beta_j^{(k)}$, for any $j = 0, \ldots, p$, are independently generated from a random walk $\beta_j^{(k)} = \beta_j^{(k-1)} + E_B$, where $E_B \sim \mathcal{N}\left(0; \sigma^2_{EB}\right)$. The proposed $\boldsymbol{\beta}^{(k)}$ is accepted in block if $u_B \leq \min\{1; A_B\}$, where u_B is a random number generated from the Uniform distribution $\mathcal{U}$ (0; 1) and the acceptance probability ratio A_B is:
$$A_B = \frac{P(R|\boldsymbol{\beta}^{(\mathbf{k})}, \boldsymbol{\gamma}^{(\mathbf{k-1})}, Y_i, W_i) p\left(\beta^{(k)}\right)}{P(R|\boldsymbol{\beta}^{(\mathbf{k-1})}, \boldsymbol{\gamma}^{(\mathbf{k-1})}, Y_i, W_i) p\left(\beta^{(k-1)}\right)}; \tag{7}$$

2. The parameters $\gamma_j^{(k)}$, for any $j = 0, \ldots, q$, are independently generated from a random walk $\gamma_j^{(k)} = \gamma_j^{(k-1)} + E_G$, where $E_G \sim \mathcal{N}\left(0; \sigma^2_{EG}\right)$. The proposed $\boldsymbol{\gamma}^{(k)}$ is accepted in block if $u_G \leq \min\{1; A_G\}$, where u_G is a random number generated from the Uniform distribution $\mathcal{U}$ (0; 1) and the acceptance probability ratio A_G is:
$$A_G = \frac{P(R|\boldsymbol{\beta}^{(\mathbf{k})}, \boldsymbol{\gamma}^{(\mathbf{k})}, Y_i, W_i) p\left(\gamma^{(k)}\right)}{P(R|\boldsymbol{\beta}^{(\mathbf{k})}, \boldsymbol{\gamma}^{(\mathbf{k-1})}, Y_i, W_i) p\left(\gamma^{(k-1)}\right)}. \tag{8}$$

At the end of a number N (large enough) of iterations we obtain N-dimensional samples of the parameter's values that will be used to estimates each β_j and γ_j through the posterior means.

It should be noted that in the case of the CUB models two of the main difficulties that have to be addressed with the Bayesian approach in the context of mixture models, are not to be considered. The first hindrance is the estimation of the number of components of the mixture that here is fixed and equal to two. Another basic feature of a mixture model is that it is invariant under permutations of the components of the mixture. In Bayesian framework this feature (exchangeability) may be very cumbersome since it generally implies that parameters are not marginally identifiable. In fact if an exchangeable prior is used on the parameters, all the posterior marginals on the parameters are identical and then it is not possible to distinguish between e.g. "first" and "second" component of the mixture. This identifiability problem is called "label switching" (see e.g. Früwirth-Schnatter, 2006). For the mixture defined by (2) and (3) no label switching question is present due to the fact that the Uniform parameter m is a known constant. In fact, also choosing an exchangeable prior on (β, γ) – as in our case – the posterior marginal of β will be distinguish from that of γ, as it can be easily observed looking at formulas (4)–(6).

4 A Simulation Study

In the first step of our preliminary study, many Monte Carlo simulations are performed on the simple case of a CUB(0,0) model, i.e. a CUB without covariates, to check the statistical performances of the algorithm. Following the scheme of the simulation study reported in D'Elia (2003), we selected some different models whose parameters values vary over the parametric space. These models show probability distributions very different in location, variability and skewness aspects. We use independent Normal priors $\mathcal{N}(0; 10)$ for the parameters β_0 and γ_0. We ran our MCMC algorithm (implemented in Digital Visual FORTRAN) for 100,000 iterations after 50,000 of burn-in and, for any model, we computed the finite bias of the posterior means based on 500 replications of the estimation procedure. Table 1 shows the results for $m = 7$ and $n = 70, 210, 700$.

We can notice that in general the bias decreases as n increases and for $n \geq 210$ it is generally limited (around 10^{-2}). The worst performances are mainly associated with the π estimator in correspondence with low values of n and of the parameter itself ($\pi < 0.4$). Comparing the finite sample bias property of Bayes and ML estimator (see D'Elia, 2003) we may observe that we find negative bias for π in most of the cases, while the bias of ML estimators is always positive. For ξ parameter the bias behaviour seems to be not so regular for both the kind of the estimators.

Many diagnostic tools are available to assess the convergence of an MCMC algorithm. Among them a few informal checks are based on graphical techniques, such as the plots of simulated values or of the ergodic means. The plot of the ergodic or running means (the posterior means updated at each iteration) provide a rough indication of stationary behaviour of the Markov chain after the burn-in iterations. The plots of the traces (the sequence of values generated at each iteration) are a valid instrument to check the mixing of the chain. A good mixing induces a fast convergence of the algorithm. For the sake of example Fig. 1 shows the behaviour over 32,000 iterations of the traces and the running means (recorded every 320 iterations) for one of the 500 replications of the case ($\pi = 0.7, \xi = 0.3$) with $n = 210$. They seem to indicate that the convergence of our algorithm is good and very clear.

Table 1 Mean and bias of the bayesian estimators based on 500 replications of the MCMC procedure

		n = 70		n = 210		n = 700	
	True values	Mean	Bias	Mean	Bias	Mean	Bias
π	0.1	0.0565	−0.0435	0.0579	−0.0421	0.0802	−0.0198
ξ	0.1	0.3041	0.2041	0.1922	0.0922	0.0939	−0.0061
π	0.1	0.0526	−0.0474	0.0450	−0.0550	0.0556	−0.0444
ξ	0.4	0.4665	0.0665	0.4637	0.0637	0.4129	0.0129
π	0.2	0.0995	−0.1005	0.1307	−0.0693	0.1877	−0.0123
ξ	0.2	0.2856	0.0856	0.2157	0.0157	0.1955	−0.0045
π	0.4	0.3622	−0.0378	0.4191	0.0191	0.3989	−0.0011
ξ	0.1	0.0854	−0.0146	0.1091	0.0091	0.1005	0.0005
π	0.4	0.2561	−0.1439	0.3622	−0.0378	0.3978	−0.0022
ξ	0.5	0.4971	−0.0029	0.4956	−0.0044	0.5013	0.0013
π	0.7	0.6973	−0.0027	0.6983	−0.0017	0.6995	−0.0005
ξ	0.3	0.2946	−0.0054	0.2985	−0.0015	0.2999	−0.0001
π	0.3	0.1980	−0.1020	0.2711	−0.0289	0.2939	−0.0061
ξ	0.8	0.7702	−0.0298	0.8109	0.0109	0.8002	0.0002
π	0.7	0.6990	−0.0010	0.7048	0.0048	0.70120	0.00120
ξ	0.6	0.5976	−0.0024	0.6005	0.0005	0.60003	0.00003
π	0.9	0.9087	0.0087	0.9003	0.0003	0.9009	0.00091
ξ	0.9	0.9022	0.0022	0.9002	0.0002	0.9000	−0.00001

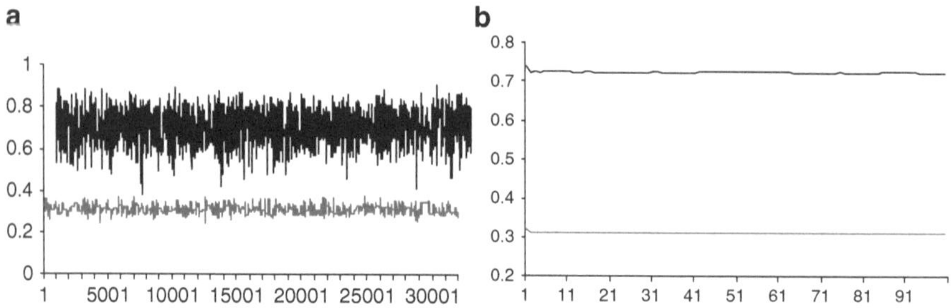

Fig. 1 Diagnostic plots for the MCMC convergence of the case $\pi = 0.7$ and $\xi = 0.3$, $n = 210$: (**a**) traces; (**b**) running means

5 An Application to Real Data

The MCMC algorithm for the CUB model with covariates was applied on a real data set concerning the students' opinions on the Orientation services of the University of Naples Federico II in years 2007 and 2008. By means of a questionnaire various items have been investigated and each student was asked to give a score for expressing his/her satisfaction with different aspect of the orientation service. For each respondents the data set contains the judgments for each item ranging from

Table 2 Posterior means of different CUB$(p, 0)$ models for *advertisement*

Models		
CUB(0,0)	$\widehat{\pi} = 0.7783$	$\widehat{\xi} = 0.3556$
CUB(1,0)	$\widehat{\beta_0} = 6.1433$	
log(Age)	$\widehat{\beta_1} = -1.5911$	$\widehat{\xi} = 0.3577$
CUB(2,0)	$\widehat{\beta_0} = 7.6359$	
log(Age)	$\widehat{\beta_1} = -1.9749$	$\widehat{\xi} = 0.3565$
Gender	$\widehat{\beta_2} = -0.5314$	
CUB(3,0)	$\widehat{\beta_0} = 5.4889$	
log(Age)	$\widehat{\beta_1} = -1.1682$	
Gender	$\widehat{\beta_2} = -0.6921$	$\widehat{\xi} = 0.3563$
Change	$\widehat{\beta_3} = -0.7623$	
CUB(4,0)	$\widehat{\beta_0} = 5.0778$	
log(Age)	$\widehat{\beta_1} = -1.1539$	
Gender	$\widehat{\beta_2} = -0.4525$	$\widehat{\xi} = 0.3556$
Change	$\widehat{\beta_3} = -0.4544$	
FT	$\widehat{\beta_4} = 0.0274$	

Table 3 Comparison of different student's profiles and corresponding parameters

Profiles	Age	Gender	Change	$\widehat{\xi}$	$\widehat{\pi_i}$	$P(R < 3)$
A	20	Woman	Yes	0.3563	0.6306	0.2367
B	20	Woman	No	0.3563	0.7854	0.1896
C	20	Man	Yes	0.3563	0.7733	0.1933
D	20	Man	No	0.3563	0.8797	0.1610
E	30	Woman	Yes	0.3563	0.5153	0.2718
F	30	Woman	No	0.3563	0.6950	0.2171
G	30	Man	Yes	0.3563	0.6799	0.2217
H	30	Man	No	0.3563	0.8199	0.1791

1 = *completely unsatisfied* to 7 = *completely satisfied* ($m = 7$) and some students' personal information such as Age, Gender, Change of original enrollment, Full Time students (FT). In Corduas et al. (2009) the data set has been extensively analyzed adopting the classical inferential procedures to estimate CUB$(0, q)$ parameters for different values of q. In the sequel we focus our attention to the analysis of the item on *advertisement of the service* since the lowest value of $\widehat{\pi}$ has been associated to it ($\widehat{\pi} = 0.78$ when all the other items have values of $\widehat{\pi}$ greater than 0.87). Our aim here is to identify which kind of students shows the greater *uncertainty* answering to this item. Then, using 2007 data set collecting n = 3,511 students' answers and their individual covariates, we exploit CUB$(p, 0)$ model for different values of p by MCMC algorithm. Using the same covariates adopted in Corduas et al. (2009), we focus our attention on the CUB$(3, 0)$ model (see Table 2) which is the best one, since the covariate FT seems not to be relevant in explaining π.

Some different profiles corresponding to the 2^3 combination of two levels for each covariate are derived from the estimated CUB$(3, 0)$ model and reported in

Table 3. We can observe that the profile that presents the greater uncertainty, i.e. the lower value of π in evaluating advertisement, is that of 30 years old women who change their original enrollment. The higher uncertainty implies a higher probability to give low evaluation ($R < 3$) as we can see looking at the last column in Table 3. Notice that $\widehat{\xi} = 0.3563$ is constant for all profiles since no covariates for ξ are present in the CUB(3, 0) model.

6 Conclusion

In this paper we adopt the Bayesian approach to the statistical analysis of a special mixture model for ordinal data. We show how it may be performed via MCMC simulation. The algorithm here introduced is extremely straightforward and it does not involve the usual problems of the MCMC methods in the standard mixtures context, or of the simulation algorithms in the classical maximum likelihood inference. Finally, through a simulation study we show that our MCMC sampler provide a good posterior inference. An application of a real data set is also studied. An advantage of the Bayesian approach is that expert knowledge may also be embedded into the model; previous studies may provide additional information on the parameters that may be expressed through the prior distributions. This topic is not discussed here because, up to now, we have adopted non-informative priors.

Future issues of the Bayesian analysis of the CUB models will regard sensitivity analysis and the implementation of model choice and variable selection.

Acknowledgements The paper has been prepared within a MIUR grant (code 2008WKHJP-KPRIN2008-PUC number E61J10000020001) for the project: "Modelli per variabili latenti basati su dati ordinali: metodi statistici ed evidenze empiriche" (research Unit University of Naples Federico II).

References

Bini, M., Monari, P., Piccolo, D., & Salmaso, L. (Eds.). (2009). *Statistical methods for the evaluation of educational services and quality of products* (Contribution to statistics). Berlin: Springer.

Corduas, M., Iannario, M., & Piccolo, D. (2009). A class of statistical models for evaluating services and performances. In M. Bini, et al. (Eds.), *Statistical methods for the evaluation of educational services and quality of products* (Contribution to statistics). Berlin: Springer.

D'Elia, A. (2003). Finite sample performance of the E-M algorithm for ranks data modelling. *Statistica, LXIII*, 41–51.

D'Elia, A., & Piccolo, D. (2005). A mixture model for preferences data analysis. *Computational Statistics & Data Analysis, 49*, 917–934.

Früwirth-Schnatter, S. (2006). *Finite mixture and markov switching models* (Springer series in statistics). New York: Springer.

Iannario, M. (2010). On the identifiability of a mixture model for ordinal data. *Metron, LXVIII*, 87.

MacLachlan, G., & Peel, D. (2000). *Finite mixture models* (Wiley series in probability and statistics). New York: Wiley

Piccolo, D. (2006). Observed information matrix for MUB models. *Quaderni di Statistica, 8*, 33–78.

Robustness Versus Consistency in Ill-Posed Classification and Regression Problems

Robert Hable and Andreas Christmann

Abstract It is well-known from parametric statistics that there can be a goal conflict between efficiency and robustness. However, in so-called ill-posed problems, there is even a goal conflict between consistency and robustness. This particularly applies to certain nonparametric statistical problems such as nonparametric classification and regression problems which are often ill-posed. As an example in statistical machine learning, support vector machines are considered.

1 Introduction

There are a number of properties which should be fulfilled by a statistical procedure. First of all, it should be *consistent*, i.e., it should converge in probability to the true value for increasing sample sizes. Another crucial property is *robustness*. Though there are many different notions of robustness, the common idea is that small model violations (particularly caused by small errors in the data) should not change the results too much. It is well-known from parametric statistics that there can be a goal conflict between efficiency and robustness. In this case one has to pay by a loss of efficiency in order to obtain more reliable results. However, in many nonparametric statistical problems, there is even a goal conflict between consistency and robustness. That is, a statistical procedure which is (in a sense) robust *cannot* always converge to the true value for increasing sample sizes. This is the case for so-called ill-posed problems. It is well-known in the machine learning theory that many nonparametric statistical problems are ill-posed. In particular, this is often true for nonparametric classification and regression problems. The rest of the paper is organized as follows: Sect. 2 introduces the setup and recalls a mathematically

R. Hable (✉) · A. Christmann
Department of Mathematics, University of Bayreuth, D-95440, Bayreuth, Germany
e-mail: Robert.Hable@uni-bayreuth.de; Andreas.Christmann@uni-bayreuth.de

A. Giusti et al. (eds.), *Classification and Data Mining*, Studies in Classification, Data Analysis, and Knowledge Organization, DOI 10.1007/978-3-642-28894-4_4,

rigorous definition of ill-posedness given by Dey and Ruymgaart (1999). Section 3 investigates the goal conflict between robustness and consistency in ill-posed problems. In Sect. 4, this is illustrated in case of support vector machines. This article brings together notions and facts which are common in different fields, namely robust statistics and machine learning.

2 Ill-Posed Statistical Problems

Many statistical estimation problems can be formalized in the following way: Let $\mathscr{P}$ be a set of probability measures on the Borel-σ-algebra of a complete separable metric space $\mathscr{Z}$ and let $\mathscr{F}$ be another complete separable metric space.[1] It is assumed that one element $P_0 \in \mathscr{P}$ is the true probability measure and the task is to estimate the value $T(P_0)$ of the functional

$$T \;:\; \mathscr{P} \to \mathscr{F}.$$

In parametric statistics, we typically have $\mathscr{P} = \{P_\theta \,|\, \theta \in \Theta\}$, $\mathscr{F} = \Theta \subset \mathbb{R}^k$ and $T(P_\theta) = \theta$ for all $\theta \in \Theta$. For a simple example of nonparametric estimation, let $\mathscr{P}$ be the set of all probability measures on $\mathbb{R}$ with finite mean and define $T(P) = \int x\, P(dx)$ for all $P \in \mathscr{P}$. Then, the task would be to estimate the mean.

The notion of an ill-posed problem originates from solving (deterministic) operator equations. Among other things, a problem is ill-posed if the solution is not stable with respect to small changes. In the above formalized estimation problem, ill-posedness has been connected with the statistical notion of qualitative robustness by Dey and Ruymgaart (1999) where the following definition was given:

Definition 1. The problem of estimating $T : \mathscr{P} \to \mathscr{F}$ is *well-posed* if T is continuous in the following sense: $\lim_{n\to\infty} T(P_n) = T(P_0)$ for every sequence of probability measures $(P_n)_{n\in\mathbb{N}} \subset \mathscr{P}$ which converges weakly to some $P_0 \in \mathscr{P}$. The problem of estimating $T : \mathscr{P} \to \mathscr{F}$ is *ill-posed* if T is <u>not</u> continuous.

3 Robustness Versus Consistency

A sequence of estimators

$$T_n \;:\; \mathscr{Z} \to \mathscr{F}, \qquad (z_1, \ldots, z_n) \mapsto T_n(z_1, \ldots, z_n), \qquad n \in \mathbb{N}, \tag{1}$$

is *consistent in* $\mathscr{P}$ for the problem of estimating $T : \mathscr{P} \to \mathscr{F}$ if, for every $P \in \mathscr{P}$,

[1] If nothing else is stated, we always use the Borel-σ-algebras.

$$T_n \xrightarrow{P} T(P) \qquad \text{for } n \to \infty. \tag{2}$$

Let d_{Pro} denote the Prokhorov metric on the set of all probability measures on the metric space $\mathscr{Z}$. That is

$$d_{\mathrm{Pro}}(\mathrm{P}_1, \mathrm{P}_2) = \inf\left\{\varepsilon \in (0,\infty) \;\middle|\; \mathrm{P}_1(B) < \mathrm{P}_2(B^{\varepsilon}) + \varepsilon \;\; \forall\, B \in \mathscr{A}\right\}$$

where $B^{\varepsilon} = \{z \in \mathscr{Z} \mid \inf_{z' \in B} d(z,z') < \varepsilon\}$ and d denotes the metric on $\mathscr{Z}$.

According to Hampel (1968, 1971) and Cuevas (1988, Definition 1), a sequence of estimators $(T_n)_{n\in\mathbb{N}}$ is called *qualitatively robust in* $\mathscr{P}$ if, for every $P \in \mathscr{P}$ and every $\rho > 0$, there is an $\varepsilon > 0$ such that, for every $Q \in \mathscr{P}$,

$$d_{\mathrm{Pro}}(Q,P) < \varepsilon \quad \Rightarrow \quad \sup_{n\in\mathbb{N}} d_{\mathrm{Pro}}\big(T_n(Q^n), T_n(P^n)\big) < \rho.$$

Accordingly, the interpretation of qualitative robustness is: if the distribution which generates the data changes slightly, then the distribution of the estimator should also change only slightly – uniformly in the sample size. By use of the Prokhorov metric, the notion of qualitative robustness covers two kinds of small errors in the data: small errors in many data points z_i and large errors in a small fraction of the data set.

Theorem 1 immediately follows from Hampel (1971, Lemma 3) for parametric statistics and Cuevas (1988, Theorem 1) for the general setting described above:

Theorem 1. *If the problem of estimating T is ill-posed, then no sequence of estimators T_n, $n \in \mathbb{N}$, can simultaneously be consistent and qualitatively robust in $\mathscr{P}$.*

This is for example also the case in nonparametric density estimation; see Cuevas (1988, Sect. 2). Though the fact that many problems are ill-posed is well-known in machine learning theory, the implications concerning robustness have hardly received any attention: A procedure which is universally consistent (i.e. consistent for *all* probability measures) cannot be stable in the sense of qualitative robustness.

In statistical machine learning theory, it is common to consider so-called *risk-consistency* instead of the consistency property defined in (2). Let $\mathscr{X}$ be an input space (with a σ-algebra $\mathscr{A}$) and $\mathscr{Y} \subset \mathbb{R}$ an output space; e.g., $\mathscr{Y} = \{-1,+1\}$ for binary classification or $\mathscr{Y} = \mathbb{R}$ for regression purposes. The quality of a (measurable) predictor $f : x \mapsto f(x)$ is measured by the risk

$$\mathscr{R}_{L,P}(f) = \int L\big(y, f(x)\big)\, P\big(d(x,y)\big)$$

where $L : \mathscr{Y} \times \mathbb{R} \to [0,\infty)$ is a (measurable) loss function. Let $\mathscr{L}_0(\mathscr{X})$ be the set of all measurable functions $f : \mathscr{X} \to \mathbb{R}$. A learning algorithm

$$T_n : (\mathscr{X} \times \mathscr{Y})^n \to \mathscr{L}_0(\mathscr{X}), \qquad D_n \mapsto T_n(D_n)$$

maps a sample D_n to a predictor $f_{D_n} : \mathscr{X} \to \mathbb{R}$. That is, T_n can be seen as an estimator which takes its values in the function space $\mathscr{L}_0(\mathscr{X})$ and $f_{D_n} = T_n(D_n)$ denotes the value of T_n in D_n. Here, we assume that the sample $D_n = \big((x_1, y_1), \dots, (x_n, y_n)\big)$ stems from i.i.d. random variables $(X_1, Y_1), \dots, (X_n, Y_n)$ with unknown distribution P. The learning algorithm $T_n : D_n \mapsto f_{D_n}$ is called *risk-consistent in* $\mathscr{P}$ if, for every $P \in \mathscr{P}$,

$$\mathscr{R}_{L,P}(f_{D_n}) \longrightarrow \inf_{f \in \mathscr{L}_0(\mathscr{X})} \mathscr{R}_{L,P}(f) \qquad (n \to \infty)$$

in probability. Accordingly, we call a learning algorithm T_n *qualitatively risk-robust in* $\mathscr{P}$ if, for every $P \in \mathscr{P}$ and every $\rho > 0$, there is an $\varepsilon > 0$ such that, for every $Q \in \mathscr{P}$,

$$d_{\text{Pro}}(Q, P) < \varepsilon \quad \Rightarrow \quad \sup_{n \in \mathbb{N}} d_{\text{Pro}}\big(\mathscr{R}_{L,P} \circ T_n(Q^n), \mathscr{R}_{L,P} \circ T_n(P^n)\big) < \rho. \quad (3)$$

That is, if the distribution P which generates the data changes slightly to Q, then the distribution of the risk $\mathscr{R}_{L,P} \circ T_n(P^n)$ only changes slightly to $\mathscr{R}_{L,P} \circ T_n(Q^n)$– uniformly in the sample size.[2]

As risk-consistency and qualitative risk-robustness are weaker properties than the ones used in Theorem 1, one may hope that there might be a learning algorithm which enjoys both of these weaker properties. However, at least for regression problems, this is not the case as the following theorem shows.

Theorem 2. *Let* $\mathscr{X} = [0, 1]$, $\mathscr{Y} = \mathbb{R}$ *and L be a distance-based loss function, i.e., there is a measurable function* $\varphi : [0, \infty) \to [0, \infty)$ *such that, for every* $(y, t) \in \mathscr{Y} \times \mathbb{R}$, $L(y, t) = \varphi(|y - t|)$. *Assume that* $\varphi(0) = 0$, *that* φ *is non-decreasing, and that* $\lim_{s \to \infty} \varphi(s) = \infty$. *Let* $\mathscr{P}$ *be the set of all probability measures P on* $\mathscr{X} \times \mathscr{Y}$ *such that* $\inf_{f \in \mathscr{L}_0(\mathscr{X})} \mathscr{R}_{L,P}(f) < \infty$. *Then, no learning algorithm* T_n, $n \in \mathbb{N}$, *defined in (1) can simultaneously be risk-consistent and qualitatively risk-robust in* $\mathscr{P}$.

As it is not possible in many nonparametric problems to have a consistent and qualitatively robust estimator/learning algorithm, one has to weaken at least one of the two desired properties. The proof of Theorem 2 shows that the incompatibility of consistency and qualitative robustness comes from the fact that the usual definition of qualitative robustness not only requires a continuity property but even equicontinuity over all possible sample sizes and this conflicts with universal consistency. However, in applications, one is usually faced with a sample of a fixed finite size so that robustness for fixed sample sizes may also be satisfactory. The following definition of "finite sample qualitative robustness" relaxes equicontinuity to continuity. This offers a possibility to get around the conflict between universal

[2]Note that it is not appropriate to consider $\mathscr{R}_{L,Q}$ instead of $\mathscr{R}_{L,P}$ in (3) because we have to evaluate the risk with respect to the true distribution P and not with respect to the erroneous Q.

consistency and qualitative robustness. However, it has to be noted that this notion of robustness is not an asymptotic one so that it is not sufficient if an asymptotic performance criterion is applied as usual for well-posed parametric problems. It is only useful for finite sample considerations.

Section 4 presents support vector machines as an example of a learning algorithm which is simultaneously universally consistent and finite sample qualitatively robust under some mild conditions.

Definition 2. An estimator T_n is called *finite sample qualitatively robust in* $\mathscr{P}$ if, for every sample size n, for every $P \in \mathscr{P}$, and every $\rho > 0$, there is an $\varepsilon_n > 0$ such that, for every $Q \in \mathscr{P}$,

$$d_{\mathrm{Pro}}(Q, P) < \varepsilon_n \quad \Rightarrow \quad d_{\mathrm{Pro}}\big(T_n(Q^n), T_n(P^n)\big) < \rho.$$

Similarly, define *finite sample qualitative risk-robustness*.

4 Example: Support Vector Machines

As an example, we consider nonparametric classification and regression where we have i.i.d. observations $(x_1, y_2), \ldots, (x_n, y_n) \in \mathscr{X} \times \mathscr{Y}$ and we want to predict the value y of an unobserved output variable Y based on the observed value x of an input variable X. That is, we want to find a "good" predictor $f : x \mapsto f(x)$. For this purpose, support vector machines attracts attention since a decade; see e.g. Vapnik (1998), Schölkopf and Smola (2002) and Steinwart and Christmann (2008).

In order to define support vector machines, a convex loss function $L : \mathscr{X} \times \mathscr{Y} \times \mathbb{R} \to [0, \infty)$ is used where $L(x, y, f(x))$ measures the quality of a prediction $f(x)$ if x is the value of the input variable X and y is the value of the output variable Y. Next, H is a Hilbert space – more precisely, a reproducing kernel Hilbert space – which consists of functions $f : \mathscr{X} \to \mathbb{R}$. Then, support vector machines are given by the estimator $T_n : (\mathscr{X} \times \mathscr{Y})^n \to H$

$$\big((x_1, y_1), \ldots, (x_n, y_n)\big) \mapsto \arg\min_{f \in H} \frac{1}{n} \sum_{i=1}^{n} L\big(x_i, y_i, f(x_i)\big) + \lambda \|f\|_H^2$$

in order to obtain good approximations of a minimizer of

$$\mathscr{L}_0(\mathscr{X}) \to \mathbb{R}, \qquad f \mapsto \mathscr{R}_{L,P}(f) = \int L\big(x, y, f(x)\big)\, P\big(d(x, y)\big).$$

The real number $\lambda \in (0, \infty)$ is a regularization parameter which prevents from overfitting. Let $\mathscr{P}$ be the set of all probability measures on $\mathscr{X} \times \mathscr{Y}$.

It was shown in Hable and Christmann (2011, Theorem 3.1) that the sequence of SVM-estimators T_n, $n \in \mathbb{N}$, is qualitatively robust in $\mathscr{P}$ under some mild conditions for any *fixed* regularization parameter $\lambda \in (0, \infty)$. However, T_n defined in this way is *not* consistent in $\mathscr{P}$. In order to obtain consistency, the fixed regularization parameter λ has to be replaced by a sequence of regularization parameters $(\lambda_n)_{n \in \mathbb{N}} \subset (0, \infty)$ which converges (not too fast) to 0 for increasing sample sizes n; see e.g. Steinwart (2002). It is also shown in Hable and Christmann (2011, Proposition 3.2) that, in the latter case, the resulting estimator using λ_n is *not* qualitatively robust. As described above, this is not a particular shortcoming of support vector machines but an unavoidable consequence of the ill-posedness in the sense of Dey and Ruymgaart (1999) of the underlying statistical problem.

However, Hable and Christmann (2011, Theorem 3.1) shows that, under some mild conditions on L and H, support vector machines are finite sample qualitatively (risk-)robust in $\mathscr{P}$ for every sequence $(\lambda_n)_{n \in \mathbb{N}} \subset (0, \infty)$. Hence, suitable choices of L, H and $(\lambda_n)_{n \in \mathbb{N}}$ guaranty that support vector machines are simultaneously (risk-) consistent and finite sample qualitatively (risk-)robust. Furthermore, it is known that support vector machines have a bounded influence function and a bounded maxbias; see e.g. Steinwart and Christmann (2008, Sect. 10).

5 Conclusions

There are a number of criteria for robustness (e.g. influence function, maxbias, breakdown point) in statistics but this article only refers to qualitative robustness. It is pointed out that there is a goal conflict between consistency and qualitative robustness in many nonparametric statistical problems such as classification and regression. This is somewhat contrary to a result from Poggio et al. (2004) and Mukherjee et al. (2006) which in some sense says that, for the method of empirical risk minimization, universal consistency is equivalent to their notion of stability even though empirical risk minimization typically is ill-posed. This shows that, in case of an ill-posed empirical risk minimization problem, no estimator (or learning procedure) can be both qualitatively robust and stable. This indicates that the notion of stability (common in machine learning theory) and the classical notion of qualitative robustness are quite conflicting even though stability is sometimes considered as some kind of a robustness property in machine learning theory.

Appendix

Proof (Theorem 2). In order to prove Theorem 2, we assume that T_n, $n \in \mathbb{N}$, is a risk-consistent learning algorithm and we show that T_n, $n \in \mathbb{N}$, is not qualitatively risk-robust. According to the assumptions on φ, for every $m \in \mathbb{N}$, there is a $c_m \in [0, \infty)$ such that $\varphi(c_m) \geq m$. For every $t \in \mathbb{R}$, let δ_t denote the Dirac-measure at t; let $U_{[0,1]}$ denote the uniform distribution on $\mathscr{X} = [0, 1]$. Define

$$P\big(d(x,y)\big) := \delta_0(dy)U_{[0,1]}(dx) \quad \text{and} \quad Q_m\big(d(x,y)\big) := \delta_{g_m(x)}(dy)U_{[0,1]}(dx)$$

for $g_m : [0,1] \to \mathbb{R}, \ x \mapsto \big(-mc_m x + 4c_m\big)I_{[0,4/m]}(x)$ for every $m \in \mathbb{N}$. Note that, for every Borel-measurable set $B \subset \mathscr{X} \times \mathscr{Y}$,

$$Q_m\left(\left\{(x,y) \in \mathscr{X} \times \mathscr{Y} \,\middle|\, x > \frac{4}{m}\right\} \cap B\right) = P\left(\left\{(x,y) \in \mathscr{X} \times \mathscr{Y} \,\middle|\, x > \frac{4}{m}\right\} \cap B\right). \tag{4}$$

Obviously, $\inf_{f \in \mathscr{L}_0(\mathscr{X})} \mathscr{R}_{L,P}(f) = 0$ and $\inf_{f \in \mathscr{L}_0(\mathscr{X})} \mathscr{R}_{L,Q_m}(f) = 0$ for every $m \in \mathbb{N}$. Then, risk-consistency implies: for every $m \in \mathbb{N}$, there is an $n_m \in \mathbb{N}$ such that, for every $n \geq n_m$,

$$Q_m^n\left(\left\{D_n \in (\mathscr{X} \times \mathscr{Y})^n \,\middle|\, \mathscr{R}_{L,Q_m}(f_{D_n}) < \frac{1}{3}\right\}\right) \geq \frac{2}{3} \tag{5}$$

$$P^n\left(\left\{D_n \in (\mathscr{X} \times \mathscr{Y})^n \,\middle|\, \mathscr{R}_{L,P}(f_{D_n}) < \frac{1}{3}\right\}\right) \geq \frac{2}{3}. \tag{6}$$

For every $m, n \in \mathbb{N}$, define $B_m^{(n)} := \big\{D_n \in (\mathscr{X} \times \mathscr{Y})^n \,\big|\, \mathscr{R}_{L,Q_m}(f_{D_n}) < \frac{1}{3}\big\}$ and

$$A_m(D_n) := \left\{x \in \mathscr{X} \,\middle|\, x \leq \frac{2}{m},\, f_{D_n}(x) \leq c_m\right\} \qquad \forall\, D_n \in (\mathscr{X} \times \mathscr{Y})^n.$$

Note that the definitions imply

$$g_m(x) - f_{D_n}(x) \geq 2c_m - c_m = c_m \geq 0 \qquad \forall\, x \in A_m(D_n). \tag{7}$$

Hence, for every $m \in \mathbb{N}$, $n \geq n_m$, and $D_n \in B_m^{(n)}$,

$$\frac{1}{3} > \mathscr{R}_{L,Q_m}(f_{D_n}) \geq \int_{A_m(D_n)} \int_{\mathbb{R}} \varphi\big(|y - f_{D_n}(x)|\big)\, \delta_{g_m(x)}(dy)\, U_{[0,1]}(dx) =$$
$$= \int_{A_m(D_n)} \varphi\big(|g_m(x) - f_{D_n}(x)|\big)\, U_{[0,1]}(dx) \overset{(7)}{\geq} \varphi(c_m) \cdot U_{[0,1]}\big(A_m(D_n)\big) \geq$$
$$\geq m \cdot U_{[0,1]}\big(A_m(D_n)\big) \tag{8}$$

Next, it follows for every $m \in \mathbb{N}$, $n \geq n_m$, and $D_n \in B_m^{(n)}$ that

$$U_{[0,1]}\big(\{x \in \mathscr{X} \mid f_{D_n}(x) > c_m\}\big) \geq U_{[0,1]}\left(\left\{x \in \mathscr{X} \,\middle|\, x \leq \frac{2}{m},\, f_{D_n}(x) > c_m\right\}\right)$$
$$= U_{[0,1]}\left(\left\{x \in \mathscr{X} \,\middle|\, x \leq \frac{2}{m}\right\}\right) - U_{[0,1]}\big(A_m(D_n)\big) \overset{(8)}{\geq} \frac{2}{m} - \frac{1}{3m} > \frac{1}{m} \tag{9}$$

and, therefore,

$$\begin{aligned}
\mathscr{R}_{L,P}\big(f_{D_n}\big) &\geq \int_{\{x\in\mathscr{X}\,|\,f_{D_n}(x)>c_m\}} \int_{\mathbb{R}} \varphi\big(|y-f_{D_n}(x)|\big)\,\delta_0(dy)\,U_{[0,1]}(dx) = \\
&= \int_{\{x\in\mathscr{X}\,|\,f_{D_n}(x)>c_m\}} \varphi\big(|f_{D_n}(x)|\big)\,U_{[0,1]}(dx) \geq \\
&\geq m\cdot U_{[0,1]}\big(\{x\in\mathscr{X}\,|\,f_{D_n}(x)>c_m\}\big) \overset{(9)}{\geq} 1. \qquad (10)
\end{aligned}$$

Define $C := [1,\infty)$. Then, for every $m\in\mathbb{N}$ and $n\geq n_m$,

$$\begin{aligned}
\big[\mathscr{R}_{L,P}\circ T_n(Q_m^n)\big](C) &= Q_m^n\Big(\big\{D_n\in(\mathscr{X}\times\mathscr{Y})^n\,\big|\,\mathscr{R}_{L,P}(f_{D_n})\geq 1\big\}\Big) \geq \\
&\overset{(10)}{\geq} Q_m^n\big(B_m^{(n)}\big) \overset{(5)}{\geq} \frac{2}{3} = \frac{1}{3}+\frac{1}{3} \\
&\overset{(6)}{\geq} P^n\left(\left\{D_n\in(\mathscr{X}\times\mathscr{Y})^n\,\middle|\,\mathscr{R}_{L,P}(f_{D_n})\geq\frac{1}{3}\right\}\right)+\frac{1}{3} \\
&\geq \big[\mathscr{R}_{L,P}\circ T_n(P^n)\big](C^{\frac{1}{3}})+\frac{1}{3}
\end{aligned}$$

where $C^{\frac{1}{3}} = \{z\in\mathbb{R}|\,\inf_{z'\in\mathbb{R}}|z-z'|<\frac{1}{3}\}$ as in the definition of d_{Pro}. This implies

$$d_{\mathrm{Pro}}\big(\mathscr{R}_{L,P}\circ T_n(Q_m^n),\mathscr{R}_{L,P}\circ T_n(P^n)\big) \geq \frac{1}{3} \qquad \forall\, n\geq n_m \quad \forall\, m\in\mathbb{N}. \qquad (11)$$

However, for every $m\in\mathbb{N}$ and every measurable $B\subset\mathscr{X}\times\mathscr{Y}$, we have

$$\begin{aligned}
Q_m(B) &= Q_m\left(\left\{(x,y)\in\mathscr{X}\times\mathscr{Y}\,\middle|\,x\leq\frac{4}{m}\right\}\cap B\right) \\
&\quad + Q_m\left(\left\{(x,y)\in\mathscr{X}\times\mathscr{Y}\,\middle|\,x>\frac{4}{m}\right\}\cap B\right) \\
&\leq \frac{4}{m}+Q_m\left(\left\{(x,y)\in\mathscr{X}\times\mathscr{Y}\,\middle|\,x>\frac{4}{m}\right\}\cap B\right) = \\
&= \frac{4}{m}+P\left(\left\{(x,y)\in\mathscr{X}\times\mathscr{Y}\,\middle|\,x>\frac{4}{m}\right\}\cap B\right) \leq \frac{4}{m}+P\left(B^{\frac{4}{m}}\right)
\end{aligned}$$

and, therefore,

$$d_{\mathrm{Pro}}\big(Q_m,P\big) \leq \frac{4}{m} \qquad \forall\, m\in\mathbb{N}. \qquad (12)$$

Inequalities (11) and (12) imply that T_n, $n\in\mathbb{N}$, is *not* qualitatively risk-robust.

References

Cuevas, A. (1988). Qualitative robustness in abstract inference. *Journal of Statistical Planning and Inference, 18*, 277–289.

Dey, A. K., & Ruymgaart, F. H. (1999). Direct density estimation as an ill-posed inverse estimation problem. *Statistica Neerlandica, 53*(3), 309–326.

Hable, R., & Christmann, A. (2011). On qualitative robustness of support vector machines. *Journal of Multivariate Analysis, 102*, 993–1007.

Hampel, F. R. (1968). *Contributions to the theory of robust estimation*. Ph.D. thesis, University of California, Berkeley.

Hampel, F. R. (1971). A general qualitative definition of robustness. *Annals of Mathematical Statistics, 42*, 1887–1896.

Mukherjee, S., Niyogi, P., Poggio, T., & Rifkin, R. (2006). Learning theory: Stability is sufficient for generalization and necessary and sufficient for consistency of empirical risk minimization. *Advances in Computational Mathematics, 25*, 161–193.

Poggio, T., Rifkin, R., Mukherjee, S., & Niyogi, P. (2004). General conditions for predictivity in learning theory. *Nature, 428*, 419–422.

Schölkopf, B., & Smola, A. J. (2002). *Learning with kernels*. Cambridge: MIT.

Steinwart, I. (2002). Support vector machines are universally consistent. *Journal of Complexity, 18*, 768–791.

Steinwart, I., & Christmann, A. (2008). *Support vector machines*. New York: Springer.

Vapnik, V. N. (1998). *Statistical learning theory*. New York: Wiley.

Issues on Clustering and Data Gridding

Jukka Heikkonen, Domenico Perrotta, Marco Riani, and Francesca Torti

Abstract This contribution addresses clustering issues in presence of densely populated data points with high degree of overlapping. In order to avoid the disturbing effects of high dense areas we suggest a technique that selects a point in each cell of a grid defined along the Principal Component axes of the data. The selected sub-sample removes the high density areas while preserving the general structure of the data. Once the clustering on the gridded data is produced, it is easy to classify the rest of the data with reliable and stable results. The good performance of the approach is shown on a complex dataset coming from international trade data.

1 Introduction

In this paper we address clustering issues in presence of data consisting of an unknown number of groups with high degree of overlapping and presenting both high and low density regions which invalidate the hypothesis of ellipticity.

J. Heikkonen (✉)
Department of Information Technology, University of Turku, Turku, Finland
e-mail: jukka.heikkonen@utu.fi

D. Perrotta
EC Joint Research Centre, Ispra site, Ispra, Italy
e-mail: domenico.perrotta@ec.europa.eu

M. Riani
University of Parma, Parma, Italy
e-mail: mriani@unipr.it

F. Torti
University of Milano Bicocca, Milan, Italy
e-mail: francesca.torti@unimib.it

A. Giusti et al. (eds.), *Classification and Data Mining*, Studies in Classification, Data Analysis, and Knowledge Organization, DOI 10.1007/978-3-642-28894-4_5,

To find the clusters on these data, the model complexity selection issue becomes difficult. Typically model complexity selection is based on the maximum likelihood formulation of the model with respect of the data and an additional cost function that penalises too complex models, i.e. the ones having too many parameters needed to capture the main characteristics of the data (Schwarz, 1978; Rissanen, 1986; Bishop, 2006). When the model complexity selection is formulated as a probabilistic problem, in addition to well known disturbing effects of noise and outliers, the presence of dense and dispersed groups of points causes additional challenges. Because of the likelihood formulation, the dense groups dictate the parameter values of the model and the groups with less points may not be properly detected. This has some similarity to sampling theory where the goal is to have representative samples from the population of interest (Cochran, 1977). Often some elements are hard to get or very costly to be obtained and one has to select the correct sampling strategy to obtain representative population statistics. For instance, in stratified sampling, the population is first divided by some meaningful rules into as homogeneous groups as possible. These groups (strata) should be mutually exclusive meaning that one element should be assigned only to one and only one group (stratum). When properly used, stratified sampling reduces sampling error, as is its goal.

In our clustering case we are interested in recognizing also those clusters that only consist of few data points. In order to achieve this goal, we propose a sampling approach that tries to avoid the disturbing effects of the dense populated data points through a data gridding technique based on Principal Component Analysis (PCA). This technique consists in defining a grid along the Principal Component (PC) axes of the data and selecting one point in each cell of the grid. Our goal is to have through the balanced data points correct model complexity for the given data and to avoid the domination of dense populated data points over the dispersed ones. The performance of the proposed data gridding technique (Sect. 2) is evaluated with two versions of Gaussian Mixture Models (GMMs) and the Forward Search (FS) on data coming from the international trade of the European Union (EU) (Sect. 3). We show that the gridded data preserve the general structure of the original dataset, which is then well captured by three clustering methods. We will see that once the clustering on the gridded data is produced, it is easier to classify the rest of the data with results which are more reliable and stable than those obtained on the original data.

To illustrate the procedure and the above problems we use the dataset in the left panel of Fig. 1. The variables are the volume and value of the monthly imports of a fishery product in the EU in a period of 3 years. One of the 27 EU Members States is producing trade flows which deviate from the main dense cluster. This situation, with a concentration of points towards the origin of the axes where the clusters intersect, is typical of international trade data. Perrotta and Torti (2009) made an exploratory analysis of this dataset and Riani et al. (2008) treated the case as regression problem. In this paper the emphasis is on inferring automatically the number and the shape of the groups. The dataset is included in the MATLAB FSDA toolbox Riani et al. (2012), which can be downloaded at http://www.riani.it/MATLAB.htm.

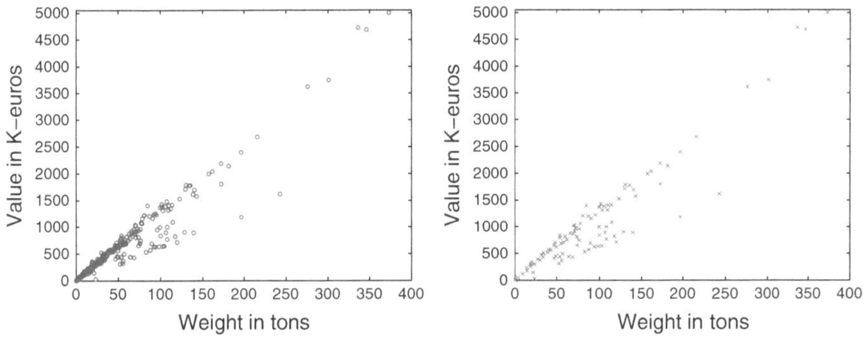

Fig. 1 Fishery data (*left plot*, 677 observations) and gridded data with 80 cells along the first principle component axis (*right plot*, 95 observations)

2 Gridding Approach

In our gridding approach the original variables are normalized to zero mean and unit variances to avoid the dominance of the scaling of variables in PCA. After scaling when the PC axes are defined, the data points are projected to this domain to have their PCs. Taking the maximum and minimum coordinates of the PCs we can define a grid of a predefined number of cells along each PC axis. In our case, when we have 2-dimensional data, the grid is also 2-dimensional and defined by the 2 PC axes. The same approach can be extended to higher dimensions. Note that the grid cells do not necessarily have equal width in all PC axes directions and a single cell can cover zero, one or more data points of the original data. Especially where the data are densely populated, the grid cells corresponding to a dense group of points include multiple original values. For each cell the goal is to search for one representative point from the original data. This is done by taking the median of points belonging to the cell and finding the closest point to the calculated median. As a result we obtain either none or one point for each cell of the grid. With the new reduced subset we can perform a desired analysis, for example estimating the Gaussian Mixture Model and the proper number of clusters over the balanced dataset.

The right panel of Fig. 1 shows the result of the gridding approach when 80 cells in each PC axis direction is applied to the original data of 677 observations. As can be observed, the gridded result represents rather well the original data and the number of points in dense and dispersed groups is better balanced.

3 Example Results

The GMM models used are Model-Based Clustering/Normal Mixture Modeling (Fraley, 1998) and its robust version Robust Trimmed Clustering (Garcia-Escudero et al., 2008). For the runs we used their well known R implementations MCLUST and TCLUST. Both methods are based on a finite mixture of distributions where each mixture component corresponds to a different group. A common reference model for the components is the multivariate Gaussian distribution. In MCLUST the standard approach for estimating the mixture consists of using the EM algorithm and the BIC to select the number of components. Each observation is assigned to the cluster to which it is most likely to belong. The TCLUST approach is defined through the search of k centers $m_1, \ldots, m_k$ and k shape matrices $U_1, \ldots, U_k$ solving the double minimization problem:

$$\arg\min_{\mathbf{Y}} \min_{\substack{m_1, \ldots, m_k \\ U_1, \ldots, U_k}} \sum_{j=1,\ldots,k} (x_i - m_j)' U_j^{-1} (x_i - m_j) \qquad i = 1, \ldots, n \tag{1}$$

where $\mathbf{Y}$ ranges on the class of subsets of size $[n(1-\alpha)]$ within the sample $\{x_1, \ldots, x_n\}$. Note that in this approach we allow for a proportion α of observations, hopefully the most outlying ones, to be left unassigned. In order to chose k, the authors suggest using the so called Classification trimmed Likelihood curves (Garcia-Escudero et al., 2011).

The third method that we consider is based on the Forward Search of Atkinson et al. (2004). This approach was originally introduced for detecting subsets and masked outliers and for estimating their effect on the models fitted to the data. This method produces a sequence of subsets of increasing size through a dynamic process that leaves outliers in the last subsets. By monitoring the trajectory of the values of the minimum Mahalanobis distance among observations outside the current subset, it is possible to detect towards the end of the search peaks that correspond to the presence of outliers. Besides, by monitoring the same statistic for searches initialised from many different randomly chosen subsets, it is possible to reveal the presence of multiple populations as separated peaks that can occur at any position along the search depending on the size and structure of the groups (Atkinson and Riani, 2007). However, problems may occur in presence of high density areas. For example, the left panel of Fig. 1 shows that more than 50 % of the data are concentrated near the origin of the axes and, thus, the random start trajectories of minimum Mahalanobis distance degenerate into the same search path in the very first steps of the FS, as shown in Fig. 2. This behaviour, which is caused by the dense population near the origin of the axes, makes the information produced by the random start FS difficult or even impossible to interpret. The detection of the first cluster is shown in the zoom of Fig. 1, where the envelopes based on about 130 size sub-sample are exceeded by the Minimum Mahalanobis distance trajectories (Riani et al., 2009). The same procedure can be repeated iteratively for the data not yet assigned to an homogeneous subgroup. However, in this case this procedure results in an excessive number of subgroups.

In presence of highly dense areas, similar difficulties arise with other classical statistical clustering methods such as K-means clustering or GMMs which, however, compared to the FS have less or even no instruments to accurately identify the parts

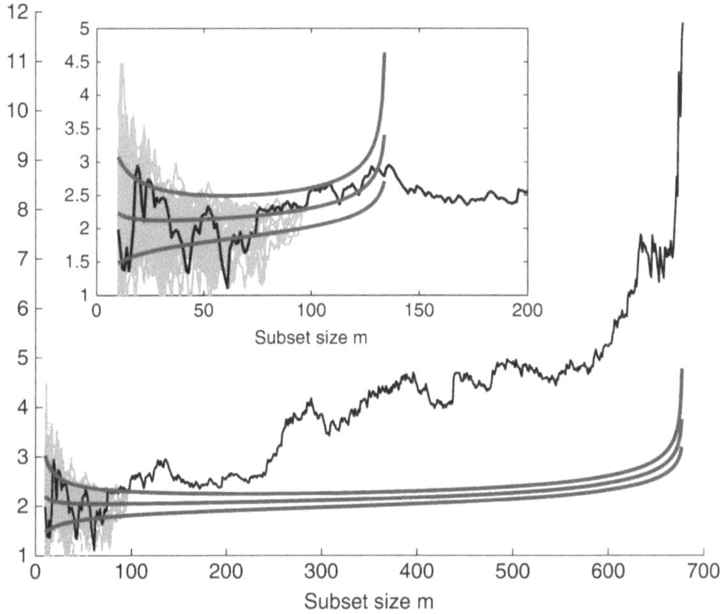

Fig. 2 Fishery dataset: 200 FS random starts and zoom on the initial part of the search (with superimposed envelopes)

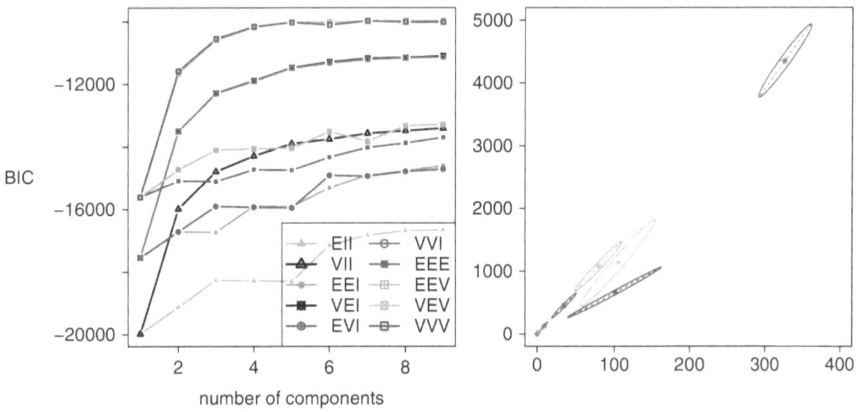

Fig. 3 Fishery dataset: BIC selection of the best MCLUST model (*left panel*) and ellipses associated with MCLUST classification (*right panel*)

of the data that are causing these issues. The left panel of Fig. 3 shows the BIC trajectories as a function of the number of groups k when using MCLUST. The highest value of BIC is obtained for $k = 8$. However, judging from the right-hand panel of the figure, which shows the ellipses associated to the eight components,

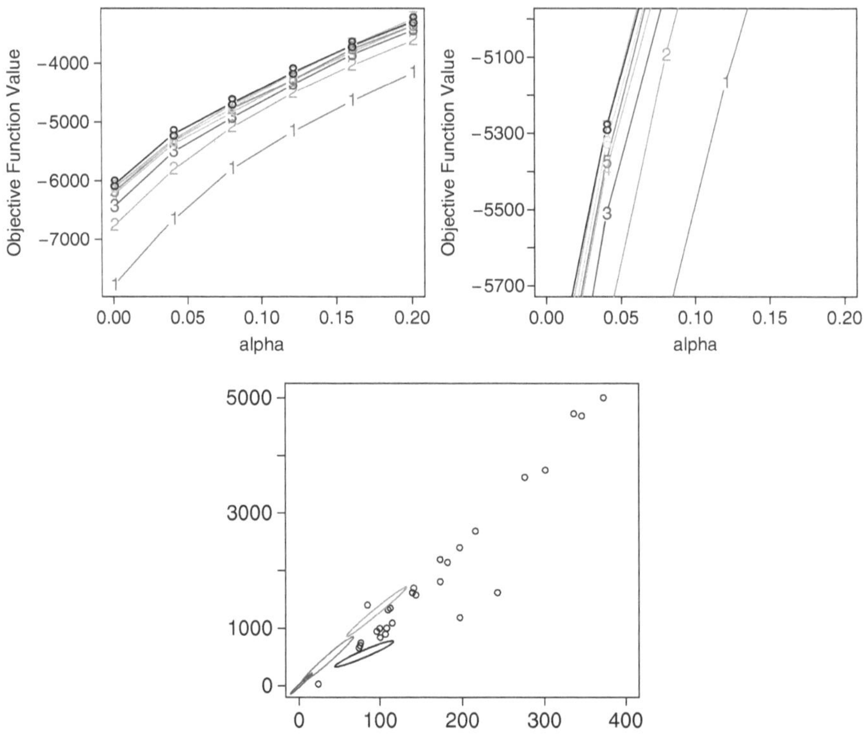

Fig. 4 Fishery dataset: Classification Trimmed Likelihood Curves for TCLUST (*upper left*), a zoom on an area of the classification plot (*upper right*) and ellipses associated with the TCLUST classification for $k = 4$ and a trimming proportion alpha $= 0.04$ (*bottom*). The unclassified data are plotted with *circles*

this number of clusters seems to be excessive. In fact the four components covering the dense area seem to be sub-samples of the same group.

Let us now consider the Gallego's algorithm (Garcia-Escudero et al., 2010), implemented in the TCLUST function in R-software, which is more sophisticated than MCLUST due to the possibility of data trimming. Based on the Trimmed Likelihood Curves shown in the top panels of Fig. 4, one should observe the correct number of clusters (say k) and the corresponding trimming level (alpha). However, the interpretation of these panels is not clear, in the sense that it is not obvious how to choose the smallest value k and alpha such that no big change in the classification trimmed likelihood curves are found when increasing from k to $k + 1$. Since here we wanted to stay on a few number of clusters and a relative small trimming, we decided in a rather arbitrary way that $k = 4$ and alpha $= 0.04$ were somehow reasonable. The ellipses associated with the four clusters produced by TCLUST method are drawn in the bottom panel of Fig. 4 together with the 4 % unclassified units shown with circles.

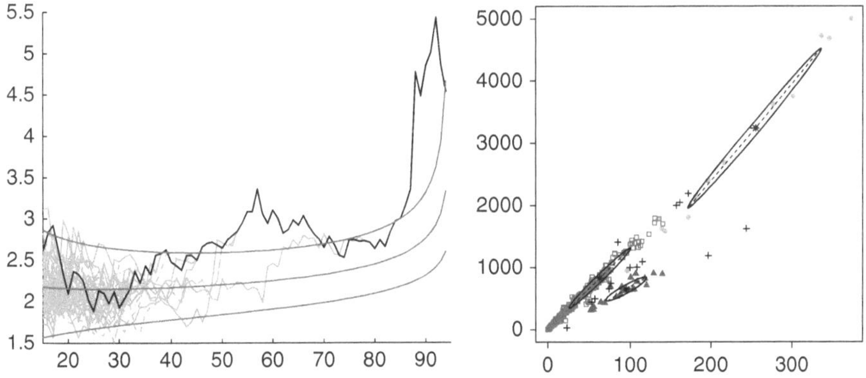

Fig. 5 Minimum deletion residual trajectories from 200 FS random starts on the gridded data (*left panel*). The final FS classification of the Fishery dataset based on centroids found on the gridded data with the FS (*right panel*)

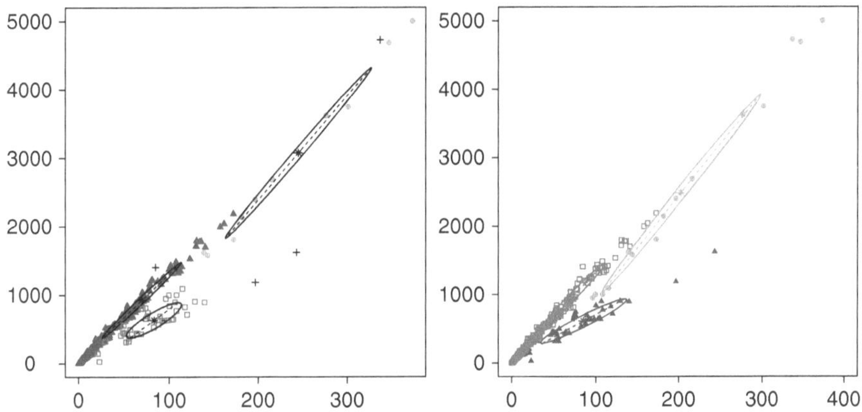

Fig. 6 Final classification of the Fishery dataset based on centroids found on the gridded data with TCLUST (*left panel*) and MCLUST (*right panel*)

When we apply the three methods to the gridded data, we obtained a more meaningful number of components as with the original data. First of all in all cases the estimated number of clusters was always three. We then classify the rest of the data based on the smallest Mahalanobis distance computed using the centroids and the variance-covariance matrices of the groups found on the gridded data. The final data classification for the three clustering methods is shown in Figs. 5 and 6. The units which remain unassigned in the FS and TCLUST are represented with the plus symbol. Compared to the previous clustering without gridding, all these new clusters are much more meaningful thanks to the data balancing of the gridding technique.

4 Conclusions

In this paper we have shown how different model-based-clustering methods are bad-performing in presence of densely populated data points with high degree of overlapping. We have therefore proposed to precede each clustering method with a technique that selects a point in each cell of a grid defined along the Principal Component axes of the data, in order to identify a sub-sample that preserves the general structure of the data, on which to apply a clustering technique.

References

Atkinson, A. C., & Riani, M. (2007). Exploratory tools for clustering multivariate data. *Computational Statistics and Data Analysis, 52*, 272–285.

Atkinson, A. C., Riani, M., & Cerioli, A. (2004). *Exploring multivariate data with the forward search*. New York: Springer.

Bishop, C. M. (2006). *Pattern recognition and machine learning*. New York: Springer.

Cochran, W. G. (1977). *Robust sampling techniques* (3rd ed.). New York: Wiley.

Fraley, C. (1998). Algorithms for model-based Gaussian hierarchical clustering. *SIAM Journal on Scientific Computing, 20*, 270–281.

Garcia-Escudero, L. A., Gordaliza, A., Matran, C., & Mayo-Iscar, A. (2008). A general trimming approach to robust cluster analysis. *Annals of Statistics, 36*, 1324–1345.

Garcia-Escudero, L. A., Gordaliza, A., Matran, C., & Mayo-Iscar, A. (2010). A review of robust clustering methods. *Advances in Data Analysis and Classification, 4*, 89–109.

Garcia-Escudero, L. A., Gordaliza, A., Matran, C., & Mayo-Iscar, A. (2011). Exploring the number of groups in robust model based clustering. *Statistics and Computing, 21*(4), 585–599.

Perrotta, D., & Torti, F. (2009). Detecting price outliers in European trade data with the forward search. In N. C. Lauro, F. Palumbo, & M. Greenacre (Eds.), *Data analysis and classification: From exploration to confirmation* (Springer studies in classification, data analysis, and knowledge organization, pp. 415–423). Berlin: Springer.

Riani, M., Cerioli, A., Atkinson, A. C., Perrotta, D., & Torti, F. (2008). Fitting robust mixtures of regression lines to European trade data. In: F. Fogelman-Soulie, et al. (Eds.), *Mining massive datasets for security applications*. Amsterdam: IOS Press.

Riani, M., Atkinson, A. C., & Cerioli, A. (2009). Finding an unknown number of multivariate outliers. *Journal of the Royal Statistical Society, Series B – Statistical Methodology, 71*, 447–466.

Riani, M., Perrotta, D. and Torti, F. (2012). FSDA: A MATLAB toolbox for robust analysis and interactive data exploration. In: *Chemometrics and Intelligent Laboratory Systems, 116*, 17–32.

Rissanen, J. (1986). Stochastic complexity and modeling. *Annals of Statistics, 14*, 1080–1100.

Schwarz, G. E. (1978). Estimating the dimension of a model. *Annals of Statistics, 6*(2), 461–464.

Dynamic Data Analysis of Evolving Association Patterns

Alfonso Iodice D'Enza and Francesco Palumbo

Abstract Dealing with large amounts of data or data flows, it can be convenient or necessary to process them in different 'pieces'; if the data in question refer to different occasions or positions in time or space, a comparative analysis of data stratified in batches can be suitable. The present approach combines clustering and factorial techniques to study the association structure of binary attributes over homogeneous subsets of data; moreover, it seeks to update the result as new statistical units are processed in order to monitor and describe the evolutionary patterns of association.

1 Introduction

The application framework of the present paper is the analysis of large date sets described by several binary attributes and stratified in different subsets of statistical units due to either the amount of data and a time/space reference of the data in question. Actually, when dealing with large amounts of data or data flows, it can be convenient or necessary to process them in different 'pieces'; if the data in question refer to different occasions or positions in time or space, a comparative analysis of data stratified in batches can be suitable.

This paper presents an approach that, through the combination of clustering and factorial techniques, aims to study the evolution of the association structure of binary attributes over homogeneous subsets of data; moreover, it seeks to update

A.I. D'Enza (✉)
Università di Cassino, Cassino, Italy
e-mail: iodicede@unicas.it

F. Palumbo
Università degli Studi di Napoli Federico II, Naples, Italy
e-mail: fpalumbo@unina.it

A. Giusti et al. (eds.), *Classification and Data Mining*, Studies in Classification, Data Analysis, and Knowledge Organization, DOI 10.1007/978-3-642-28894-4_6,

the result as new statistical units are processed in order to monitor and describe the evolutionary patterns of association.

A typical real world example of such data structures is market basket analysis (MBA) where each statistical unit is a *transaction* and the binary attributes indicate whether a product is purchased or not: here the aim is to study and monitor the attributes association structure over time. Further examples of such a kind of data structure are available in finance, environmental and social sciences.

The association study of a large number of attributes and the comparison among the different solutions obtained for each data batch represent a two-fold problem. Two different but not independent aspects have to be considered in the analysis: the former can be suitably faced by factorial techniques; the latter, the comparison among different solutions obtained for each data batch, remains the main issue in the analysis. A straightforward approach for a one-to-one comparison is to perform a supplementary projection of a data batch on the factorial structure resulting from a previous data batch; a slightly more sophisticated approach is to update the obtained solution progressively when new data batches comes in (Iodice D'Enza and Greenacre, 2010).

In this contribution the proposal is to introduce a latent categorical variable which is determined and updated at each incoming batch; in other words this variable is determined according to the association structure and represents the 'link' among the solutions. The latent categorical variable is endogenously determined by the procedure. The procedure consistency is assured by the fact that both the factorial technique and the determination of the latent variable satisfy the same criterion. To determine the latent categorical variable, a good solution consists in grouping statistical units into homogeneous groups in order to get a set of profiles that are representative of similar units.

In the literature different proposals aim to explore the relationship structure characterizing a data set through the combination of clustering procedures and factorial techniques. Dealing with continuous variables, an example of such a combined approach is *tandem analysis* proposed by Arabie and Hubert (1994): following this approach, a principal component analysis (PCA) is first applied on data and a clustering procedure is then performed on the statistical unit scores of a reduced number of components. A sequential application of PCA and clustering may not reveal the group structure underlying data, or it can even mask it: procedures suitably combining clustering with factorial analysis (FA) techniques have been proposed. Vichi and Kiers (2001) propose a combination of principal component analysis with k-means clustering method. In the framework of categorical data, another interesting approach combining clustering and multiple correspondence analysis (MCA) (Greenacre, 2007) is proposed by Hwang et al. (2006). Similarly, yet dealing with binary data, Palumbo and Iodice D'Enza (2010) propose a suitable dimension reduction and clustering.

The paper is structured as follows: in Sect. 2 the problem is introduced and formalized; furthermore the interpretation of the maximization criterion is provided. In Sect. 3 the whole procedure is described in detail. The last section shows an example of application on a large and sparse real world data set.

2 Problem Statement

In this section we provide a description of the proposed approach to study the structure of associations in binary high-dimensional data.

Let n and p be respectively the number of statistical units and the number of binary attributes Z_j $(j = 1, \dots, p)$; let K be the number of groups of homogeneous statistical units. Assigning a single statistical unit to one of K groups is a single occurrence of a multinomial experiment with K possible outcomes. The group k, $k = 1, \dots, K$, is coded via the indicator variable I_k, where $I_k = 1$ if the statistical unit is assigned to the kth group, $I_k = 0$ otherwise. Considering n trials, the random vector $X = (X_1, X_2, \dots, X_K)$ follows a multinomial distribution with parameters $(n; \pi_1, \pi_2, \dots, \pi_K)$, with $\pi_k = Pr(I_k = 1)$: n is the only known parameter. Assuming the parameters $(\pi_1, \pi_2, \dots, \pi_K)$ to be known, the optimal criterion for the allocation of the n statistical units into the K groups is to randomly assign units to groups proportionally to the corresponding π_k parameters.

In this contribution the aim is to estimate π_k (that is to optimally assign units to groups) in order to maximize the heterogeneity among groups with respect to the Z_j attributes. Each of the attributes Z_j is Bernoulli distributed (with z indicating success and $\bar{z}$ failure) distributed with parameter π_Z. According to the same criterion, new statistical units are processed in order to update both the clustering solution and the binary attribute quantifications.

We illustrate the criterion to maximize in the simple case of one single binary attribute Z, and its generalization to the p attributes case afterwards. Consider X as a qualitative variable with $k = 1, \dots, K$ categories and Z a binary variable with attributes $\{z, \bar{z}\}$.

Let $\mathbf{F}$ be the cross-classification table (represented above) with general element f_{kh} being the co-occurrence number of the categories k and h, with $h = 1, 2$; the row margin f_{k+} is the number of occurrences of the category $k (k = 1, \dots, K)$ and the column margin f_{+h} is the number of occurrences of the category $h (h = 1, 2)$, with $f_{++} = n$ being the grand total of the table. The qualitative variance, or heterogeneity, of X can be defined by the Gini index

$$G(X) = 1 - \sum_{k=1}^{K} \left(\frac{f_{k+}}{n} \right)^2 = 1 - \sum_{k=1}^{K} \frac{f_{k+}^2}{n^2}. \tag{1}$$

Within the category of z (the same occurs for $\bar{z}$) the variation is

$$G\,(X \mid z) = 1 - \sum_{k=1}^{K} \frac{f_{k1}^2}{f_{+1}^2}. \tag{2}$$

The variation of X within the categories of the variable Z is obtained by averaging $G\,(X \mid z)$ and $G\,(X \mid \bar{z})$ and it is denoted by $G\,(X \mid Z)$, formally

$$G(X \mid Z) = \sum_{h=1}^{2} \frac{f_{+h}}{n} \left(1 - \sum_{k=1}^{K} \frac{f_{kh}^2}{f_{+h}^2} \right) = 1 - \frac{1}{n} \sum_{k=1}^{K} \sum_{h=1}^{2} \frac{f_{kh}^2}{f_{+h}}. \tag{3}$$

Then the variation of X explained by the categories of Z is

$$G(X) - G(X \mid Z) = 1 - \sum_{k=1}^{K} \frac{f_{k+}^2}{n^2} - \left(1 - \frac{1}{n}\sum_{k=1}^{K}\sum_{h=1}^{2} \frac{f_{kh}^2}{f_{+h}}\right)$$
$$= \frac{1}{n}\sum_{k=1}^{K}\sum_{h=1}^{2} \frac{f_{kh}^2}{f_{+h}} - \frac{1}{n}\sum_{k=1}^{K} \frac{f_{k+}^2}{n}. \quad (4)$$

It is worth noting that the quantity in Eq. (4) can also be expressed in terms of proportional prediction that refers to the situation when statistical units are randomly assigned to the group k with probability $\frac{f_{k+}}{n}$. When additional information is provided by a variable Z, the proportional prediction becomes $\frac{f_{kh}}{f_{+h}}$, $h = 1, 2$. The average proportion of correct prediction with additional information is $\sum_{k=1}^{K} \frac{f_{k+}^2}{n^2}$ and the average proportion of correct prediction without additional information is $\sum_{k=1}^{K}\sum_{h=1}^{2} \frac{f_{kh}^2}{f_{+h}}$, (Mirkin, 2001).

The groups heterogeneity corresponds to the qualitative variance between the K levels of the X variable, then with n statistical units described by p binary attributes $Z_1, Z_2, \ldots, Z_j, \ldots, Z_p$, the quantity to maximize is

$$\sum_{j=1}^{p} \left(G(X) - G(X \mid Z_j)\right). \quad (5)$$

The expression (5) represents the sum of variances explained by each of the attributes Z_j and it is the generalization of expression (4) to the p-attributes case.

3 The Procedure

This section illustrates the procedure to alternatively determine the factorial structure that better synthesizes the multiple associations among attributes and the latent variable that is the link of the solutions of each data batch. It is worth to note that the subsequent batches must be described by the same set of attributes. It keeps the between groups heterogeneity maximized as new data batches are analysed and it consists of three phases:

1. Analysis of the starting batch: an iterative factorial clustering procedure is used to obtain the starting solution as proposed by Palumbo and Iodice D'Enza (2010);
2. Processing of upcoming batches;
3. Update of the solution according to new data.

The latter two phases are repeated for each new data batch to analyse.

The following algebraic notation will be used:

- $\mathbf{Z}(n \times 2p)$ disjunctive coded binary data matrix, with two columns per attribute (presence-absence);

Table 1 Example of two-variable cross-classification table

		Z		
		z	$\bar{z}$	
X	1	f_{11}	f_{12}	f_{1+}
	$\vdots$	$\vdots$	$\vdots$	$\vdots$
	K	f_{K1}	f_{K2}	f_{K+}
		f_{+1}	f_{+2}	n

- $\mathbf{z}_j (j = 1, \ldots, 2p)$ the general column vector of $\mathbf{Z}$;
- $\mathbf{X}(n \times K)$ a binary matrix that assigns each statistical unit to one of the K groups;
- $\mathbb{F} = \mathbf{X}^\mathsf{T}\mathbf{Z} = \left[\mathbf{F}_1, \mathbf{F}_2, \ldots, \mathbf{F}_j, \ldots, \mathbf{F}_p\right]$ is a block matrix, the jth block corresponds to the cross-tabulation of the categorical variable X with the attribute Z_j, (as described in Table 1).

A two-step iterative procedure is used to analyse the starting data batch; then new data are processed and the corresponding data structures are updated.

In algebraic terms, the criterion in expression (5) to maximize corresponds to

$$\begin{aligned} &\operatorname{tr}\left[\tfrac{1}{n}\mathbb{F}(\Delta)^{-1}\mathbb{F}^\mathsf{T} - \tfrac{p}{n^2}\left(\mathbf{X}^\mathsf{T}\mathbf{1}\mathbf{1}^\mathsf{T}\mathbf{X}\right)\right] \\ \equiv &\operatorname{tr}\left[\tfrac{1}{n}\mathbf{X}^\mathsf{T}\mathbf{Z}(\Delta)^{-1}\mathbf{Z}^\mathsf{T}\mathbf{X} - \tfrac{p}{n^2}\left(\mathbf{X}^\mathsf{T}\mathbf{1}\mathbf{1}^\mathsf{T}\mathbf{X}\right)\right] \end{aligned} \tag{6}$$

where $\Delta = diag(\mathbf{Z}^\mathsf{T}\mathbf{Z})$ is the matrix of the diagonal elements of $\mathbf{Z}^\mathsf{T}\mathbf{Z}$ and $\mathbf{1}$ is a n-dimensional vector of ones.

The solution to the problem lies in maximizing the trace of the above matrix, and the least square solution consists in the eigen-analysis of

$$\frac{1}{n}\left[\mathbf{X}^\mathsf{T}\mathbf{Z}(\Delta)^{-1}\mathbf{Z}^\mathsf{T}\mathbf{X} - \frac{p}{n}\left(\mathbf{X}^\mathsf{T}\mathbf{1}\mathbf{1}^\mathsf{T}\mathbf{X}\right)\right]\mathbf{U} = \Lambda\mathbf{U}. \tag{7}$$

A direct solution is not admissible, since both the optimal $\mathbf{X}$ and $\mathbf{U}$ must be determined: the orthonormal basis $\mathbf{U}$ depends on $\mathbf{X}$, on the other hand, the allocation matrix $\mathbf{X}$ is defined on the sub-space with basis $\mathbf{U}$. The determination of $\mathbf{X}$ and $\mathbf{U}$ maximizing the criterion (6) requires an iterative procedure that runs over the following steps:

- *step 0*: pseudo-random generation of matrix $\mathbf{X}$;
- *step 1*: eigenvalue decomposition of the matrix resulting from expression (7), obtaining the matrix

$$\Psi = \left(\mathbf{Z}(\Delta)^{-1}\mathbf{Z}^\mathsf{T} - \frac{p}{n}\mathbf{1}\mathbf{1}^\mathsf{T}\right)\mathbf{X}\mathbf{U}\Lambda^{\frac{1}{2}}; \tag{8}$$

- *step 2*: update matrix $\mathbf{X}$ according to a Euclidean squared distance-based non-hierarchical clustering algorithm (k-means) on the projected statistical units (Ψ matrix).

While the quantity in (6) increases with respect to the previous iteration, steps 1 and 2 are iterated. It is not possible to embed the optimal quantities in a single target function, then it is not possible to formally demonstrate the convergence of the iterative procedure. In fact, the convergence is testified by empirical studies on real and synthetic data sets. Results are not presented here for sake of space. Although the procedure may reach local maxima, as in K-means-like algorithms, the random multiple starts strategy can be used to identify global maxima.

From a geometrical point of view, given a clustering solution $\mathbf{X}$, the columns of matrix $\mathbf{U}$ define the orthogonal sub-space that better separates the projection of the centroids stored in the rows of the matrix $\mathbb{F} = \mathbf{X}^{\mathsf{T}}\mathbf{Z}$.

Let $\mathbf{Z}^{+}$ be a $(n^{+} \times 2p)$ matrix of a new data batch consisting of n^{+} statistical units. The new data are projected on the factorial plan defined by the orthonormal basis $\mathbf{U}$ according to the following formula

$$\Psi^{+} = \left(\mathbf{Z}^{+}(\Delta)^{-1}\mathbf{Z}^{\mathsf{T}} - \frac{p}{n}\mathbf{1}^{+}\mathbf{1}^{\mathsf{T}}\right)\mathbf{X}\mathbf{U}\Lambda^{\frac{1}{2}}. \tag{9}$$

The data points in Ψ^{+} are then assigned to the closest of the K centroids defined by the starting solution: then a $(n^{+} \times K)$ allocation matrix $\mathbf{X}^{+}$ for the new data is obtained. The update process requires the orthonormal basis $\mathbf{U}$ to be updated, too. Thus, all the following quantities are updated according to the new available information:

- $n^{*} = n + n^{+}$ is the total number of statistical units;
- $\mathbb{F}^{*} = \mathbb{F} + \mathbb{F}^{+}$, with $\mathbb{F}^{+} = \mathbf{X}^{+\mathsf{T}}\mathbf{Z}^{+}$;
- $\Delta^{*} = \Delta^{+} + \Delta$, where $\Delta^{+} = diag(\mathbf{Z}^{+\mathsf{T}}\mathbf{Z}^{+})$ is the diagonal matrix of the attribute occurrences.

The updated orthonormal basis $\mathbf{U}^{*}$ is obtained via the eigen-analysis of the following quantity

$$\frac{1}{n^{*}}\left[\mathbb{F}^{*}(\Delta^{*})^{-1}\mathbb{F}^{*\mathsf{T}} - \frac{p}{n^{*}}\left(\mathbf{f}^{*}\mathbf{f}^{*\mathsf{T}}\right)\right]\mathbf{U}^{*} = \Lambda^{*}\mathbf{U}^{*} \tag{10}$$

where $\mathbf{f}^{*}$ is the row-margin vector of the $\mathbb{F}^{*}$ matrix.

4 Example on Real Data

In this section the proposed procedure is applied to the 'retail' data set (Brijs et al., 1999). The retail market basket data set is supplied by a anonymous Belgian retail supermarket store. The data are collected over three non-consecutive periods, for a time range of approximately 5 months. The total amount of receipts (statistical units) being collected equals n = 88,163, whereas the number of products (binary attributes) p = 28,549. The data set is very sparse, and the analysis aim is to study

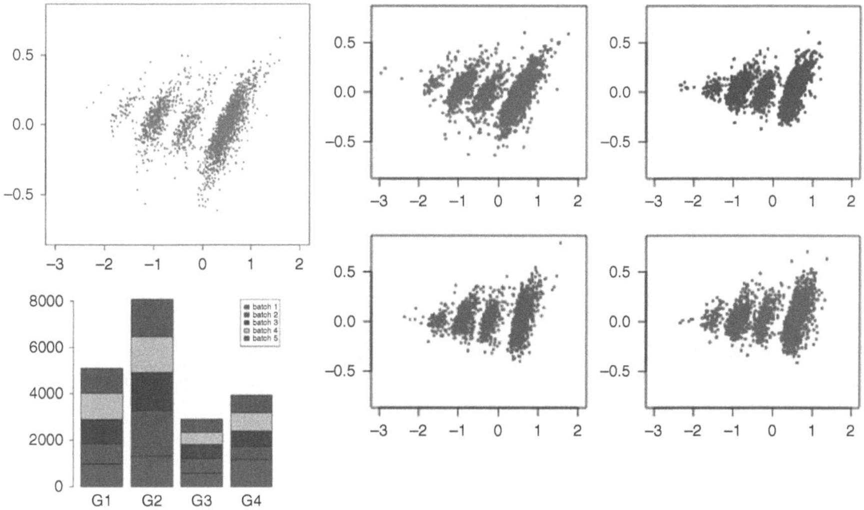

Fig. 1 Statistical units and cluster activity: starting (*top-left*) versus the four upcoming data batches (*right*, from the *top* to the *bottom*)

the association structures: in a pre-processing phase the attributes occurring in less than 1 % of statistical units are discarded, as well as the receipts with less than three products. In summary, we analyze 60 attributes observed on 20,000 statistical units. The data set is then split in five batches of 4,000 statistical units each.

The splitting criterion is due to the fact that the statistical units have been progressively recorded over time, but since we had no information on the exact date of each record, we decided to consider five batches of the same size. We consider the first 4,000 statistical units as the starting batch, then the solution is updated according to further data batches analyzed. We assume that the number of groups underlying the statistical units is $K = 4$. The top-left window in Fig. 1 shows the statistical units in the starting batch; in the bottom-left side the upcoming batch statistical units are represented. The top-right window represents the activity of the clusters: the stacked bars refer to the number of units assigned to each cluster for each of the subsequent batch. All the displays in Fig. 1 show a substantial stability in the buying behaviors of the supermarket store customers: in fact, the displays of the upcoming batches slightly differ from the starting one. This aspect is confirmed by the clusters activity: the amount of statistical units assigned to each cluster is almost proportional to the starting size of the cluster.

To appreciate the changes in the attribute associations, refer to Fig. 2. The factorial representation in Fig. 2 represents a common visualization support of the attribute associations for each of the subsequent batch. In particular, such common support is obtained by performing a multi-way multidimensional scaling (Borg and Groenen, 2005) on the chi-square distances characterizing the attributes. Figure 2 displays the trajectories of the attribute points batch after batch. The longer the

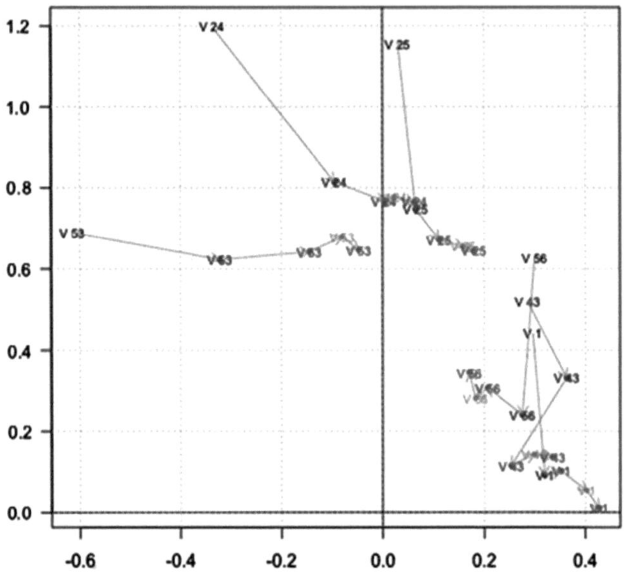

Fig. 2 Attributes representation: plot of 10 % of the longest trajectories

trajectory, the larger the change in the association structure of the corresponding attribute. In order to increase the readability of the display we plotted only the 10 % of the longest trajectories, so that the most changing attributes are highlighted. Each trajectory is represented by an arrow pointing towards the subsequent position.

The results confirm that, in general, the buying behavior of customers does not radically change from a month to another. However, some product sales do change over time and a common graphical display of the attributes (Fig. 2) turns out to be helpful in quickly identifying which attributes changed more. The proposed strategy leads to display the multiple association structure of attributes. Results are updated adaptively as new data batches are processed. Furthermore, the procedure does not require all the data batches are permanently stored in memory, but the last one.

References

Arabie, P., & Hubert, L. (1994). Cluster analysis in marketing research. *IEEE Transactions on Automatic Control, 19*, 716–723.

Borg, I., & Groenen, P. (2005). *Modern multidimensional scaling*. New York: Springer.

Brijs, T., Swinnen, G., Vanhoof, K., & Wets, G. (1999). Using association rules for product assortment decisions: A case study. In *Proceedings of the fifth ACM SIGKDD international conference on knowledge discovery and data mining*, San Diego, California, United States (pp. 254–260). New York: ACM.

Greenacre, M. J. (2007) *Correspondence analysis in practice* (2nd ed.). Boca Raton: Chapman and Hall/CRC.

Hwang, H., Dillon, W. R., & Takane, Y. (2006). An extension of multiple correspondence analysis for identifying heterogenous subgroups of respondents. *Psychometrika, 71*, 161–171.

Iodice D'Enza, A., & Greenacre, M.J. (2010). Multiple correspondence analysis for the quantification and visualization of large categorical data sets. In Proceedings of SIS09 Statistical Methods for the Analysis of Large Data-Sets, Pescara. Padova: CLEUP.

Mirkin B. (2001). Eleven ways to look at the Chi-squared coefficient for contingency tables. *The American Statistician, 55*(2), 111–120.

Palumbo F., & Iodice D'Enza A. (2010). A two-step iterative procedure for clustering of binary sequences. In: *Data analysis And classification* (pp. 50–60). Berlin: Springer.

Vichi M., & Kiers H. (2001). Factorial k-means analysis for two way data. *Computational Statistics and Data Analysis, 37*(1), 49–64.

Classification of Data Chunks Using Proximal Vector Machines and Singular Value Decomposition

Antonio Irpino, Mario Rosario Guarracino, and Rosanna Verde

Abstract Data production grows at an unprecedented increasing rate in every research and technical field. Furthermore, with the explosion of sensors networks and proprietary/legacy classifiers, like those used by banks for assessing the credit approvals, the data production and modeling is done locally, where only the local classifiers are available. In order to find a global classification rule, the ensemble classification paradigm proposes several methods of aggregation. In this paper, starting from a set of classifiers obtained by using a recently developed classification technique, known as Regularized Generalized Eigenvalues Classifier, we present a novel way of aggregating linear classification models using the Singular Value Decomposition. Using artificial datasets, we compare the developed algorithm with a voting scheme, showing that the proposed strategy allows a reduction in computational cost with a classification accuracy that well compares with the original method.

1 Introduction

Classification refers to the capability of a system to learn from examples how to discriminate cases in two or more given classes. The system learns from a set of cases, usually referred as the *training set*. Each case is described by a set of

A. Irpino (✉) · R. Verde
Dipartimento di Studi Europei e Mediterranei, Seconda Università degli Studi di Napoli, Caserta, Italy
e-mail: antonio.irpino@unina2.it; rosanna.verde@unina2.it

M.R. Guarracino
High Performance Computing and Networking, National Research Council of Italy, Naples, Italy
Center for Applied Optimization, University of Florida, Gainesville, FL, USA
e-mail: mariog@ufl.edu

A. Giusti et al. (eds.), *Classification and Data Mining*, Studies in Classification, Data Analysis, and Knowledge Organization, DOI 10.1007/978-3-642-28894-4_7,

variables and a class label. For each new case, the trained system predicts its class label. If we limit to only two classes, the problem is called *binary classification*; in all other cases it is named *n-class* or *multiclass classification*. Support Vector Machines (SVM) (Vapnik, 1995) are among the most used techniques in supervised learning. Recently, the Regularized Generalized Eigenvalue Classifier (ReGEC) extension has been proposed to solve binary and multiclass classification problems (Irpino et al., 2010).

As data production grows at an unprecedented increasing rate in every technological and research field, and we witness the explosion of ubiquitous sensors networks, it becomes impossible to store data for later processing. A plausible solution for analyzing such data is the use and development of incremental or distributed algorithms that can handle streams of data. In the case of classification algorithms, the aforementioned strategies furnish partial models of classification that subsequently need to be fused (Sinha et al., 2008). Let us consider the classification problem in a distributed computational paradigm, characterized by a set of R sensors collecting data that are labeled according to K classes. Each sensor can only process and communicate a limited amount of data, due to its memory, computational and energy consumption characteristics. Therefore, only resulting models are exchanged among sensors, while data are processed locally and then discarded. The situation is reminiscent of privacy preserving systems, as those used in banks and other financial institutes, processing data and communicating only their classification models. An example could be an insurance company and a bank which exchange their models for identifying good and bad customers, maintaining privacy on personal data. In such a scenario, to merge the classification models, we need to set up an ensemble classification system, since it is not possible to share training sets, but only classification models. When classification models are based on linear functions, for separating (e.g. SVM) or representing (e.g. ReGEC) classes, we propose a novel technique for merging such models. Let Ω be a dataset of N labeled data (i.e. classified into $K > 1$ classes), partitioned into R data chunks, each one of cardinality N_r, and described by P explicative variables. We propose a strategy for merging the classification models for each class in each chunk using the Singular Value Decomposition. This is actually a form of classifiers fusion, that, to our knowledge has never been described before in literature.

In the present work, we detail the computational advantages of such strategy: it needs less computational resources to reach a classification performance, in terms of accuracy, comparable with the original algorithms. The paper is organized as follows. In Sect. 2 we introduce the main classification algorithm based on vector machines. In Sect. 3 we propose a method, based on Singular Value Decomposition, for fusing all the local models into a single one. In Sect. 4, we show the performance and compare the proposed strategy in terms of accuracy and execution time on artificial datasets.

2 ReGEC and Vector Machines Based Classification Methods

SVM are powerful classification and regression techniques, but their computational and storage requirements rapidly increase with the number of training vectors, putting many problems of practical interest away from their reach. The core of an SVM is a quadratic programming problem, separating support vectors from the rest of the training data.

In case of two linearly separable classes, SVM finds a hyperplane that separates the elements belonging to two different classes. The separating hyperplane is usually chosen to maximize the margin between the two classes. The margin can be defined as the maximum distance between two parallel boundary hyperplanes $x'w - \gamma = \pm 1$ that leave all cases of the two classes on different sides. The classification hyperplane $x'w - \gamma = 0$ is midway from the boundary planes. The points that are closest to the hyperplane are called *support vectors*, and are the only points needed to train the classifier.

In the Proximal Support Vector Machines (PSVM) classification (Fung and Mangasarian, 2001; Suykens et al., 2002), two parallel planes are generated so that each plane is the closest to the points of one of the two classes and as far apart as possible from the points of the other class. The classifying plane is again midway between the parallel proximal planes. PSVM find two parallel planes $x'w - \gamma = \pm 1$ such that the points of the two classes are clustered around these two planes, and the reciprocal of the two-norm of the distance of two planes in the (w, γ) space of $\Re^{P+1}$ is minimum.

Mangasarian and Wild (2004) propose to generalize previous technique and to classify two classes of points using two non parallel planes, each the closest to one set of points, and the furthest from the other. Let A and B be the matrices containing on each row a training point, and $x'w - \gamma = 0$ be a hyperplane in $\Re^P$. In order to satisfy the previous condition for all points in A, the plane can be obtained by solving the following optimization problem:

$$\min_{w,\gamma \neq 0} \frac{\|Aw - e\gamma\|^2}{\|Bw - e\gamma\|^2}. \tag{1}$$

The hyperplane for cases in B can be obtained by minimizing the inverse of the objective function in (1). Now, let

$$G = [A \quad -e]'[A \quad -e], \quad H = [B \quad -e]'[B \quad -e], \quad z = [w' \ \ \gamma]', \tag{2}$$

where $[A \quad -e]$ is the matrix obtained from A adding the column vector $-e$, composed of -1s, of proper dimension. Using (2), Eq. (1) becomes:

$$\min_{z \in R^P} \frac{z'Gz}{z'Hz}. \tag{3}$$

The expression (3) is the Raleigh quotient of the generalized eigenvalue problem $Gx = \lambda Hx$. The stationary points are obtained at, and only at, the eigenvectors of (3), and the value of the objective function (1) is given by the corresponding eigenvalues. When H is positive definite, the Raleigh quotient is bounded and it ranges over the interval determined by minimum and maximum eigenvalues (Parlett, 1998). H is positive definite under the assumption that the columns of $[B \; -e]$ are linearly independent.

The inverse of the objective function in (3) has the same eigenvectors and reciprocal eigenvalues. Let $z_{min} = [w_1' \quad \gamma_1]'$ and $z_{max} = [w_2' \quad \gamma_2]'$ be the eigenvectors related to the smallest and largest eigenvalues, respectively. Then, $x'w_1 - \gamma_1 = 0$ is the closest hyperplane to the set of points in A and the furthest from those in B, and $x'w_2 - \gamma_2 = 0$ is the closest hyperplane to the set of points in B and the furthest from those in A. This method uses a single hyperplane to describe each class. It is worth noting that the hyperplanes in general are not parallel, as for PSVM. Since a point is assigned to a class based on its distance from the corresponding hyperplane, the method can also classify problems that are not linearly separable.

Mangasarian and Wild (2004) propose to use Tikhonov regularization applied to a two-fold problem. The use of the proposed regularization requires the solution of two separate generalized eigenvalue problems. Guarracino et al. (2007) propose a new regularization that only needs the solution of one generalized eigenvalue problem.

3 ReGEC for Data Chunks: SVD Based Ensemble Classifier

Suppose to have a training set T of N elements labeled by K class identifiers, described by P explicative variables that is partitioned into R chunks. Each chunk r contains N_r training units with K unique class labels described by the same variables. For each chunk r ($r = 1, \ldots, R$), the local ReGEC computes the z_k^r ($k = 1, \ldots, K$) models. Each model z_k^r is a hyperplane described by a linear equation with $P + 1$ coefficients as follows:

$$z_k^r \; : \; w_{k1}^r x_1 + w_{k2}^r x_2 + \ldots + w_{kP}^r x_P + \gamma_k^r = 0. \tag{4}$$

For each class, we have a set of R models. In order to obtain a single classifier z_k we propose to average the R models to obtain a new hyperplane that is closest to all of them. In other words, we search for a hyperplane that maximizes the sum of the cosines of the angles among all the local models. Consider the normal vector n_k^r with P components associated to each hyperplane:

$$\mathbf{n}_k^r = [w_{k1}^r \; w_{k2}^r \; \ldots \; w_{kP}^r]. \tag{5}$$

To have a similar contribution from all planes, we normalize all hyperplanes obtaining the following normalized models $\hat{z}_k^r$:

$$\hat{z}_k^r \; : \; \hat{w}_{k1}^r x_1 + \hat{w}_{k2}^r x_2 + \ldots + \hat{w}_{kP}^r x_P + \hat{\gamma}_k^r = 0, \quad r = 1, \ldots, R. \tag{6}$$

where

$$\hat{w}^r_{kj} = \frac{w^r_{kj}}{\|\mathbf{n}^r_k\|}(j = 1, \ldots, p), \hat{\gamma}^r_k = \frac{\gamma^r_k}{\|\mathbf{n}^r_k\|}.$$

After the normalization, we set up a matrix $\widehat{\mathbf{W}}_k$ where each row contains the normal vector:

$$\widehat{\mathbf{W}}_k = \begin{bmatrix} \hat{w}^1_{k1} & \ldots & \hat{w}^1_{kP} \\ \ldots & \ldots & \ldots \\ \hat{w}^R_{k1} & \ldots & \hat{w}^R_{kP} \end{bmatrix} \tag{7}$$

The matrix $\widehat{\mathbf{W}}_k$ is a set of R unitary row vectors. Finding the vector that maximizes the sum of cosines of the angles among all the vectors is equivalent to determining the vector with maximum correlation. To this extend, we use the Singular Value Decomposition for factorizing the matrix $\widehat{\mathbf{W}}_k$:

$$\widehat{\mathbf{W}}_k = \mathbf{U}_k \Lambda_k \mathbf{V}'_k \tag{8}$$

The left eigenvector $\mathbf{u}^1_k = [u^1_{k1} \ldots u^1_{kP}]$ associated with the largest eigenvalue λ^1_k is the normal vector of a new hyperplane that is closer (in the sense of the cosines sum) to all the R hyperplanes. Choosing a single average hyperplane is consistent with the base ReGEC algorithm, in which a single hyperplane describes each class. To compute the constant term γ^{1*}_k of the hyperplane, let us consider the column vector $\hat{\gamma}_h = [\hat{\gamma}^1_k \ldots \hat{\gamma}^R_k]'$ and the right first eigenvector $\mathbf{v}^1_k = [v^1_{k1} \ldots v^1_{kR}]$:

$$\gamma^{1*}_k = \frac{1}{\lambda^1_k} \hat{\boldsymbol{\gamma}}_k \mathbf{v}^1_k. \tag{9}$$

Then the average (normalized) hyperplane for the generic class k is the following:

$$\hat{z}^1_k \ : \ u^1_{k1} x_1 + u^1_{k2} x_2 + \ldots + u^1_{kP} x_P + \gamma^{1*}_k = 0. \tag{10}$$

In a similar way, it is possible to compute a plane for each singular value. In such a case, it is worth noting that the rows of $\widehat{\mathbf{W}}_k$ have unitary length, the matrix $\mathbf{M} = \widehat{\mathbf{W}}_k \widehat{\mathbf{W}}'_k$ has the same characteristics of a correlation matrix: its trace is invariant and equals R, thus:

$$trace\,(\Lambda_k)^2 = \sum_{j=1}^{rank(\widehat{\mathbf{W}}_k)} \left(\lambda^j_k\right)^2 = R. \tag{11}$$

If we choose to represent a class a single average hyperplane, the assignment of a test point x is assigned to the class related to the closest hyperplane as described by Guarracino et al. (2007). When more than one average model is used to represent a class, the assignment of a test point x can be obtained computing its weighted distance from all planes related to the singular values:

$$d(x, Class_k) = \sqrt{\frac{\sum_{j=1}^{rank(\widehat{\mathbf{W}}_k)} \left(\lambda^j_k\right)^2 d^2\left(x, \hat{z}^j_k\right)}{R}} \tag{12}$$

where $\hat{z}_k^j$ is the average plane associated to λ_k^j. It is clear that the contribution to the distance of a plane related to a null eigenvalue is zero.

The computational complexity of the proposed method is lower then the one of the original method. Indeed, in the original algorithm, building the matrices G and H costs $O(PN^2)$ and the solution of the generalized eigenvalue problem $O(P^3)$, with an asymptotic complexity of $O(PN^2 + P^3)$ and $P << N$. On the other hand, the training from a chunk of dimension N/R points the overall complexity is $O(PN^2/R^2 + P^3 N/R)$, with an advantage in execution time, as it will be clarified in the following numerical examples.

4 Experiments

The algorithm has been implemented with Matlab (R2007a)[1]. Results are calculated using an Intel Q9550 CPU 2.83 GHz, 4 GB RAM running Windows Vista. Matlab function *eig* for the solution of the generalized eigenvalue problem has been used as computational kernel of ReGEC.

To validate the strategy, experiments on synthetic data, with different number of chunks, are considered. The synthetic dataset consists of 300,000 points in $\Re^2$ classified into three classes (100,000 points for each class). They are generated by three bivariate normal distributions that moderately overlap, distributed as follows:

$$\mathit{Class}\ 1 \sim N\left(\begin{bmatrix}0\\2\end{bmatrix}, \begin{bmatrix}1 & -1\\-1 & 3\end{bmatrix}\right) \quad \mathit{Class}\ 2 \sim N\left(\begin{bmatrix}1\\6\end{bmatrix}, \begin{bmatrix}5 & 1\\1 & 2\end{bmatrix}\right)$$

$$\mathit{Class}\ 3 \sim N\left(\begin{bmatrix}5\\-1\end{bmatrix}, \begin{bmatrix}0.5 & 0.3\\0.3 & 5.0\end{bmatrix}\right).$$

The data for the experiments are represented in Fig. 1.

We set up five experimental conditions with 10, 50, 100, 500 and 1,000 chunks that are chosen from the dataset. We show the results of classification using a ten-fold cross validation. The chunking operation is done on the training set. Then, the global model is computed and the test set is used for classification accuracies. We have also considered a majority voting (MV) scheme in order to compare the proposed strategy with a well grounded method of fusion of classifiers: a test point is assigned to the class with the highest number of votes for all the chunks.

The chunks are chosen according to two schemes. In the first, we consider a random partition of the training set into R chunks. This partitioning schema implicitly assumes that the generator distribution of data is the same for each chunk. The second partition schema uses k-means algorithm to cluster each class of the training set into R clusters, then using each cluster as a data chunk. This partition schema assumes data distribution can be different in each chunk. In Table 1 we report the execution time of the algorithm to build the local classification models

[1](www.mathworks.com/products/matlab/)

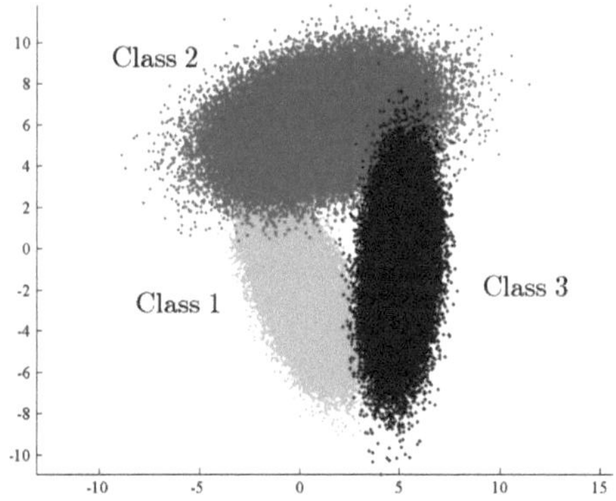

Fig. 1 The synthetic dataset

Table 1 Execution times in milliseconds with increasing data chunks

# of chunks	Mean time	Merging time
1 (Base algorithm)	–	238.9
10	15.8	0.5
50	4.1	1.3
100	2.1	3.2
500	1.2	37.6
1,000	1.1	111.1

and to average them in a global one. The execution time does not take into account the time to generate the chunks, and it is the same for both schemes.

The first row of Table 1 represents the execution time of the ReGEC algorithm on all data, and therefore there is no merging. We note that for an increasing number of chunks, the mean execution time for building a local mode decreases, as expected, and the merging operation takes longer, resulting in a lower overall execution time, with respect to the base ReGEC algorithm applied to the whole training set.

In Table 2 we report the accuracies for the random partitioning and k-means partitioning schemes used for generating the chunks. Regarding classification accuracy, when the partitioning schema is random, the accuracy is 96.4 %, and as expected remains almost constant for all experiments and for the three classification rules. Indeed, the data distribution in each chunk is the same as in the complete training set. In this case, for the prediction of new data we can use one average hyperplane for each class.

In the latter case (the K-means partitioning scheme), the distribution of data in each chunk is not the same as in the training set. Observing the accuracies, the use of a single average hyperplane for each class gives lower accuracies with respect the MV scheme. The strategy of using two average hyperplanes for each class is the best according to the accuracy and to the involved computational resources. Even if the accuracy is slightly better than the MV fusion strategy, the system needs to store only two models for each class and compute only two distances for each class (and a step for averaging them), while the voting MV the system needs to store all the

Table 2 Random and K-means partitioning schemes, 1p (one av. plane), 2p (two av. planes) and MV (Majority Voting). Accuracy of ten-fold cross validation for different number of chunks and partitioning scheme. In bold the best accuracies for each partitioning scheme and no. of chunks

	Random partition			K-means partition		
Chunks	1p SVD(%)	2p SVD(%)	MV(%)	1p SVD(%)	2p SVD(%)	MV(%)
1	96.4	–	–	96.4	–	–
10	**96.3**	**96.3**	**96.3**	90.4	**98.5**	97.0
50	**96.3**	**96.3**	**96.3**	93.8	**98.7**	98.3
100	**96.3**	**96.3**	**96.3**	94.6	**99.0**	98.7
500	96.3	**96.4**	96.3	94.7	**99.0**	98.8
1,000	96.3	**96.4**	96.3	94.8	**99.0**	98.8

hyperplanes of each chunk and to compute the corresponding distances (and a step for counting the votes).

5 Conclusions

We have presented a novel technique for merging partial classification models, based on Proximal Vector Machines. The method is based on the idea of computing a model that minimizes the distance among the partial models using singular value decomposition. Results show that, for normal distributions of data in each class, chunking data provides an accurate solution to the problem, with competitive execution time. In future, we will investigate the possibility of introducing incremental techniques to handle distributed data streams with non-stationary distributions.

References

Fung, G., & Mangasarian, O. L. (2001). Proximal support vector machine classifiers. In *Proceedings of the seventh ACM SIGKDD international conference on knowledge discovery and data mining (KDD '01)*. ACM, New York, NY, USA (pp. 77–86).

Guarracino, M. R., Cifarelli, C., Seref, O., & Pardalos, P. (2007). A classification algorithm based on generalized eigenvalue problems. *Optimization Methods and Software, 22*(1), 73–81.

Irpino, A., Guarracino, M. R., & Verde, R. (2010) Multiclass generalized eigenvalue proximal support vector Machines. In: *4th IEEE Conference on Complex, Intelligent and Software Intensive Systems (CISIS 2010)*, Krakow (pp. 25–32).

Mangasarian, O., & Wild, E. (2004). *Multisurface proximal support vector classification via generalized eigenvalues* (Tech. Rep. 04-03). Data Mining Institute.

Parlett, B. (1998). *The symmetric eigenvalue problem*. Philadelphia: SIAM.

Sinha, A., Chen, H., Danu, D. G., Kirubarajan, T., & Farooq, M. (2008). Estimation and decision fusion: A survey. *Neurocomputing, 71*(13–15), 2650–2656.

Suykens, J. A. K., Van Gestel, T., De Brabanter, J., De Moor, B., & Vandewalle, J. (2002). *Least squares support vector machines*. Singapore: World Scientific.

Vapnik, V. (1995). *The nature of statistical learning theory*. New York: Springer.

Correspondence Analysis in the Case of Outliers

Anna Langovaya, Sonja Kuhnt, and Hamdi Chouikha

Abstract Analysis of categorical data by means of Correspondence Analysis (CA) has recently become popular. The behavior of CA in the presence of outliers in the table is not sufficiently explored in the literature, especially in the case of multidimensional contingency tables. In our research we apply correspondence analysis to three-way contingency tables with outliers, generated by deviations from the independence model. Outliers in our work are chosen in such a way that they break the independence in the table, but still they are not large enough to be easily spotted without statistical analysis. We study the change in the correspondence analysis row and column coordinates caused by the outliers and perform numerical analysis of the outlier coordinates.

1 Introduction

Correspondence Analysis has proven itself to be one of the most popular and important tools in statistical analysis of data in psychology and social sciences (Blasius, 2001; Greenacre, 1984). In CA as well as in every statistical analysis, observations can appear that seem to deviate strongly from the majority of the data. Such observations are usually called outliers and may contain important information about unknown irregularities, dependencies and interactions within the data.

As concerns correspondence analysis, outliers are given by specific cell frequencies of the underlying contingency table. Situations can occur where outliers are present in the table, which are not immediately suspicious, but play a crucial role for the statistical analysis. In such cases our approach will be useful.

A. Langovaya (✉) · S. Kuhnt · H. Chouikha
TU Dortmund University, Dortmund, Germany
e-mail: langovaya@statistik.tu-dortmund.de; kuhnt@statistik.tu-dortmund.de; chouikha@statistik.uni-dortmund.de

A. Giusti et al. (eds.), *Classification and Data Mining*, Studies in Classification, Data Analysis, and Knowledge Organization, DOI 10.1007/978-3-642-28894-4_8,

In our research we apply CA to three-way contingency tables (Blasius and Greenacre, 2006; Kroonenberg, 2007) with data entries generated by deviations from the independence model. Specific dependencies are caused by outliers of moderate size. Moderate means that values of this outliers are not as large to be recognized in the table because of their size alone but still large enough to cause some irregularities in the data. In the present work, outliers are rather independence breakers of considerable size rather than merely unlikely large elements of the table. We study the change in CA row and column coordinates, caused by moderate outliers.

In Sect. 2, we introduce the notation and describe the initial steps of our analysis of three-way tables. In Sect. 3 we propose a model to generate outliers in contingency tables. Sect. 4 describes our simulation study and contains the main results of the present paper. We conclude with some final remarks in Sect. 5.

2 Notation

In this paper we will be studying the change in the CA row and column coordinates caused by outliers. Our focus is on three-dimensional tables. In the literature on CA the problem of the presence of outliers in contingency tables has not been explored. In this paper we try to figure out, how the CA behaves under the influence of moderate outliers.

Let X_1, X_2, X_3 be categorical random variables, and let $\mathscr{X} = \{1, \ldots, I\} \times \{1, \ldots, J\} \times \{1, \ldots, K\}$ be the set of their categories. In general, X_1, X_2 and X_3 are not assumed to be independent. Denote as n_{ijk} the observed frequencies of the event $(X_1 = i, X_2 = j, X_3 = k)$, where $i = 1 \leq i \leq I$, $j = 1 \leq j \leq J$, $k = 1 \leq k \leq K$. Define

$$n_{i\cdot\cdot} = \sum_j \sum_k n_{ijk}\ , \qquad n_{\cdot j\cdot} = \sum_i \sum_k n_{ijk}\ , \qquad n_{\cdot\cdot k} = \sum_i \sum_j n_{ijk}\ . \tag{1}$$

Therefore, $n_{i\cdot\cdot}, n_{\cdot j\cdot}, n_{\cdot\cdot k}$ are marginal sums corresponding to the variables X_1, X_2, X_3. Let N_{ijk} be the random variable corresponding to cell (i, j, k) of a three-dimensional contingency table. Denote the sample size by $n = \sum_{i,j,k} n_{ijk}$.

Suppose first that we are interested in testing the null hypothesis H_0: X_1, X_2, X_3 are completely independent.

Let π_{ijk}, for all $i = 1 \leq i \leq I$, $j = 1 \leq j \leq J$, $k = 1 \leq k \leq K$, denote the joint probability corresponding to the cell (i, j, k) (and the random variable N_{ijk}). Then $0 < \pi_{ijk} < 1$, $\forall i, j, k$, and

$$\sum_i \sum_j \sum_k \pi_{ijk} = \sum_i \pi_{i\cdot\cdot} = \sum_j \pi_{\cdot j\cdot} = \sum_k \pi_{\cdot\cdot k} = 1, \tag{2}$$

where we denote

Table 1 Two-dimensional slice of a three-dimensional contingency table

		X_3				Sum
X_1	X_2	1	2	...	K	
	1	n_{111}	n_{112}	n_{11k}	n_{11K}	$n_{11\cdot}$
1	$\vdots$	n_{1j1}	n_{1j2}	$\ddots$	n_{1jK}	
	J	n_{1J1}	n_{1J2}	n_{1Jk}	n_{1JK}	
	1	n_{i11}				
$\vdots$	$\vdots$		$\ddots$	n_{ijk}		$n_{ij\cdot}$
	J				n_{iJK}	
	1	n_{I11}				
I	$\vdots$			$\ddots$		
	J				n_{IJK}	$n_{IJ\cdot}$
Sum		$n_{\cdot\cdot 1}$	$\cdots$	$n_{\cdot\cdot k}$	$n_{\cdot\cdot K}$	n

$$\pi_{i\cdot\cdot} = \sum_j \sum_k \pi_{ijk}\,, \qquad \pi_{\cdot j\cdot} = \sum_i \sum_k \pi_{ijk}\,, \qquad \pi_{\cdot\cdot k} = \sum_i \sum_j \pi_{ijk}\,. \tag{3}$$

Clearly, $\pi_{i\cdot\cdot}, \pi_{\cdot j\cdot}, \pi_{\cdot\cdot k}$ are the marginal probabilities. With this notation, the null hypothesis H_0 of complete independence is equivalent to

$$\pi_{ijk} = \pi_{i\cdot\cdot}\pi_{\cdot j\cdot}\pi_{\cdot\cdot k}, \quad \forall i, j, k. \tag{4}$$

There is no canonical way of performing CA for three or more dimensional tables. Essentially, all the methods in multidimensional CA use transformations of multidimensional matrices into two-dimensional matrices, which in some cases are larger than the multidimensional ones. Subsequently, the usual CA (or CA with some restrictions or/and assumptions) is applied to the transformed matrices.

In this paper, we consider the following transformation of the initial three-dimensional table as given in Table 1.

For complete analysis of the three-way table, one should consider all combinations of such two-dimensional slices of partial dependencies of these three variables (X_1, X_2, X_3). To give the main idea, but in order to save space, we consider here just one type of such two-dimensional slices (shown above).

To apply CA, we construct from the above transformed table a matrix of standardized residuals **S**, with dimensions $(I \cdot J) \times K$ and elements

$$s_{(ij)k} = \frac{(n_{ijk}/n) - r_{(ij)}c_k}{\sqrt{r_{(ij)}c_k}}, \tag{5}$$

where $c_k = n_{\cdot\cdot k}/n$ and $r_{(ij)} = n_{ij\cdot}/n$ are the weighted marginal sums of columns and rows respectively. We define additionally two diagonal matrices $\mathbf{D_r} = diag(r_{11}, \ldots, r_{IJ})$ and $\mathbf{D_c} = diag(c_1, \ldots, c_K)$.

Suppose that the matrix **S** is such that the singular value decomposition (SVD) of this matrix: $\mathbf{S} = \mathbf{U}\Sigma\mathbf{V}^T$ is well defined, i.e. there are no rows and columns, consisting of zeros only. Then we are provided with all necessary components to derive the coordinates of rows and columns in CA-plot. The following matrices give us coordinates for symmetric CA-plots that we consider in the current paper:

- Principal coordinates of rows: $\mathbf{F} = \mathbf{D_r}^{-\frac{1}{2}}\mathbf{U}\Sigma$
- Principal coordinates of columns: $\mathbf{G} = \mathbf{D_c}^{-\frac{1}{2}}\mathbf{V}\Sigma$.

3 Outliers

A matter of particular interest now is the behavior of CA when the independence is only violated by individual cell counts. These specific cell frequencies, generated as deviation from the independence model, can be interpreted as outliers. Outliers for contingency tables in general have been considered before (Kuhnt, 2004; Shane and Simonoff, 2001; Barnett and Lewis, 1984) but rarely with constructive outlier generating model.

3.1 Generation of Outliers

In this paper outliers are understood as *specific cell frequencies* – "unusual" observations in the underlying contingency table. *Specific* – in the sense of deviation from the assumed null model. The independence model assuming a multinomial distribution is taken to be the null model. To begin with we consider the partial independence model corresponding to our transformed two-dimensional table.

The outlier generating model for the case of multinomially distributed entries could be defined as follows. For the $(N_l)_{l\in L} \sim Multinomial(n, (\pi_l)_{l\in L})$, the random variables $(N_l)_{l\in L^\star}$, with parameters $(\pi_l)_{l\in L^\star}$, where $L^\star \subset L$, are **outliers** with respect to the given model class, if for$(N_l)_{l\in L\setminus L^\star}$ with $(\pi_l)_{l\in L\setminus L^\star}$, there exists $(\tilde{\pi}_l)_{l\in L}$ from the model class, and a normalizing constant $c \in \mathbb{R}$ such, that $c(\tilde{\pi}_l)_{l\in L\setminus L^\star} = (\pi_l)_{l\in L\setminus L^\star}$, but $c(\tilde{\pi}_l)_{l\in L^\star} \neq (\pi_l)_{l\in L^\star}$ in every component.

In our research we focus on outliers of moderate size, occurring as cell counts generated from an outlier model as specified above. We explore afterwards the CA-coordinates.

4 Simulation Study

To investigate the impact that moderate outliers have on coordinates in CA-plots we have executed the following simulation study.

First, we randomly generate marginal probabilities $\pi_{i\cdot\cdot}, \pi_{\cdot j\cdot}, \pi_{\cdot\cdot k}$, based on which the probability matrix **P** for the case of independence is constructed with the joint probabilities as the product of the respective marginal probabilities: $\pi_{ijk} = \pi_{i\cdot\cdot}\pi_{\cdot j\cdot}\pi_{\cdot\cdot k}$. Hence, we don't consider a single independence model but simulate from the whole class of independence models.

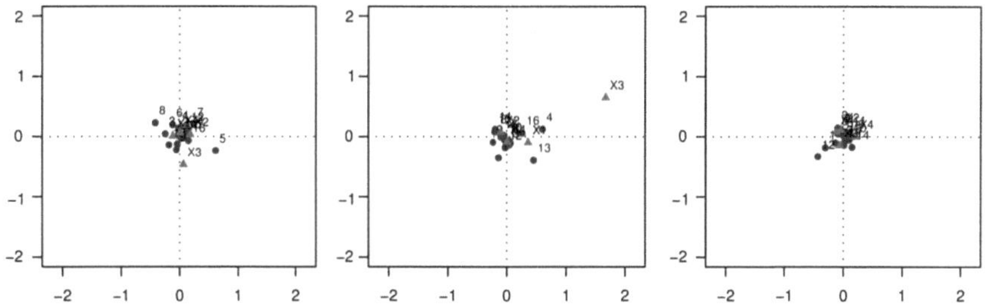

Fig. 1 Three CA-plots for the independence case

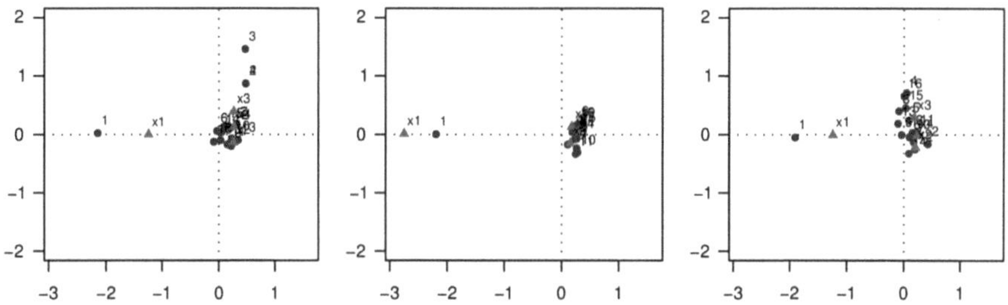

Fig. 2 Three CA-plots for the case of the outlier in the first cell

After that we simulate a matrix of observations $(\mathrm{X}^{(i,j,k)})$ for $n = 1{,}000$ observations from *Multinomial*$(n, (\pi_l)_{l=1,\cdots,IJK})$ with $I = J = K = 4$. The CA (**R**-Package: '**ca**', (Nenadić and Greenacre, 2007)) is then applied to **X** (Fig. 1). To generate an outlier in the table, one π_{ijk} is replaced by $(1,2) * \max(\mathbf{P})$. For example, for the outlier in the first cell, we replace the probability π_{111} of **P** by the value mentioned above. Afterwards all joint probabilities of the matrix **P** will be rescaled so that $\sum_{i,j,k} \pi_{ijk} = 1$.

Then CA is applied to the observation matrix $(\mathrm{X}^{(i,j,k)})$, based on this new probability matrix. Since examples (see Figs. 1 and 2)[1] already suggest that the CA-plots differ dramatically in the case of independence compared to the case of the model with an outlier, it is sensible to analyze the row and column coordinates of the CA-plot.

In Fig. 2 we can see that the coordinates of the outlier, which is placed in the first cell (i.e. first row (blue circle 1) and first column (red triangle X1) simultaneously), are located pretty far from the set of coordinates of other points (rows and columns of the underlying table). Whereas in the case of independence, all points are concentrated around the origin. The tables with an outlier (in the first cell of the table, marked with red), corresponding to the CA-plot on the Fig. 2, show that although these values (outliers) break the independence in the table, they are not particularly conspicuous (Fig. 3).

[1]The printed version of the paper contains only black-white pictures. Coloured versions of this pictres are available from the authors upon request.

Fig. 3 Three tables for the case of the outlier in the first cell

```
[[1]]
      X1 X2 X3 X4
,,1   62  0  0  1
       0  0  0  2
       0  0  1  2
       0  0  0  1

,,2    4 22  0 16
      19 45  7 32
       5 42  7 39
      15 50  6 23

,,3    4 25  1 13
      12 73  8 41
      12 60  4 29
      12 58 11 43

,,4    1 11  0  9
      10 29  4 21
       9 31  7 22
       8 19  1 11
```

```
[[2]]
      X1 X2 X3 X4
,,1   61 20 15  4
       1 64 47 39
       0 21 20 14
       3 53 51 41

,,2    0 11 10  6
       0 46 34 31
       0 20 10 11
       0 27 34 37

,,3    0  3  3  2
       0 11  3  9
       0  4  2  1
       0  5  6  5

,,4    1 11  6  7
       0 32 28 16
       0  7  7  7
       0 31 40 22
```

```
[[3]]
      X1 X2 X3 X4
,,1   54  5  3  4
       0  4  2 11
       3  2  6  6
       2  1  8  5

,,2    8 15 23 24
      13 29 22 52
      10 40 24 43
       7 18 17 33

,,3   13 22 18 28
      13 45 33 50
      10 42 39 60
       7 33 13 38

,,4    2  2  3  4
       1  5  2  1
       2  2  6  5
       1  0  3  3
```

Table 2 Numerical results for CA-coordinates for independence (left part) and with one outlier (right part)

	Independence			Outlier in the first cell		
Coord	25-quantil	75-quantil	Median	25-quantil	75-quantil	Median
xr1	−1.27	0.03	−0.54	−2.94	−1.96	−2.47
yr1	−0.56	0.91	0.21	−0.18	0.22	0.02
xr2	−0.67	0.78	0.07	−0.02	0.62	0.36
yr2	−0.68	0.94	0.19	−0.18	1.19	0.49
xr3	−0.79	0.91	0.06	−0.05	0.63	0.36
yr3	−0.88	0.77	−0.05	−0.91	0.74	−0.03
xc1	−0.86	0.86	0.00	−1.90	−1.15	−1.48
yc1	−0.83	0.90	0.03	−0.07	0.11	0.02
xc2	−0.81	0.94	0.07	0.42	0.84	0.61
yc2	−0.97	0.81	−0.10	−1.13	1.04	−0.05
xc3	−0.92	0.88	−0.01	0.39	0.81	0.60
yc3	−0.87	0.82	0.05	−1.12	1.06	0.04

These generated outliers (in the first cell in each case) are very different from their neighbors in the same row and column, but they are not the largest values in the table. Therefore they are called outliers of moderate size or moderate outliers. The fact that these outliers can be immediately recognized in CA-plots, illustrates the usefulness of Correspondence Analysis.

4.1 *CA-Coordinates*

Next we investigate changes in CA-coordinates in the case of outliers. As first step we execute 1,000 simulations for two cases: (1) independence and (2) with an outlier in the first cell. We also did up to 10^6 simulations for each case, but the tendency remains the same.

In the Table 2 the quartiles of the distribution of CA-plot coordinates (Coord) for the first four rows and columns of the transformed two-dimensional contingency

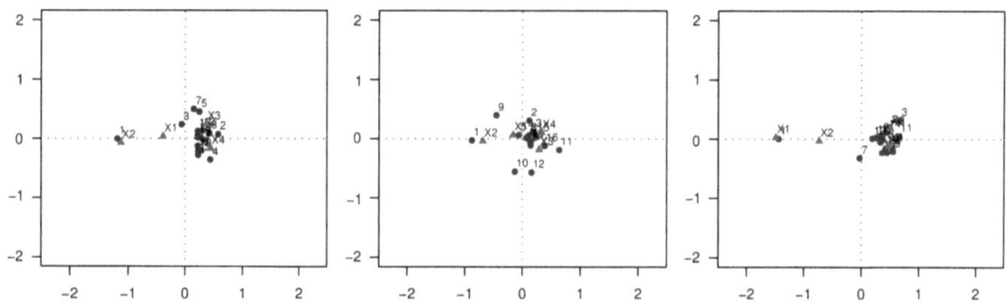

Fig. 4 CA-plots for the case of two outliers

Table 3 Numerical results for CA-coordinates for independence (left part) and with two outliers (right part)

Independence				Two outliers in the first two cells		
Coord	25-quantil	75-quantil	Median	25-quantil	75-quantil	Median
xr1	−1.34	0.02	−0.56	−2.29	−1.80	−2.06
yr1	−0.61	0.93	0.15	−0.15	0.17	0.01
xr2	−0.77	0.77	0.04	0.01	0.64	0.39
yr2	−0.67	0.95	0.17	−0.28	1.41	0.58
xr3	−0.86	0.85	−0.07	0.10	0.70	0.45
yr3	−0.81	0.81	−0.01	−0.86	0.83	−0.05
xc1	−0.81	0.87	0.07	−1.67	−0.38	−0.98
yc1	−0.81	0.89	0.08	−0.43	0.62	0.22
xc2	−0.87	0.87	−0.06	−1.52	−0.10	−0.66
yc2	−0.90	0.89	−0.04	−0.71	0.50	−0.24
xc3	−0.76	0.85	0.01	0.62	1.15	0.84
yc3	−0.93	0.85	−0.09	−1.13	1.22	0.14

table are shown, where xr$i$ - CA-coordinates of the i-th row in the first dimension and yr$i$ in the second dimension, $i = 1, 2, 3$. And xc$i$, yc$i$ are analogous but for the columns.

As we can see from this table, in the case of independence the majority of CA-coordinates are concentrated in the center for rows, as well as for columns, in both dimensions. But in the case of outliers in the first cell, that means first row and first column, CA-coordinates of these first row and first column are shifted far to the left (into the negative part of the axis). Meanwhile CA-coordinates of other rows and columns are distributed again around the origin.

We also experimented with placing several outliers in the table. The situation with two or more outliers is different from the case of one outlier.

For illustration, we describe only the case of two outliers, where both outliers are placed into the two first cells of the first row (blue circle 1 in Fig. 4) of the table, that means first (red triangle X1) and second (red triangle X2) columns respectively.

From the Table 3 for the CA-coordinates with two outliers in the table (right part of the table) we can see, that the results of analysis are now more ambiguous. For example, for the case of two outliers, rows and columns containing outliers are still shifted to the left in the first dimension, whereas other rows and columns are shifted now to the right, and nothing unusual happens with the coordinates in the second dimension. This phenomenon is less traceable, and more detailed and theoretically inferred analysis of CA-coordinates is needed, where a closer combination of CA, outlier detection methods and log-linear models (Agresti, 2002; Andersen, 1994) might be useful.

5 Conclusion

It is clearly of importance to explore further distributions of CA-coordinates. This includes confidence intervals for CA-coordinates under the null hypothesis, as well as in the case of outliers: to identify particular dependencies in the table and to suggest possible criteria for identifying hidden outliers in multi-way contingency tables by means of correspondence analysis coordinates.

However, in statistical data analysis, it is often important not only to spot outliers, but also to model interactions in the table without the outliers. For this purpose, one has to complement the CA-method with log-linear analysis.

Acknowledgements We appreciate valuable comments of Mikhail Langovoy as well as of anonymous reviewers of our article.

References

Agresti, A. (2002). *Categorical data analysis*. Hoboken: Wiley.

Andersen, E. B. (1994). *The statistical analysis of categorical data*. Berlin: Springer.

Barnett, V., & Lewis, T. (1984). *Outliers in statistical data* (Wiley Series in Probability and Mathematical Statistics. Applied Probability and Statistics 2nd ed.). Chichester: Wiley.

Blasius, J. (2001). *Korrespondenzanalyse*. München: Oldenbourg Verlag.

Blasius, J., & Greenacre, M. (2006). *Multiple correspondence analysis and related methods*. London: Chapman and Hall.

Greenacre, M. J. (1984). *Theory and applications of correspondence analysis*. London: Academic.

Kroonenberg, P. M. (2007). *Applied multiway data analysis*. Hoboken: Wiley.

Kuhnt S. (2004). Outlier identification procedures for contingency tables using maximum likelihood and L_1 estimates. *Scandinavian Journal of Statistics, 31*, 431–442.

Nenadić, O. & Greenacre, M. (2007). Correspondence analysis in **R**, with two- and three-dimensional graphics: The **ca** package. *Journal of Statistical Software, 20*(3), 1–13.

Shane, K. V., & Simonoff, J. S. (2001). A robust approach to categorical data analysis. *Comput-GraphStat, 10*, 135–157.

Variable Selection in Cluster Analysis: An Approach Based on a New Index

Isabella Morlini and Sergio Zani

Abstract In cluster analysis, the inclusion of unnecessary variables may mask the true group structure. For the selection of the best subset of variables, we suggest the use of two overall indices. The first index is a distance between two hierarchical clusterings and the second one is a similarity index obtained as the complement to one of the previous distance. Both criteria can be used for measuring the similarity between clusterings obtained with different subsets of variables. An application with a real data set regarding the economic welfare of the Italian Regions shows the benefits gained with the suggested procedure.

1 Introduction

In cluster analysis, the inclusion of 'noisy' variables may mask the recovery of the true underlying structure. In the literature, various procedures aimed at determining the best subset of variables have been proposed, both in the context of model-based and not-model-based clustering (Fowlkes et al., 1988; Gnanadesikan et al., 1995; Montanari and Lizzani, 2001; Tadesse et al., 2005; Raftery and Dean, 2006; Fraiman et al., 2008; Steinley and Brusco, 2008). In this paper we propose a new approach, based on an overall index measuring the distance between two hierarchical clusterings. This criterion is novel since it is applied directly to the whole hierarchies and may be thought of as a generalization of the measures used

I. Morlini (✉)
Department of Economics, University of Modena and Reggio Emilia, Via Berengario 51, 41100, Modena, Italy
e-mail: isabella.morlini@unimore.it

S. Zani
Department of Economics, University of Parma, Via Kennedy 6, 43100, Parma, Italy
e-mail: sergio.zani@unipr.it

A. Giusti et al. (eds.), *Classification and Data Mining*, Studies in Classification, Data Analysis, and Knowledge Organization, DOI 10.1007/978-3-642-28894-4_9,

for comparing two partitions (Rand, 1971; Fowlkes and Mallows, 1983; Hubert and Arabie, 1985). The paper is organized as follows: in Sect. 2 we define the index, we present its properties and its decomposition with reference to each stage of the hierarchy; in Sect. 3 we consider the similarity index obtained as the complement to one of the suggested distance and we deal with the adjustment for agreement due to chance; in Sect. 4 we describe the use of the index for measuring the similarity between clusterings obtained with different subsets of variables, following a forward and a backward approach; in Sect. 5 we present results on a real data set.

2 The Index and Its Properties

Suppose we have two hierarchical clusterings of the same number of objects, n. Let us consider the $N = n(n-1)/2$ pairs of objects and let us define, for each non trivial partition in k groups ($k = 2, \ldots, n-1$), a binary variable X_k with values $x_{ik} = 1$ if objects in pair $i (i = 1, \ldots, N)$ are classified in the same cluster in partition in k groups and $x_{ik} = 0$ otherwise. A binary $(N \times (n-2))$ matrix $\mathbf{X}_g$ for each clustering g $(g = 1, 2)$ may be derived, in which the columns are the binary variables X_k. A global measure of dissimilarity between the two clusterings may be defined as follows:

$$Z = \frac{\| \mathbf{X}_1 - \mathbf{X}_2 \|}{\| \mathbf{X}_1 \| + \| \mathbf{X}_2 \|}, \tag{1}$$

where $\| \mathbf{A} \| = \sum_i \sum_k \| a_{ik} \|$ is the L_1 norm of the matrix $\mathbf{A}$. In (1) the matrices involved take only binary values and the L_1 norm is equal to the square of the L_2 norm. The derivation of Z uses the Rand's idea of considering the N object pairs. However, Z is a new index since it is applied to a whole hierarchy and not only to a single partition. Z has the following properties.

- It is bounded in [0,1]. $Z = 0$ iff the two hierarchical clusterings are identical and $Z = 1$ when the clusterings have the maximum degree of dissimilarity, that is when for each partition in k groups and for each i, objects in pair i are in the same group in clustering 1 and in different groups in clustering 2 (or vice versa).
- It is a distance, since it satisfies the conditions of non negativity, identity, symmetry and triangular inequality (Zani, 1986).
- The complement to 1 of Z is a similarity measure, since it satisfies the conditions of non negativity, normalization and symmetry.
- It does not depend on the group labels since it refers to pairs of objects.
- It may be decomposed in $(n-2)$ parts related to each pair of partitions in k groups since:

$$Z = \sum_k Z_k = \sum_k \sum_i \frac{|x_{1ik} - x_{2ik}|}{\| \mathbf{X}_1 \| + \| \mathbf{X}_2 \|}. \tag{2}$$

  The plot of Z_k versus k shows the distance between the two clusterings at each stage of the procedure.

Table 1 Contingency table of the cluster membership of the N object pairs

	Second clustering ($g = 2$)		
First clustering ($g = 1$)	Pairs in the same cluster	Pairs in different clusters	Sum
Pairs in the same cluster	T_k	$P_k - T_k$	P_k
Pairs in different clusters	$Q_k - T_k$	$U_k = N + T_k - P_k - Q_k$	$N - P_k$
Sum	Q_k	$N - Q_k$	$N = n(n-1)/2$

3 The Complement of the Index

Consider the quantities in the (2×2) contingency table showing the cluster membership of the object pairs in each of the two partitions (Table 1).

Since $\| \mathbf{X}_1 \| = \sum_k Q_k$ and $\| \mathbf{X}_2 \| = \sum_k P_k$, the complement to 1 of Z is:

$$S = 1 - Z = \frac{2\sum_k T_k}{\sum_k Q_k + \sum_k P_k}. \tag{3}$$

Also the similarity index S may be decomposed in $(n-2)$ parts V_k related to each pair of partitions in k groups:

$$S = \sum_k V_k = \sum_k \frac{2T_k}{\sum_k Q_k + \sum_k P_k}. \tag{4}$$

The components V_k, however, are not similarity indices for each k since they assume values < 1 even if the two partitions in k groups are identical. For this reason, we consider the complement to 1 of each Z_k in order to obtain a single similarity index for each pair of partitions:

$$S_k = 1 - Z_k = \frac{\sum_{j=2}^{n-1} P_j + \sum_{j=2}^{n-1} Q_j - P_k - Q_k + 2T_k}{\sum_{j=2}^{n-1} P_j + \sum_{j=2}^{n-1} Q_j} = \frac{\sum_{j \neq k} P_j + \sum_{j \neq k} Q_j + 2T_k}{\sum_j P_j + \sum_j Q_j}. \tag{5}$$

A similarity index between two partitions may be adjusted for agreement due to chance (Hubert and Arabie, 1985; Albatineh et al., 2006; Warrens, 2008). With reference to formula (5) the adjusted similarity index AS_k has the form:

$$AS_k = \frac{S_k - E(S_k)}{max(S_k) - E(S_k)}. \tag{6}$$

Under the hypothesis of independence of the two partitions, the expectation of T_k in Table 1 $E(T_k) = P_k Q_k / N$. Therefore, the expectation of S_k is given by:

$$E(S_k) = \frac{\sum_{j \neq k} P_j + \sum_{j \neq k} Q_j + 2P_k Q_k / N}{\sum_j P_j + \sum_j Q_j}. \tag{7}$$

Considering $max(S_k) = 1$ and simplifying terms we obtain:

$$AS_k = \frac{2T_k - 2P_k Q_k / N}{P_k + Q_k - 2P_k Q_k / N}. \tag{8}$$

The adjusted Rand index for two partitions in k groups is given by Warrens (2008):

$$AR_k = \frac{2(NT_k - P_k Q_k)}{N(P_k + Q_k) - 2P_k Q_k}, \tag{9}$$

and so AS_k is equal to the Adjusted Rand Index.

4 Criteria for Variable Selection

Indexes Z and S can be used for variable selection in cluster analysis (Fowlkes et al., 1988; Fraiman et al., 2008; Steinley and Brusco, 2008). The inclusion of 'noisy' variables can actually degrade the ability of the clustering procedures to recover the true underlying structure (Friedman and Meulman, 2004). For a set of p variables and a certain clustering method, we suggest three different approaches, suitable for data sets with tens of variables. Variable selection in data sets containing hundreds or thousands of variables (like gene expression data) is not considered in this paper.

First we may obtain the p one dimensional clusterings with reference to each single variable and then compute the $p \times p$ similarity matrix $\mathbf{S}$. The pairs of variables reflecting the same underlying structure show high similarity. On the contrary, the noisy variables should present a similarity with the other variables near to the expected value for chance agreement. We may select a subset of variables that best explains the classification into homogeneous groups. These variables help us to better understand the multivariate structure and suggest a dimension reduction that can be used in a new data set for the same problem (Tadesse et al., 2005).

Next we may find the similarities between clusterings obtained with subsets of variables (regarding, for example, different features). This approach is helpful in showing aspects that lead to similar partitions and subsets of variables that, on the contrary, lead to different clusterings.

A third way to proceed consists in finding the similarities between the 'master' clustering obtained by considering all the variables and the clusterings obtained by eliminating each single variable in turn, in order to highlight the 'marginal' contribution of each variable to the master structure.

5 An Application to a Real Data Set

We consider the 20 Italian regions and the following 9 variables measuring different aspects of the economic wealth: X_1 = activity rate, X_2 = unemployment rate, X_3 = youth unemployment rate, X_4 = family average income, X_5 = family median income, X_6 = income Gini concentration index, X_7 = % of poor families, X_8 = % of people dissatisfied for their economic conditions, X_9 = % of families with inadequate income. We standardize variables to zero mean and unit variance before

Table 2 Values of S between pair of clusterings of the Italian regions data set

	Euclidean distance					Manhattan distance				
Method	Average	Complete	Single	Ward	Centroid	Average	Complete	Single	Ward	Centroid
Average	1	0.80	0.90	0.76	0.96	0.96	0.81	0.88	0.79	0.95
Complete	0.80	1	0.73	0.72	0.78	0.83	0.98	0.72	0.82	0.78
Single	0.90	0.73	1	0.71	0.93	0.87	0.73	0.92	0.72	0.90
Ward	0.76	0.72	0.71	1	0.75	0.78	0.73	0.68	0.79	0.74
Centroid	0.96	0.78	0.93	0.75	1	0.92	0.79	0.88	0.77	0.95
Average	0.96	0.83	0.87	0.78	0.92	1	0.84	0.87	0.83	0.94
Complete	0.81	0.98	0.73	0.73	0.79	0.84	1	0.72	0.82	0.78
Single	0.88	0.72	0.92	0.68	0.88	0.87	0.72	1	0.72	0.93
Ward	0.79	0.82	0.72	0.79	0.77	0.83	0.82	0.72	1	0.78
Centroid	0.95	0.78	0.90	0.74	0.95	0.94	0.78	0.93	0.78	1

applying hierarchical cluster analysis with different distances and different methods. We compute the S index for each pair of clusterings. Results, reported in Table 2, show that, in general, clustering remains stable varying distances or methods or both (all pairwise similarity indexes take values greater than 0.7). The fact that the clustering does not change appreciably leads to the evidence that the topologies of the trees are natural and are not simply artifacts of the algorithms. Analyzing the values of the pairwise similarities, we note that the Ward and the single linkage seem to behave a little bit differently from the other methods, while the complete linkage, the average linkage and the centroid method seem to be more similar to each other. The global measure of similarity S may be decomposed in parts related to each partition in $k = 2, \ldots, 18$ groups. As an example, Table 3 presents the values of S_k and AS_k for two pairs of clusterings. This table shows the reason why the second couple has a slightly less similarity. In these two dendrograms, 12 partitions (among the 18 ones) are exactly the same while for the first two dendrograms the identical partitions are 13. In order to determine the 'true' number of clusters, we may count the couples of clusterings in which each partition in k groups is identical. From counts reported in Table 4 we see that the partition in 2 groups remains identical in 36 clusterings. Only partition in 18 clusters has a larger count. This may be taken as evidence that partition in two groups comes naturally from data and is not driven by the algorithm. In this partition, northern and central regions are separated from southern regions.

Table 5 reports the values of S between the clustering obtained considering all variables ($\{X_i\}_{i=1,\ldots,9}$) (in the following we will refer to this tree as the overall tree) and the clusterings obtained eliminating each variable in turn. In the table, the column or row header $\{X_i\}_{i \neq j}$ indicates the subset of variables without X_j. For example, $\{X_i\}_{i \neq 1}$ is the subset $\{X_2, X_3, X_4, X_5, X_6, X_7, X_8, X_9\}$. The Euclidean distance and the average method are used for obtaining partitions.

If we eliminate X_9, the clustering remains identical. This means that X_9 has no 'marginal' contribution to the overall clusterings, given the other variables. X_8 is the variable which seems to have the major marginal influence to the overall

Table 3 Values of S_k and AS_k for two pairs of clusterings of the Italian regions data set

	Number k of clusters																	
	2	3	4	5	6	7	8	9	10	11	12	13	14	15	16	17	18	19
	Similarity between partitions obtained with Euclidean distance and the average method and partitions obtained with Euclidean distance and the centroid method																	
S_k	1	1	1	1	0.98	0.98	0.98	1	0.98	1	1	1	1	1	0.98	1	1	1
AS_k	1	1	1	1	0.86	0.75	0.86	1	0.81	1	1	1	1	1	0.66	1	1	1
	Similarity between partitions obtained with Euclidean distance and the average method and partitions obtained with Manhattan distance and the centroid method																	
S_k	1	1	1	1	0.98	0.97	0.98	1	1	1	1	0.98	1	1	0.98	1	1	0.98
AS_k	1	1	1	1	0.86	0.73	0.86	1	1	1	1	0.57	1	1	0.66	1	1	0.00

Table 4 Counts of pairs of clusterings in which each partition in k groups is identical

k	2	3	4	5	6	7	8	9	10	11	12	13	14	15	16	17	18	19
n. of pairs	36	29	8	8	7	3	1	6	6	8	12	9	17	29	21	16	45	20

Table 5 Values of S for couples of clusterings obtained with different subsets of variables

	$\{X_i\}_{i\neq1}$	$\{X_i\}_{i\neq2}$	$\{X_i\}_{i\neq3}$	$\{X_i\}_{i\neq4}$	$\{X_i\}_{i\neq5}$	$\{X_i\}_{i\neq6}$	$\{X_i\}_{i\neq7}$	$\{X_i\}_{i\neq8}$	$\{X_i\}_{i\neq9}$	$\{X_i\}$
$\{X_i\}_{i\neq1}$	1	0.96	0.83	0.85	0.88	0.86	0.93	0.84	0.89	0.89
$\{X_i\}_{i\neq2}$	0.96	1	0.86	0.88	0.86	0.87	0.95	0.85	0.91	0.91
$\{X_i\}_{i\neq3}$	0.83	0.86	1	0.83	0.82	0.87	0.84	0.81	0.89	0.89
$\{X_i\}_{i\neq4}$	0.85	0.88	0.83	1	0.97	0.86	0.86	0.80	0.89	0.89
$\{X_i\}_{i\neq5}$	0.88	0.86	0.82	0.97	1	0.85	0.85	0.80	0.89	0.89
$\{X_i\}_{i\neq6}$	0.86	0.87	0.87	0.86	0.85	1	0.85	0.84	0.94	0.94
$\{X_i\}_{i\neq7}$	0.93	0.95	0.84	0.86	0.85	0.85	1	0.85	0.88	0.88
$\{X_i\}_{i\neq8}$	0.84	0.85	0.81	0.80	0.80	0.84	0.85	1	0.86	0.86
$\{X_i\}_{i\neq9}$	0.89	0.91	0.89	0.89	0.89	0.94	0.88	0.86	1	1
$\{X_i\}$	0.89	0.91	0.89	0.89	0.89	0.94	0.88	0.86	1	1

clustering structure. The value of S between $\{X_i\}_{i\neq4}$ and $\{X_i\}_{i\neq5}$ ($S = 0.97$) shows that X_4 and X_5, as one would expect, bring the same marginal contribution. We may also consider the similarities between the clustering recovered by all variables $\{X_i\}$ and the clusterings obtained by using each single variable. The values of S are:

$$S(\{X_i\},\{X_1\}) = 0.74, \quad S(\{X_i\},\{X_2\}) = 0.66, \quad S(\{X_i\},\{X_3\}) = 0.58,$$
$$S(\{X_i\},\{X_4\}) = 0.55, \quad S(\{X_i\},\{X_5\}) = 0.76, \quad S(\{X_i\},\{X_6\}) = 0.69,$$
$$S(\{X_i\},\{X_7\}) = 0.77, \quad S(\{X_i\},\{X_8\}) = 0.52, \quad S(\{X_i\},\{X_9\}) = 0.53.$$

None of the values are particularly high and thus the clustering recovered with all variables seems to derive from a multivariate effect and not to be dominated by the univariate effect of a single variable. As shown in Fig. 1, variables X_1, X_5 and

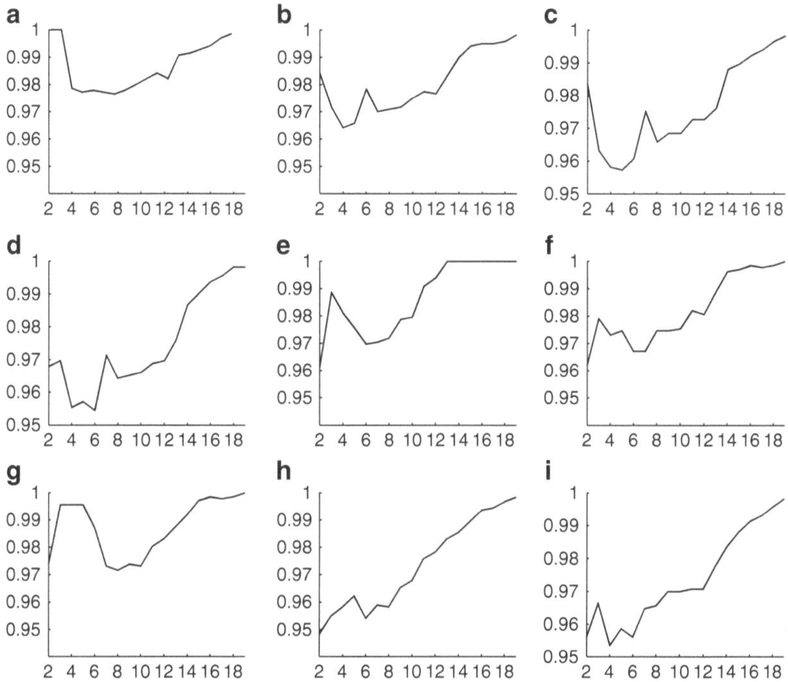

Fig. 1 Values of S_k for the partitions obtained with all variables and the partitions obtained with (**a**) X_1, (**b**) X_2, (**c**) X_3, (**d**) X_4, (**e**) X_5, (**f**) X_6, (**g**) X_7, (**h**) X_8 and (**i**) X_9

X_7 have a peak of the similarity values S_k for $k = 3$. X_1 is in perfect agreement also for $k = 2$ while X_7 for $k = 4, 5$. Variables X_4, X_6 and X_9 have a different S_k pattern, but they also have a peak for $k = 3$. On the contrary, the peak for X_2 and X_3 is for $k = 2$. Thus, in this case, the choice for the 'correct' number of clusters is somehow difficult, since both $k = 2$ and $k = 3$ seem to be good alternative. Figure 1 also shows that variables which have the smaller values in the similarity S, like X_3, X_4, X_5 and X_6, exhibit a less agreement to the overall clusters for small numbers k of groups. The patterns of S_k for these variables display smaller values for $k < 12$. For the other variables, S_k increase less rapidly, with respect to k. Finally, we study the behavior of three subsets of variables, each one related to a specific feature of the economic situation. We consider subset $\{X_1, X_2, X_3\}$, related to the demographic structure, subset $\{X_4, X_5, X_6\}$ related to the income structure and subset $\{X_7, X_8, X_9\}$, related to the relative and the perceived poverty. The similarities between the cluster trees of each subset and of all variables are: $S(\{X_i\}, \{X_{1,2,3}\}) = 0.76$, $S(\{X_i\}, \{X_{4,5,6}\}) = 0.66$, $S(\{X_i\}, \{X_{7,8,9}\}) = 0.78$. The similarities between clusterings of each subsets are: $S(\{X_{4,5,6}\}, \{X_{7,8,9}\}) = 0.59$, $S(\{X_{1,2,3}\}, \{X_{4,5,6}\}) = 0.61$, $S(\{X_{1,2,3}\}, \{X_{7,8,9}\}) = 0.62$. Here again we note that none of the three subsets reveals a clustering very similar to the clustering obtained with all the variables. All the three aspects of the economic health seem equally

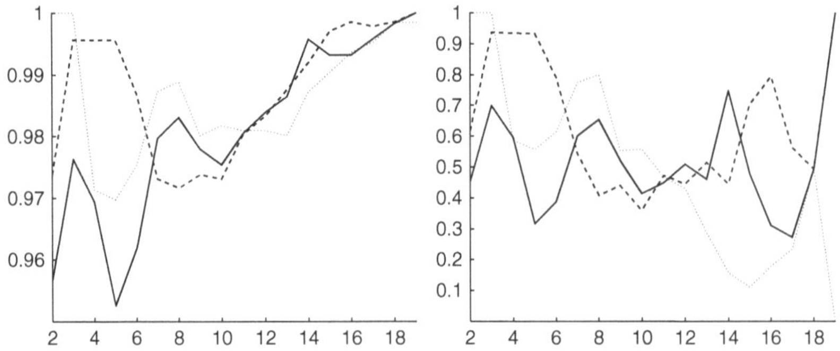

Fig. 2 Plots of S_k (*left*) and AS_k (*right*) between partitions obtained with all variables $\{X_i\}$ and subset $\{X_1, X_2, X_3\}$ (*dotted line*), subset $\{X_4, X_5, X_6\}$ (*solid line*), subset $\{X_7, X_8, X_9\}$ (*dashed line*)

to contribute to the overall clustering. Figure 2 reports the plots of S_k and AS_k. The scales in the Y-axis are different. However, the patterns of S_k and AS_k are nearly identical, for $k \leq 12$. It is a desirable property that the correction for the chance influences the values but not the configuration of the plot for small k. For large k, as one would expect, the correction for chance do influence the patterns of the index and S_k tends to one while AS_k tends to zero. We note that, for example, the configuration in two groups is largely dominated by the demographic structure, while configurations in 3, 4 and 5 clusters are mostly influenced by the perceived poverty.

References

Albatineh, A. N., Niewiadomska-Bugaj, M., & Mihalko, D. (2006). On similarity indices and correction for chance agreement. *Journal of Classification, 23*, 301–313.

Fowlkes, E. B., & Mallows, C. L. (1983). A method for comparing two hierarchical clusterings. *JASA, 78*, 553–569.

Fowlkes, E. B., Gnanadesikan, R., & Kettenring, J. R. (1988). Variable selection in clustering. *Journal of Classification, 5*, 205–228.

Fraiman, R., Justel, A., & Svarc, M. (2008). Selection of variables for cluster analysis and classification rules. *JASA, 103*, 1294–1303.

Friedman, J. H., & Meulman, J. J. (2004). Clustering objects on subset of attributes. *Journal of the Royal Statistical Society B, 66*, 815–849.

Gnanadesikan, R., Kettering, J. R., & Tsao, S. L. (1995). Weighting and selection of variables for cluster analysis. *Journal of Classification, 12*, 113–136.

Hubert, L. J., & Arabie, P. (1985). Comparing partitions. *Journal of Classification, 2*, 193–218.

Montanari, A., & Lizzani, L. (2001). A projection pursuit approach to variable selection. *Computational Statistics and Data Analysis, 35*, 463–473.

Raftery, A. E., & Dean, N. (2006). Variable selection for model based clustering. *JASA, 101*, 168–178.

Rand, W. M. (1971). Objective criteria for the evaluation of clustering methods. *JASA, 66*, 846–850.

Steinley, D., & Brusco, M. J. (2008). Selection of variables in cluster analysis: An empirical comparison of eight procedures. *Psychometrika, 73*, 125–144.

Tadesse, M. G., Sha, N., & Vannucci, M. (2005). Bayesian variable selection in clustering high-dimensional data. *JASA, 100*, 602–617.

Warrens, M. J. (2008). On the equivalence of Cohen's Kappa and the Hubert-Arabie adjusted Rand index. *Journal of Classification, 25*, 177–183.

Zani, S. (1986). Some measures for the comparison of data matrices. In *Proceedings of the XXXIII Meeting of the Italian Statistical Society* (pp. 157–169), Bari, Italy.

A Model for the Clustering of Variables Taking into Account External Data

Karin Sahmer

Abstract In this paper, a statistical model for the clustering of variables taking into account external data is proposed. This model is particularly appropriate for preference data in the presence of external information about the products. The clustering of variables around latent components (CLV method) is analysed on the basis of this model. Within the CLV method, there is one option without external data and one option taking into account external data. The criteria of both options can be expressed in function of the parameters of the postulated model. It is shown that the hierarchical algorithm finds the correct partition when the parameters of the model are known, no matter which option of CLV is used. Furthermore, the two options of CLV are compared by means of a simulation study. Both options perform well except for the case of small samples with a very large noise. Moreover, in most cases the performance of both options is equivalent.

1 Introduction

A method for the clustering of variables (CLV) was proposed by Vigneau and Qannari (2003). This method is based on a hierarchical clustering followed by a partitioning algorithm and includes several options. It can be used when variables with a negative correlation should be grouped together. It is also possible to use this method when a negative correlation between variables shows disagreement. In both cases, it is possible to take into account external data in the clustering procedure. Sahmer (2006) analysed the CLV method in the case where a high negative correlation shows a proximity of variables. The scope of the present paper is to state a model that can be used to analyse the CLV method in the

K. Sahmer (✉)
Groupe ISA, 48 boulevard Vauban, F-59046 Lille Cedex, France
e-mail: karin.sahmer@isa-lille.fr

A. Giusti et al. (eds.), *Classification and Data Mining*, Studies in Classification, Data Analysis, and Knowledge Organization, DOI 10.1007/978-3-642-28894-4_10,

case of grouping together only variables with positive correlations. An important application is the segmentation of consumers according to their liking of products, taking into account sensory data (Vigneau and Qannari, 2002).

In Sect. 2, a model for the clustering of variables taking into account external data is proposed. The corresponding covariance matrix is specified. The properties of the CLV method are analysed in Sect. 3. The analysis is based on the stated model and concerns CLV taking into account external data but also CLV without external data. Finally, the two options of CLV are compared by means of a simulation study (Sect. 4).

2 A Model for Preference Data

Vigneau and Qannari (2002) proposed their method for a segmentation of consumers according to their preferences for products, taking into account external information about the products. The following model has in mind this application but it fits for all cases where a linear relationship between external variables and the variables to be clustered can be assumed. The variables to be clustered will be called the x variables. In the case of preference data, the liking score of each consumer is one x variable. The external variables, for example the sensory descriptors, will be called the z variables.

It is assumed that there exist K groups of consumers. They will be denoted by $G^{(1)}, G^{(2)}, \ldots, G^{(K)}$. The number of consumers in group $G^{(k)}$ is denoted by $p^{(k)}$. In each group, a linear relationship between sensory descriptors and liking scores is assumed. The liking score of the jth consumer in group $G^{(k)}$ is given by:

$$x_j^{(k)} = \mathbf{z}'\boldsymbol{\beta}^{(k)} + \epsilon_j^{(k)} \tag{1}$$

where $x_j^{(k)}$ is the random variable representing the liking score of the jth consumer, $\mathbf{z} = \left(z_1, z_2, \ldots, z_q\right)'$ is the random vector corresponding to the q sensory descriptors, $\boldsymbol{\beta}^{(k)} = \left(\beta_1^{(k)}, \beta_2^{(k)}, \ldots, \beta_q^{(k)}\right)'$ is the parameter vector of group $G^{(k)}$, and $\epsilon_j^{(k)}$ corresponds to the error term. In this model, all consumers belonging to the same group have the same parameter vector $\boldsymbol{\beta}$. It is assumed that the error terms are uncorrelated with each other and with the z variables. Furthermore, an equal error variance is assumed: $\text{var}\left(\epsilon_j^{(k)}\right) = \sigma^2 \; \forall k \, \forall j$. The covariance matrix of the z variables will be denoted by $\boldsymbol{\Sigma}_z$.

Under this model, the covariance matrix of $\mathbf{x}^{(k)}$ (the random vector of the x variables belonging to group $G^{(k)}$) is given by:

$$\boldsymbol{\Sigma}_x{}^{(k)} = \mathbf{1}_{p^{(k)}}\mathbf{1}_{p^{(k)}}{}'\boldsymbol{\beta}^{(k)\prime}\boldsymbol{\Sigma}_z\boldsymbol{\beta}^{(k)} + \sigma^2\mathbf{I} \tag{2}$$

where $\mathbf{1}_{p^{(k)}}$ is the vector consisting of $p^{(k)}$ ones and $\mathbf{I}$ is the $p^{(k)}$-dimensional identity matrix. The matrix of covariances of x variables belonging to different groups $G^{(k)}$ and $G^{(m)}$ is given by:

$$\boldsymbol{\Sigma}_x{}^{(k,m)} = \mathbf{1}_{p^{(k)}} \mathbf{1}_{p^{(m)}}{}' \boldsymbol{\beta}^{(k)\prime} \boldsymbol{\Sigma}_z \boldsymbol{\beta}^{(m)} \tag{3}$$

while the matrix of covariances between $\mathbf{x}^{(k)}$ and $\mathbf{z}$ is equal to:

$$\boldsymbol{\Sigma}_{xz}{}^{(k)} = \mathbf{1}_{p^{(k)}} \boldsymbol{\beta}^{(k)\prime} \boldsymbol{\Sigma}_z \tag{4}$$

The relationship between the z variables and the x variables is reflected in the covariance matrix of the x variables. So a clustering of the x variables without explicitly taking into account the z variables does take them into account implicitly.

3 Properties of CLV

In the following, the two options of CLV (with or without external data) will be analysed. Vigneau and Qannari (2003) consider the clustering on observed data. In order to analyse the clustering based on the covariance matrix stated in the model of Sect. 2, the notation of the clustering criterion has been slightly modified. The following analysis concerns only the first part of the CLV approach, the hierarchical clustering. The partitioning algorithm within the CLV method is not considered here. A distinction has to be made between the real groups $G^{(k)}$ and the groups formed by the algorithm. The latter will be called the clusters $C_1, C_2, \ldots, C_{K^*}$ where K^* is the current number of clusters in a given step of the algorithm. According to the stated model, there exists a linear relationship between the variables to be clustered and some external variables. So the appropriate clustering method would be the option of CLV taking into account external data. This option will first be analysed (Sect. 3.1). Afterwards (Sect. 3.2), it will be analysed what happens if, in spite of the existence of external data, the option without external data is used.

3.1 Properties of CLV Taking into Account External Data

When taking into account external data, the criterion $\tilde{S}$ has to be maximized:

$$\tilde{S} = \sum_{k=1}^{K^*} \sum_{j \text{ with } x_j \in C_k} \operatorname{cov}\left(x_j, c_k\right) \tag{5}$$

under the constraints $c_k = \mathbf{z}' \mathbf{a}_k$ and $\mathbf{a}'_k \mathbf{a}_k = 1$. The variable c_k is called the latent component of cluster C_k and is constrained to be a linear combination of the

z variables. Define $\tilde{\mathbf{a}}_k = \left(\mathrm{cov}\left(z_1, \bar{x}_k\right), \ldots, \mathrm{cov}\left(z_q, \bar{x}_k\right)\right)'$ and $\mathbf{a}_k = \frac{\tilde{\mathbf{a}}_k}{\sqrt{\tilde{\mathbf{a}}_k'\tilde{\mathbf{a}}_k}}$ where $\bar{x}_k$ is the mean of all variables in cluster C_k. In each step, merge the two clusters resulting in the smallest decrease $\Delta\tilde{S}$ of criterion $\tilde{S}$.

Merging two subgroups of the same real group. Consider a cluster A that is a subgroup of group $G^{(k)}$. The mean of the variables belonging to A is given by $\bar{x}_A = \frac{1}{p_A}\sum_{j\in A}\left(\mathbf{z}'\boldsymbol{\beta}^{(k)} + \epsilon_j^{(k)}\right) = \mathbf{z}'\boldsymbol{\beta}^{(k)} + \bar{\epsilon}_A$ where $\bar{\epsilon}_A$ is the average of the error terms corresponding to the variables in cluster A. The vector of covariances between $\bar{x}_A$ and $\mathbf{z}$ is given by:

$$\mathrm{cov}\left(\bar{x}_A, \mathbf{z}\right) = \mathrm{cov}\left(\mathbf{z}'\boldsymbol{\beta}^{(k)} + \bar{\epsilon}_A,\ \mathbf{z}\right) = \boldsymbol{\Sigma}_z\boldsymbol{\beta}^{(k)}. \tag{6}$$

This vector is the same for all subgroups of group $G^{(k)}$. If a subgroup A of group $G^{(k)}$ is one of the current clusters, the vector $\tilde{\mathbf{a}}$ used to calculate the latent component is equal to $\boldsymbol{\Sigma}_z\boldsymbol{\beta}^{(k)}$, and its latent component is equal to

$$c_A = \mathbf{z}'\,\frac{\boldsymbol{\Sigma}_z\boldsymbol{\beta}^{(k)}}{\sqrt{\boldsymbol{\beta}^{(k)\prime}\boldsymbol{\Sigma}_z\boldsymbol{\Sigma}_z\boldsymbol{\beta}^{(k)}}}. \tag{7}$$

The covariance of a variable belonging to group $G^{(k)}$ with this latent component c_A is equal to:

$$\begin{aligned}\mathrm{cov}\left(x_j^{(k)}, c_A\right) &= \mathrm{cov}\left(\mathbf{z}'\boldsymbol{\beta}^{(k)} + \epsilon_j^{(k)},\ \mathbf{z}'\,\frac{\boldsymbol{\Sigma}_z\boldsymbol{\beta}^{(k)}}{\sqrt{\boldsymbol{\beta}^{(k)\prime}\boldsymbol{\Sigma}_z\boldsymbol{\Sigma}_z\boldsymbol{\beta}^{(k)}}}\right)\\ &= \sqrt{\boldsymbol{\beta}^{(k)\prime}\boldsymbol{\Sigma}_z\boldsymbol{\Sigma}_z\boldsymbol{\beta}^{(k)}}.\end{aligned} \tag{8}$$

So the part of cluster criterion $\tilde{S}$ corresponding to cluster A can be written as:

$$\tilde{S}_A = \sum_{j \text{ with } x_j\in A} \mathrm{cov}\left(x_j^{(k)}, c_A\right) = p_A\sqrt{\boldsymbol{\beta}^{(k)\prime}\boldsymbol{\Sigma}_z\boldsymbol{\Sigma}_z\boldsymbol{\beta}^{(k)}} \tag{9}$$

where p_A is the number of variables belonging to cluster A. Finally, when merging two subgroups A and B of the same real group $G^{(k)}$, the decrease $\Delta\tilde{S}$ in criterion $\tilde{S}$ is equal to

$$\begin{aligned}\Delta\tilde{S} &= \tilde{S}_A + \tilde{S}_B - \tilde{S}_{A\cup B}\\ &= p_A\sqrt{\boldsymbol{\beta}^{(k)\prime}\boldsymbol{\Sigma}_z\boldsymbol{\Sigma}_z\boldsymbol{\beta}^{(k)}} + p_B\sqrt{\boldsymbol{\beta}^{(k)\prime}\boldsymbol{\Sigma}_z\boldsymbol{\Sigma}_z\boldsymbol{\beta}^{(k)}} - (p_A+p_B)\sqrt{\boldsymbol{\beta}^{(k)\prime}\boldsymbol{\Sigma}_z\boldsymbol{\Sigma}_z\boldsymbol{\beta}^{(k)}}\\ &= 0\end{aligned} \tag{10}$$

Merging subgroups of different groups. Let A be a subgroup of the real group $G^{(k)}$ and let B be a subgroup of the real group $G^{(m)}$. Adapting the calculations of the preceding paragraph we find:

$$\Delta\tilde{S} = p_A\sqrt{\boldsymbol{\beta}^{(k)'}\boldsymbol{\Sigma}_z\boldsymbol{\Sigma}_z\boldsymbol{\beta}^{(k)}} + p_B\sqrt{\boldsymbol{\beta}^{(m)'}\boldsymbol{\Sigma}_z\boldsymbol{\Sigma}_z\boldsymbol{\beta}^{(m)}} - \sqrt{\left(p_A\boldsymbol{\beta}^{(k)} + p_B\boldsymbol{\beta}^{(m)}\right)'\boldsymbol{\Sigma}_z\boldsymbol{\Sigma}_z\left(p_A\boldsymbol{\beta}^{(k)} + p_B\boldsymbol{\beta}^{(m)}\right)}. \qquad (11)$$

Here, $\Delta\tilde{S} = 0$ when there is a group with no linear relationship between the z variables and the x variables, that is when $\boldsymbol{\beta}^{(k)} = 0$ or $\boldsymbol{\beta}^{(m)} = 0$. Criterion $\Delta\tilde{S}$ is also zero when the two parameter vectors are collinear and of the same direction ($\boldsymbol{\beta}^{(k)} = d\,\boldsymbol{\beta}^{(m)}$ for a positive constant d). Otherwise, $\Delta\tilde{S} > 0$.

If we consider two groups with collinear parameter vectors of the same direction as only one group, the hierarchical algorithm of CLV taking into account external data will first merge variables or subgroups belonging to the same real group before merging variables or subgroups belonging to different groups. So, the correct partition is found. The only condition is that, in each group, there must be a linear relationship between z variables and x variables.

3.2 *Properties of CLV Without External Data*

The hierarchical algorithm of CLV without external data is equivalent to the Ward algorithm on centered and transposed data. The cluster criterion to be maximized can be expressed as:

$$S = \sum_{k=1}^{K^*} p_k \operatorname{var}(\bar{x}_k) \qquad (12)$$

where p_k is the number of variables in cluster C_k and $\bar{x}_k$ is the average of all variables in cluster C_k. When merging two clusters A and B, there is a decrease ΔS in criterion S. Let $\boldsymbol{\Sigma}_A$ and $\boldsymbol{\Sigma}_B$ be the covariance matrices of clusters A and B, and let $\boldsymbol{\Sigma}_{AB}$ be the matrix of covariances between the variables belonging to cluster A and the variables belonging to cluster B. It is possible to write ΔS as a function of these matrices:

$$\begin{aligned}\Delta S &= p_A \operatorname{var}(\bar{x}_A) + p_B \operatorname{var}(\bar{x}_B) - (p_A + p_B)\operatorname{var}(\bar{x}_{A\cup B}) \\ &= \frac{1}{p_A + p_B}\left[\frac{p_B}{p_A}\mathbf{1}'\boldsymbol{\Sigma}_A\mathbf{1} + \frac{p_A}{p_B}\mathbf{1}'\boldsymbol{\Sigma}_B\mathbf{1} - 2\times\mathbf{1}'\boldsymbol{\Sigma}_{AB}\mathbf{1}\right] \qquad (13)\end{aligned}$$

Table 1 A summary of the main results. Value of the clustering criteria in different cases

	$\Delta\tilde{S}$	ΔS
Merging subgroups of the same real group	0	σ^2
Merging subgroups of groups with collinear parameter vectors of same direction	0	$> \sigma^2$
Merging subgroups of different groups	> 0	$> \sigma^2$

since

$$\operatorname{var}(\bar{x}_A) = \operatorname{var}\left(\frac{1}{p_A}\sum_{j \text{ with } x_j \in A} x_j\right) = \frac{1}{p_A^2}\mathbf{1}'\boldsymbol{\Sigma}_A\mathbf{1} \tag{14}$$

and

$$\operatorname{var}(\bar{x}_{A\cup B}) = \frac{1}{(p_A + p_B)^2}\left(\mathbf{1}'\boldsymbol{\Sigma}_A\mathbf{1} + \mathbf{1}'\boldsymbol{\Sigma}_B\mathbf{1} + 2 \times \mathbf{1}'\boldsymbol{\Sigma}_{AB}\mathbf{1}\right). \tag{15}$$

If the postulated model is true, we can express the covariance matrices in formula (13) using the model parameters (see Sect. 2).

When merging subgroups of the same real group $G^{(k)}$, we obtain:

$$\begin{aligned}\Delta S &= \frac{1}{p_A + p_B}\left[\frac{p_B}{p_A}\left(p_A\boldsymbol{\beta}^{(k)\prime}\boldsymbol{\Sigma}_z\boldsymbol{\beta}^{(k)}p_A + p_A\sigma^2\right)\right.\\ &\quad \left. + \frac{p_A}{p_B}\left(p_B\boldsymbol{\beta}^{(k)\prime}\boldsymbol{\Sigma}_z\boldsymbol{\beta}^{(k)}p_B + p_B\sigma^2\right) - 2 \times p_A\boldsymbol{\beta}^{(k)\prime}\boldsymbol{\Sigma}_z\boldsymbol{\beta}^{(k)}p_B\right]\\ &= \sigma^2 \end{aligned} \tag{16}$$

So when merging two subgroups of a same real group, the criterion ΔS is equal to the error variance σ^2. For the merging of subgroups of two different real groups $G^{(k)}$ and $G^{(m)}$, we obtain:

$$\Delta S = \frac{p_A p_B}{p_A + p_B}\left(\boldsymbol{\beta}^{(k)} - \boldsymbol{\beta}^{(m)}\right)'\boldsymbol{\Sigma}_z\left(\boldsymbol{\beta}^{(k)} - \boldsymbol{\beta}^{(m)}\right) + \sigma^2 \tag{17}$$

This expression is equal to σ^2 when $\boldsymbol{\beta}^{(k)} = \boldsymbol{\beta}^{(m)}$ (merging two subgroups of the same real group). It also may be equal to σ^2 when two z-variables are perfectly correlated, that is, when $\boldsymbol{\Sigma}_z$ is positive semi definite but not positive definite. In all other cases, ΔS is larger than σ^2. So the clustering with criterion S will result in the correct partition.

3.3 Conclusion

The main results of the analysis of the CLV method are summarised in Table 1. Both options of CLV find the correct partition since the cluster criterion is larger when merging variables belonging to two different groups than when merging subgroups of the same real group.

Table 2 Parameters used for simulation

Parameter	Values
Number of observations	– Small samples: 8 observations; – Medium samples: 20 observations; – Large samples: 100 observations
Number of groups	Two groups or three groups
Number of x variables	For data sets with two groups: 40 variables or 120 variables; for data sets with three groups: 60 variables or 180 variables
Group membership	– Equal group size; – Unequal group size
Parameter vectors $\boldsymbol{\beta}$	– For each z variable, there is exactly one group with a non-zero β; – Opposed groups: For each z variable, there is one group with a positive β and one group with a negative β
Importance of noise	– Small noise: 99% of the variance of a x variable is explained by the model; – Medium noise: 80% of the variance of a x variable is explained by the model; – Large noise: 50% of the variance of a x variable is explained by the model

4 Simulation Study

The analysis described above is based on the known covariance matrix. In reality, this covariance matrix has to be estimated. A simulation study has been performed in order to compare the two options of CLV when the clustering is based on a sample covariance matrix. The simulated data correspond to the model stated in Sect. 2. Each simulated **Z** data set was simulated with ten uncorrelated variables of variance 1. Some other details about the simulated data sets are listed in Table 2. There are 144 possible combinations of simulation parameters. For each combination, 100 data sets were simulated, resulting in 14,400 simulated data sets. For each data set, both options of CLV were used. The dendrogram was cut at the correct number of groups, and the obtained partition was compared to the correct partition.

Both options of CLV always found the correct partition for data sets with 100 observations. Also in the case of small noise, the two options perform very well. Only for one data set (out of 1,600 data sets) with small noise and eight observations, none of the two options of CLV finds the correct partition. For all other data sets with a small noise, the correct partition is found by both options. Table 3 shows the results for the other data sets. For both options of CLV, the percentage of data sets for which the correct partition is found, is indicated. For small data sets (eight observations) with a medium noise, CLV without taking into account external data found the correct partition in 76 % of the cases while CLV taking into account external data found this partition in only 68 % of the cases. When there is a medium noise and 20 observations, both options perform very well (almost 100 % of correct partitions). In the case of a large noise, none of the options obtains good results (only 16 % of correct partitions) when there are only eight observations. With 20 observations, both options attain a percentage of correct partitions above 60 %.

Table 3 Percentage of correct partitions

Sample size	Medium noise		Large noise	
8 observations	CLV with external data:	68.4	CLV with external data:	16.4
	CLV without external data:	76.0	CLV without external data:	15.9
20 observations	CLV with external data:	99.6	CLV with external data:	63.6
	CLV without external data:	99.9	CLV without external data:	62.1

5 Conclusion and Perspectives

A model has been stated for the clustering of variables taking into account external data. The CLV method for the case of grouping together only variables with positive correlations has been analysed on the basis of this model. Both options of CLV (with or without external data) have been considered. It has been shown that the hierarchical algorithm finds the correct partition when the parameters of the model are known, no matter which option of CLV is used. A simulation study has confirmed that the performance of the two options is comparable. CLV without external data is equivalent to the Ward algorithm (after centering and transposing the data matrix). So it is not necessary to take into account the external data when the aim is to find the correct partition. For practical purposes, it can hence be better to use the clustering without external data. After cutting the dendrogram, a linear regression can be done in each cluster in order to determine the relationship between z variables and x variables.

This paper only considers the performance of CLV concerning the detection of the correct partition. It is possible that the estimation of the parameter vectors $\boldsymbol{\beta}$ is better when taking into account the external data in the clustering process. Some more work has to be done in order to analyse this parameter estimation.

References

Sahmer, K. (2006). *Propriétés et extensions de la classification de variables autour de composantes latentes. Application en évaluation sensorielle*. Ph.D. thesis, Rennes, France/Dortmund, Germany.

Vigneau, E., & Qannari, E. M. (2002). Segmentation of consumers taking account of external data. A clustering of variables approach. *Food Quality and Preference, 13*(7–8), 515–521.

Vigneau, E., & Qannari, E. M. (2003). Clustering of variables around latent components. *Communications in Statistics – Simulation and Computation, 32*(4), 1131–1150.

Calibration with Spatial Data Constraints

Ivan Arcangelo Sciascia

Abstract We describe an approach that combines the calibrated estimation and the spatial data analysis. In particular we want to describe the possibility of using calibrated estimators when spatial constraints arise in the estimation process with respect to some information that were considered available instead. We describe some possible constraints that could emerge during the estimation procedure and we develop an example of a constrained situation where the constraints are on auxiliary information available and on the density of the units in the spatial domain considered.

1 Introduction

Modern remote sensing systems allow to have a significant amount of data to estimating processes on land areas. These systems coupled with Geographic Information System (GIS) support the design of surveys on spatial domains in disciplines such as biology and forestry science. In this work remote sensing is coupled to the calibrated estimation proposed by Deville and Sarndal (1992) and the application to forest resources is inspired by the work of Opsomer et al. (2007a) where they face the characteristics of the design-based and model-assisted estimators. The work of Opsomer et al. (2007a) stimulated the debate among other authors: Christman (2007), Little (2007) and Ruppert (2007) in the comments to the paper and in the rejoinder by the same authors (Opsomer et al., 2007b). In particular Little in his comments describes concerns about the possibility of using together the peculiarities of design-based and model-assisted estimation paradigms also described in Little (2004).

I.A. Sciascia (✉)
Dipartimento di Scienze Economico-Sociali e Matematico-Statistiche,
Università di Torino, Turin, Italy
e-mail: ivan.sciascia@unito.it

A. Giusti et al. (eds.), *Classification and Data Mining*, Studies in Classification, Data Analysis, and Knowledge Organization, DOI 10.1007/978-3-642-28894-4_11,

Little and the commentaries makes it possible to bring out the essential aspects that are the *uncertainties* of the estimation process that could become *constraints*. The present work aims to spread the debate on design-based estimation for spatial data to develop a note of what are the uncertainties at the beginning of the estimation process that can become constraints for the estimation process.

2 Constraints

The uncertainties for the estimation process are the following: sample design and its peculiarities should be taken into account in the estimation process, the lack of answers that can generate inefficiency estimates, measurement errors and measures disturbances that can cause inefficiency of the estimates, population models and their characteristics in the case of choice of linear or nonlinear models. These uncertainties are budgeted in the design of the survey to lead to unbiased and consistent estimates of quantities such as averages or totals. Another condition that the researcher must address is instead one in which the uncertainties are transformed into constraints during the estimation process. These constraints include: the limitation of information supply, the X_i vectors which describe the auxiliary information available through remote sensing, low density units in the sample, high rate of non-response, high rate of response errors. If you experience any of these constraints during the estimation process the efficiency of the estimators utilized could decrease.

We consider a survey for spatial domains using a calibrated estimator and we define the following notations: a finite population $U = \{1,2,\ldots,N\}$ and a partition of U in d subdomains U_i composed by N_i units, with $i = 1, 2, \ldots, d$, and $\bigcup_{i=1}^{d} U_i = U$ and $\sum_{i=1}^{d} N_i = N$.

We use the calibration estimator to estimate the total of a continuous variable y $\hat{Y}_c = \sum_{j \in U_i} y_{ij}$:

$$\hat{Y}_c = \sum_{i=1}^{n} c_i y_i \tag{1}$$

where c_i is the calibrated weight of ith observation according to constrained optimization.

The constrained optimization for the calibrated estimator is:

$$c_i : min \sum_{i=1}^{n} G_i(c_i, w_i) \tag{2}$$

$$subject\ to \sum_{i=1}^{n} c_i x_i = X. \tag{3}$$

$G_i(c_i, w_i)$ is the generic distance function among c_i and w_i; considering the Euclidean distance among c_i and w_i:

$$G_i(c_i, w_i) = (c_i - w_i)^2 / w_i \tag{4}$$

resolving the optimization problem the general regression estimator (GREG) is obtained:

$$\hat{Y}_c = \sum_{i=1}^{n} c_i y_i = \hat{Y} + \hat{\beta}_c (X - \hat{X}) \tag{5}$$

where β_c is:

$$\hat{\beta}_c = \sum_{i=1}^{n} w_i x_i y_i / \sum_{i=1}^{n} w_i x_i^2. \tag{6}$$

2.1 Constraint on Auxiliary Information

We assume that for calibration estimation with Eq. (5) a matrix $\mathbf{X_N}$ is known for every auxiliary variable considered in the survey. The auxiliary information constraint can reduce the number of the auxiliary variables previous considered and change calibration estimator performances.

2.2 Constraint on Missing Responses

Another constraint we could face is the missing responses on survey. So we could not utilize the response of one or more of the observed units $i = 1, 2, \ldots, d$ with consequent differences in the formula $\sum_{i=1}^{d} N_i = N$ and in the calibrated estimator $\hat{Y}_c = \sum_{i=1}^{n} c_i y_i = \hat{Y} + \hat{\beta}_c (X - \hat{X})$.

2.3 Constraint on the Density of the Units in the Subset Grid

This constraint concerns the presence of a low number of units in the estimation subdomain so to assure the reliability of stratum/units' weights. If during the estimation process it is not possible to respect the stratum proportion we are in presence of a density constraint that we can formalize in a condition:

$$\sharp(U_i) \leq \alpha \tag{7}$$

where $\sharp(U_i)$ is the number of units included in the subdomain space unit U_i and α is a threshold.

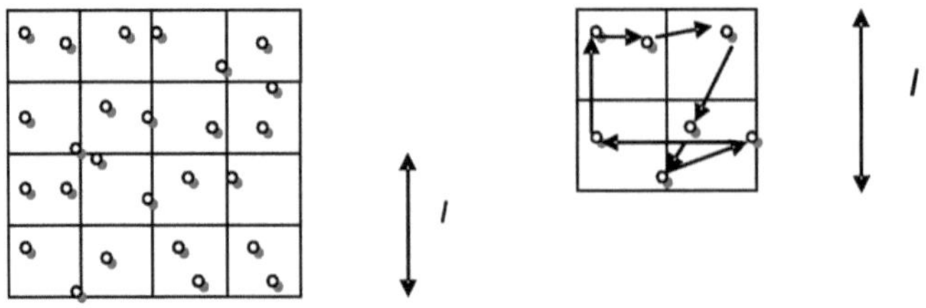

Fig. 1 Simulated sampling points

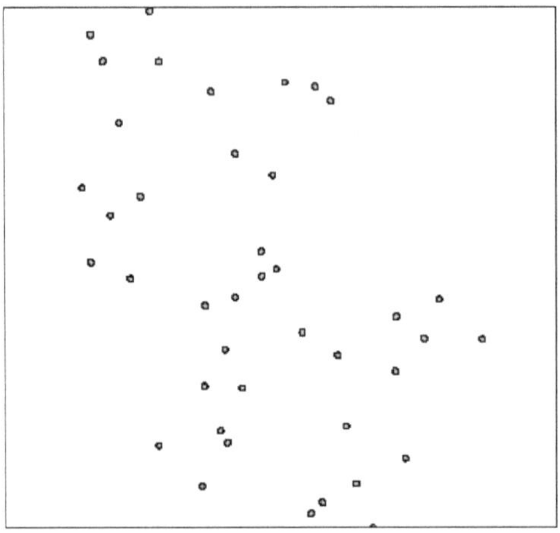

Fig. 2 Ventina sampling points

3 Simulation

Imagine applying two of these three constraints in a survey for spatial domains. We can apply the following simulation to forest data.

Figure 1 describes a simulated example of sampling point distribution with constraint as in Eq. (8); the population grid on the left and the coarse grid on the right with the dimension I. Figure 2 describes a real distribution of sampling points utilized in the forest study of Garbarino et al. (2009). Imagine to apply a two-phase systematic design:

- In the first stage a coarse grid is extracted: size l^2;
- In the second stage a fine grid is extracted from the previous: size $\frac{l^2}{4}$.

The sampling design provides: (a) from remote sensing system we get various x_i of auxiliary variables and (b) the forest area has high density of sampling units (threes or cluster of threes), but the following constraints are arisen: (1) only an auxiliary variable x_i is available, (2) since the area of forest has changed (environment and anthropogenic disturbances) the forest area has low density actually.

We measure the second stage grid variables y_i, x_i on a territorial sample being known the population total X. We estimate the total calibration estimator $\hat{Y}_c$ given the defined constraints. If the number of units in the grid at the second level is lower than stratum proportion of the stratified sampling design we consider the possibility of developing an algorithm that corrects the calibrated estimator, which, in this case, would considers only α units. The estimate would therefore be inefficient and inconsistent with the sampling design. It is possible, however, to carry out the sampling heuristic going back to the grid of the first level and considering a function with the neighborhood of units deployed in space.

The adjusted sampling algorithm is as follows:

1. Selection in the first stage of the grid size l^2;
2. selection in the second stage of the one of the four grid size $l^2/4$ included in the first stage grid size:
   (a) If the stratum proportion is satisfied we proceed with the calibrated estimation using the measurements of y_i, x_i on the units that fall in the grid;
   (b) If the constraint $\sharp(U_i) < \alpha$ is active we develop a correction term that depends on the location of units that fall within the first stage grid.

Considering the simulated distribution of Fig. 1 we defined a grid of sparse data where the units meet the following space constraint:

$$\sharp(z_i) \leq 2 \tag{8}$$

where $\sharp(z_i)$ is the number of units included in the minimum space unit z_i. Consider a finite population of N units where associated with the ith unit are the study variable y_i and the auxiliary variable x_i.

Our correction to $\hat{Y}_c$ involves the distance between the second stage sampled unit and its nearest neighbour in the first stage sampling unit in the following way:

$$\hat{Y}_{cs} = \hat{Y}_c + f(d_{ij}) \tag{9}$$

where $\hat{Y}_{cs}$ is the adjusted calibration estimator for sparse data, $f(d_{ij})$ is a function of distance between the sampled unit i and its nearest neighbour j; $i \in \frac{l^2}{4} grid$ and $j \notin \frac{l^2}{4} grid$, $j \in l^2 grid$.

Note that, if the distance is approximated by a straight line, then it can be computed by using the coordinates of the units (x_i, y_j), (x_i, y_j).

$$d_{ij} = \sqrt{(x_i - x_j)^2 + (y_i - y_j)^2} \tag{10}$$

In this simulation example a sample obtained is composed by two units, the first sampled, the second chosen according to distance. Further work should focus on the explanation of the distance function as a correction term, according to the information available in the constrained situation.

4 Concluding Remarks

Inspired by research on the sampling of forest resources in this work we used the design-based estimation procedure in the presence of constraints. The estimated bound for calibrated spatial data as developed in this argument is applicable to various natural resource estimation problems. The procedure is apt to estimate the two-phase systematic sampling where in phase one the information on an intensive sample grid is extracted and the field visit in the phase two consists in the measurement of interest and ancillary variables on a subset of the phase one grid. The phase one can be divided into two stages: in the first stage the sampling grid is chosen and in the second stage a subset of the sampled grids are chosen. Once the field measurements have been done in the phase two the estimations of population totals can be calculated for the overall spatial population according to the design-based weights. The ancillary information improve the efficiency of the estimators. Supplying the calibration technique the auxiliary information is considered for the survey inference through a parametric linear model and remote sensing and GIS give the availability of a great amount of data to reduce survey costs and improve the precision of the estimates of the survey on natural resources.

Although there are these advantages some constraints could be considered: constraints on auxiliary information, constraints on missing responses and constraints in the density of the units in the subset grid are faced when they are active during the estimation procedure. In this work we suggested some initial consideration on a possible strategy of estimation in the presence of two of the three constraints previous described and we developed a simulation example. We would like to test this strategy with further studies on forest resources.

References

Christman, M. C. (2007). Comment. *Journal of the American Statistical Association, 102*(478), 411–412.

Deville, J. C., & Sarndal, C. E. (1992). Calibration estimators in survey sampling. *Journal of the American Statistical Association, 87*(418), 376–382.

Garbarino, M., Weisberg, P. J., & Motta, R. (2009). Interacting effects of physical environment and anthropogenic disturbances on the structure of European larch (*Larix decidua* Mill.) forest. *Forest Ecology and Management, 257*, 1794–1802.

Little, R. J. (2004). To model or not to model? Competing modes of inference for finite population sampling. *Journal of the American Statistical Association, 99*(466), 546–556.

Little, R. J. (2007). Comment. *Journal of the American Statistical Association, 102*(478), 412–415.

Opsomer, J. D., Breidt, F. J., Moisen, G. G., & Kauermann, G. (2007a). Model-assisted estimation of forest resources with generalized additive models. *Journal of the American Statistical Association, 102*(478), 400–409.

Opsomer, J. D., Breidt, F. J., Moisen, G. G., & Kauermann, G. (2007b). Rejoinder. *Journal of the American Statistical Association, 102*(478), 415–416.

Ruppert, D. (2007). Comment. *Journal of the American Statistical Association, 102*(478), 409–411.

Part II
Data Mining

Clustering Data Streams by On-Line Proximity Updating

Antonio Balzanella, Yves Lechevallier, and Rosanna Verde

Abstract In this paper, we introduce a new clustering strategy for temporally ordered data streams, which is able to discover groups of homogeneous streams performing a single pass on data. It is a two steps approach where an on-line algorithm computes statistics about the dissimilarities among data and then, an off-line algorithm computes the final partition of the streams. The effectiveness of the proposal is evaluated through tests on real data.

1 Introduction

In recent years a wide number of domains is generating temporally ordered, fast changing, potentially unbounded data streams. Some examples are daily fluctuations of stock market, fault diagnosis, web data, network traffic monitoring, electricity consumptions, remote sensors data.

In such cases, due to the arrival of new data at a very high data rate and to the need of providing fast answers to queries on data, it is needed to move from the analysis of data bases to the analysis of data streams. The latter is based on satisfying several constraints which make traditional data mining techniques unusable. In particular: (1) Data elements are on-line collected, (2) Data are potentially unbounded in size, (3) Data after processing are discarded or archived and become not easily available anymore.

A. Balzanella (✉) · R. Verde
Seconda Universitá degli Studi di Napoli, Via del Setificio 81100, Caserta, Italy
e-mail: antonio.balzanella@gmail.com; rosanna.verde@unina2.it

Y. Lechevallier
INRIA, 78153, Le Chesnay cedex, France
e-mail: Yves.Lechevallier@inria.fr

A. Giusti et al. (eds.), *Classification and Data Mining*, Studies in Classification, Data Analysis, and Knowledge Organization, DOI 10.1007/978-3-642-28894-4_12,

As a consequence, strategies for data stream analysis should meet the following design criteria (Ganguly et al., 2009): (1) Time required for processing the incoming observations should be small and constant, (2) Memory requirements have to be reduced with reference to the amount of data to process, (3) Algorithms have to perform only one scan of the data (4) The knowledge about data should be available at any point in time or on user demand.

In this paper, we focus on the clustering of data streams. In this framework, clustering is used to deal with two different challenges. The first one is related to analyze a single univariate or multivariate data stream to discover a partitioning of the observations it is composed of. The second one is based on processing data streams generated by a set of sources (let us think about sensor networks) to discover a partitioning of the sources selves. Our focus is on the second one, which is usually referred as clustering of streams.

If we consider a data stream as a continuously growing time series, clustering of streams shares several topics with the clustering of time series.

In most of cases, clustering methods for time series adapt algorithms for static data to time series, according to two main approaches (Liao, 2005). The first one is based on introducing a dissimilarity function for time series comparison in a conventional clustering algorithm; the second kind of approaches is made by a preliminary step where the raw data are processed by some dimensionality reduction or modeling technique and by a second step which is the running of a clustering algorithm on the results of the first step.

However, the development of clustering algorithms in data stream framework needs further considerations: (1) algorithms should be able to provide suitable data summaries since observations are discarded after the processing; (2) methods should keep into account the evolution of the data over time; (3) algorithms should support the possibility to recover the clustering structure of user defined time periods rather than providing the partitioning of the whole recording period.

The previous requirements imply that adapting conventional clustering algorithms to data stream processing is not sufficient but it is necessary to develop new appropriate methods.

With this aim, we introduce a new strategy for clustering temporally ordered data streams which satisfies the requirements of data stream processing framework and which supports the adoption of user chosen distance measures and the preliminary processing of raw data through some dimensionality reduction technique.

Is is based on discovering a global partitioning of the streams starting from the clustering of local batches of data. The clustering of incoming data batches provides, as output, a set of locally representative profiles of data and allows to update a suitable distance matrix which records the proximities among the streams. The latter, allows to get the partition of the streams by means of an off-line clustering algorithm that is run on the on-line updated proximity matrix.

The paper is organized as follows. Section 2 describes some of the main existing proposals for data streams clustering. Section 3, introduces the details of our strategy. Section 4 presents the evaluation of the algorithm on real data. Finally, Sect. 5 ends the paper with some conclusions and perspectives.

2 State of Art

The data stream clustering problem has been widely dealt in recent years. The most of proposals aim at performing clustering of observations in univariate or multivariate data streams (a wide review is available in Kavitha and Punithavalli 2010). The clustering of multiple data streams is, instead, a more recent challenge. Interesting proposals have been introduced in Beringer and Hullermeier (2006), Dai et al. (2006), Rodriguess and Pedroso (2008) and Balzanella et al. (2011). The first one is an extension to the data stream framework of the *k-means* algorithm performed on time series. Basically, the idea is to split parallel arriving streams into non overlapping windows and to process the data of each window performing, at first, a Discrete Fourier Transform to reduce the dimensionality of data, and then, a *k-means* algorithm on the coefficients of the transformation. The main drawback of this strategy is the inability to deal with evolving data streams. This is because the final data partition only depends on the data of the most recent window.

The second proposal is based on performing a dimensionality reduction of the incoming streams by means of a wavelet transform or a piecewise linear regression and, then, on a suitable clustering strategy on the coefficients of the stream transformation. Although this method is able to deal with evolving data streams, its main drawback is that the approach used for summarization is only based on storing compressed streams.

The third mentioned approach is a top-down strategy named Online Divisive-Agglomerative Clustering (ODAC) where a hierarchy is built according to the correlation among the streams. The proposed divisive approach incrementally updates the distance among the streams and executes a procedure for splitting and aggregating the clusters on the basis of the comparison between the diameter of each cluster to a threshold obtained using Hoeffding bounds. In order to deal with evolution in data, ODAC provides a criterion to aggregate the leafs still based on the clusters diameters and Hoeffding bounds.

Finally, a further proposal deals with the clustering of multiple data streams through a strategy which includes the on-line updating of a co-association matrix using the output of the clustering of local batches of data and the off-line clustering of the co-association matrix by spectral clustering algorithm, for discovering the final partition of the streams.

3 Clustering Data Streams Through the On-Line Clustering of Data Batches

Let $S = \{Y_1, \ldots, Y_i, \ldots, Y_n\}$ be a set of n streams $Y_i = [(y_1, t_1), \ldots, (y_j, t_j), \ldots, (y_\infty, t_\infty)]$ made by real valued ordered observations on a discrete time grid $T = \{t_1, \ldots, t_j, \ldots t_\infty\} \in \Re$. A time window w_f, with $f = 1, \ldots, \infty$, is an ordered subset of T having size s. Each time window w_f frames a subset Y_i^w of Y_i called subsequence where $Y_i^w = \{y_j, \dots, y_{j+s}\}$.

The objective is to find a partition P of S into C clusters such that each stream Y_i belongs to a cluster C_k with $k = 1, \dots, C$ and $\bigcap_{k=1}^{C} C_k = \phi$. Streams are allocated to each cluster C_k with the aim to minimize the dissimilarity within each cluster and to maximize the dissimilarity between clusters.

In order to get a partition P, the incoming parallel streams are, at first, split into non overlapping batches $[Y_1^w, \dots, Y_i^w, \dots, Y_n^w]$ by means of windows of fixed size s. On each batch of data we run a Dynamic Clustering Algorithm (DCA) extended to complex data Diday (1971) and De Carvalho et al. (2004).

The DCA looks for a local partitioning $P^w = C_1^w \cup \dots \cup C_\kappa^w \cup \dots \cup C_K^w$ into K clusters of the current batch of data and an associated set of prototypes $B^w = (b_1^w, \dots, b_\kappa^w, \dots, b_K^w)$ which are the synthesis of the streams behavior in time localized area.

Given a suitable dissimilarity measure $d(\cdot)$, DCA optimizes the following criterion:

$$\Delta(P^w, B^w) = \sum_{k=1}^{K} \sum_{Y_i^w \in C_\kappa^w} d(Y_i^w, b_\kappa^w) \tag{1}$$

The partition P^w and the related prototypes B^w are obtained by the iteration, until convergence, of a representation step, where prototypes are computed, and an allocation step where the subsequences are allocated to the clusters.

3.1 Dissimilarity Updating

The local partition P^w of the streams is the basis for the online updating of the dissimilarities among the streams. In particular, every time a new incoming batch of data is processed by DCA, we update a matrix $A_w = [a_w(i,m)]$ (with $i, m = 1, \dots, n$) in order to record, in each cell, the status of the proximity between a couple of streams.

In our previous proposal Balzanella et al. (2011), we introduced an updating approach based on the co-associations of the streams in the clusters so that each cell $a_w(i,m)$ collects the number of times each couple of streams is allocated to the same cluster of the local partitions P^w. According to this schema, A_w is a proximity graph which can be processed through a spectral clustering algorithm in order to provide the final partition of the streams.

Unlikely to our previously proposal we update the matrix A_w so to consider the internal homogeneity of the clusters and their separation with the aim to improve the performance of the method.

Let $W_\kappa^w = \frac{\sum_{Y_i^w \in C_\kappa^w} d(Y_i^w; b_\kappa^w)}{|C_\kappa^w|}$ be the average distance of the subsequences Y_i^w in a cluster C_κ^w to the correspondent prototype b_κ^w and $D_{i,\kappa} = d(Y_i^w; b_\kappa^w)$ with $Y_i^w \notin C_\kappa^w$, be the distance of a subsequence Y_i^w to the prototype b_κ^w of a cluster C_κ^w to which Y_i^w does not belong.

The cell $a_w(i,m)$ is updated according to the following rule:

$$\begin{cases} a(i,m) = W_\kappa^w, & \text{if } Y_i^w, Y_m^w \in C_\kappa^w \\ a(i,m) = D_{i,\kappa}, & \text{if } (Y_i^w \notin C_\kappa^w) \cap (Y_m^w \in C_\kappa^w) \end{cases} \tag{2}$$

The online updating strategy can be summarized as follows:

---

for Each window w_f **do**
  $(P^w, B^w) = DCA([Y_1^w, \ldots, Y_i^w, \ldots, Y_n^w], K)$
  for $i = 1 : n$ **do**
    for $m = 1 : n$ **do**
      if $Y_i^w, Y_m^w \in C_\kappa^w$ **then**
        $a(i,m) = a(i,m) + W_\kappa^w$
      end if
      if $Y_i^w \in C_\kappa^w$ and $Y_m^w \notin C_\kappa^w$ **then**
        $a(i,m) = a(i,m) + D_{i,\kappa}$
      end if
    end for
  end for
end for

Such updating strategy allows to get a measure of the distance between each pair of streams using only the distance of each subsequence to the prototypes. Note that these are already computed in the DCA clustering, thus no further computation is needed.

An interesting feature of the distance matrix A_w is that it satisfies the additive property. Given two time stamp t_1, t_2 where the distances are available through the matrices A_{t_1} and A_{t_2}, it is possible to recover the updates in $t_2 - t_1$ by the difference $A_{t_2} - A_{t_1}$.

Since in data stream analysis computational constraints impose that it is not possible to store the status of the distance matrix A_w at each updating, we can use the additive property in order to introduce a tilted windows schema for storing recent information at a fine scale and long-term information at a coarse scale. Especially we advice a logarithmic time frame where let us suppose that the most recent matrix stores the proximities until the current time, the remaining slots for recording the proximity information, are, for example, the last quarter (15 min), the next two quarters (ago), 4 quarters, 8 quarters, 16 quarters, and so on, growing at an exponential rate.

This can be realized by the simple deletion of the non required matrices allowing to keep the computational constraints under control.

3.2 Off-Line Partitioning

On user demand or according to a suitable temporal calendar, it is possible to get a partition P of streams of S over a time interval. We can make use of the additive property of the distance matrix A_w in order to get the proximities updates over the queried time interval and then, to process the resulting matrix through a proper clustering algorithm.

We propose to use dynamical clustering algorithm applied to a dissimilarity table *DCLUST* proposed in Diday and Noirhomme-Fraiture (2008) since it is consistent with the criterion optimized for the local clustering by DCA.

The aim of the DCLUST is to partition a set of objects into a fixed number of homogeneous classes on the basis of the proximities between pairs of objects. The optimized criterion is based on the sum of the dissimilarities between elements belonging to the same cluster:

$$\Delta(P,G) = \sum_{k=1}^{C} \sum_{Y_i \in C_k} d(Y_i, g_k) \tag{3}$$

where the prototype $g_k \in G$ of the cluster C_k corresponds to the stream Y_{m^*}: $m^* = argmin(\sum_{Y_i \in C_k} d(Y_i, Y_m))$ with $Y_m \in C_k$. Note that the distances in the algorithm are the values available in the matrix A_w that is, $d(Y_i, Y_m) = a(i,m)$.

The DCLUST algorithm is as follows:

1. *Initialization*: The initial vector of prototypes, G contains random elements of S
2. *Allocation step*: A stream Y_i is allocated to the cluster C_k if and only if $k = \text{argmin}\, d(Y_i, g_k)$ with $k = 1, \ldots, C$
3. *Representation step*: For each $k = 1, \ldots, C$, the prototype g_k representing the class C_k is the stream $Y_{m^*} \in C_k$

The steps 2 and 3 are repeated until convergence.

4 Main Results

In order to evaluate the performance of the proposed strategy we have performed several tests on real datasets. Moreover, we have made a comparison with the well known k-means algorithm applied on stocked data.

We have chosen two datasets in the evaluation process: The first one is made by 76 highly evolving time series, downloaded from Yahoo finance, which represent the daily closing price of random chosen stocks. Each time series is made by 4,000 observation. The second one is made by 179 highly evolving time series which collect daily electricity supply at several locations in Australia. Each time series is made by 3,288 recordings.

We have considered several common indexes to assess the effectiveness of the proposal (see Maulik and Bandyopadhyay 2002). Calinski-Harabasz Index(CH), Davies-Bouldin (DB) Index and Silhouette Width Criterion(SW), are used as internal validity criteria for evaluating the compactness of clusters and their separation. The Rand index (RI) and the Adjusted Rand index (ARI) are used to measure the consensus between the partition obtained by our proposal and the partition obtained using the k-means.

In order to perform the testing, we need to set the following input parameters for the proposed procedure: (1) the size s of each temporal window, (2) the number of clusters K of each local partition P_w, (3) the final number of cluster C to get the partition of S. For the k-means we only need to set the number of clusters C.

The Euclidean distance is used as dissimilarity function in both the procedures after that the raw data have been standardized in order to account for different scales which could dominate the clustering. According to this choice, DCA algorithm on the windows data, becomes a *k-means* where the prototypes are the average of the data in a cluster.

The parameter C has been set, for the first and second datasets, running the *k-means* algorithm using $C = 2, \ldots, 8$. For each value of C we have computed the total within deviance. We have chosen $C = 4$ for the first dataset and $C = 3$ for the second dataset, since these are the values which provide the highest improvement of the clusters homogeneity.

By evaluating, through the mentioned indexes, the partitioning quality for several values of s we can state that the choice of the windows size does not impact on the clusters homogeneity. As a consequence, the choice of the value of such parameter, can be performed according to the kind of required summarization. For example, if we need to detect a set of prototypes for each week of data, we choose a value of the window size which frames the observations in a week.

In our tests, we have used windows made by 30 observations for the first two datasets and 50 for the third one.

The third required input parameter K does not strongly impact on the clustering quality. We have tested this by evaluating the behavior of the Calinski-Harabasz Index and of the Davies-Bouldin Index according to $k = 2, \ldots, k = 10$.

In Table 1 we show the main results of the evaluated indexes:

Table 1 External and internal validity indices

	On-line clustering					k-means clustering		
Dataset	DB	CH	SW	RI	ARI	DB	CH	SW
Power supply	2.104	26.353	0.227	0.95	0.88	2.172	26.504	0.229
Financial data	1.793	15.291	0.307	0.91	0.86	1.754	15.594	0.321

Looking at the values of the internal validity indexes, computed for our proposal and for the k-means on stocked data, it emerges that the homogeneity of the clusters and their separation, is quite similar.

Moreover, the value of the Rand Index and of the Adjusted Rand Index, highlights the strength of the consensus between the obtained partitions.

5 Conclusions

In this paper we have introduced a new strategy for clustering of data streams. It is able to provide a set of summaries of local behaviors by means of the DCA of batches of data and it allows to set at query time the temporal interval over which to provide the partition of the streams. To manage the storage requirements, a tilted windows schema has still been introduced such that recent information are stored at a finer level of detail while older information at a lower detail level. The performance of the algorithm have still compared to a standard algorithm for stocked data.

References

Balzanella, A., Lechevallier, Y., & Verde, R. (2011). Clustering multiple data streams. In S. Ingrassia, R. Rocci, & M. Vichi (Eds.), *New perspectives in statistical modeling and data analysis*. Berlin: Springer.

Beringer, J., & Hullermeier, E. (2006). Online clustering of parallel data streams. *Data and Knowledge Engineering, 58*(2), 180–204.

Dai, B.-R., Huang, J.-W., Yeh, M.-Y., & Chen, M.-S. (2006). Adaptive clustering for multiple evolving streams. *IEEE Transactions on Knowledge and Data Engineering, 18*(9), 1166–1180.

De Carvalho, F., Lechevallier, Y., & Verde, R. (2004). Clustering methods in symbolic data analysis. In D. Banks, et al. (Eds.), *Classification, clustering, and data mining applications* (Studies in classification, data analysis, and knowledge organization, pp. 299–317). Berlin: Springer.

Diday, E. (1971). La methode des Nuees dynamiques. *Revue de Statistique Appliquee, 19*(2), 19–34.

Diday, E., & Noirhomme-Fraiture, M. (2008). *Symbolic data analysis and the SODAS software*. Chichester/Hoboken: Wiley.

Ganguly, A. R., Gama, J., Omitaomu, O. A., Gaber, M. M., & Vatsavai, R. R. (2009). *Knowledge discovery from sensor data*. Boca Raton: CRC.

Kavitha, V., & Punithavalli, M. (2010). Clustering time series data stream – A literature survey. *International Journal of Computer Science and Information Security, 8*(1), 289–294.

Liao, T. W. (2005). Clustering of time series data. A survey. *Pattern Recognition, 38*(11), 1857–1874.

Maulik, U., & Bandyopadhyay, S. (2002). Performance evaluation of some clustering algorithms and validity indices. *IEEE Transactions on Pattern Analysis and Machine Intelligence, 24*(12), 1650–1654.

Rodriguess, P. P., & Pedroso, J. P. (2008). Hierarchical clustering of time series data streams. *IEEE Transactions on Knowledge and Data Engineering, 20*(5), 615–627.

Summarizing and Detecting Structural Drifts from Multiple Data Streams

Antonio Balzanella and Rosanna Verde

Abstract In recent years the analysis of data streams has received a lot of attention. This is motivated by the increase of the number of applications which generate huge amounts of high speed temporal data. Let us think to sensor networks, computer networks, manufactures. Data streams are usually highly evolving, thus mining changes in data is a challenging task. In this paper we will deal with the structural drift detection problem where the aim is to discover and to describe changes in proximity relations among multiple data streams. We will introduce a new strategy whose effectiveness is shown through an application on simulated data.

1 Introduction

A growing number of applicative fields is generating huge amount of temporal data. Some examples are rfids, sensors and web logs across industries including manufacturing, financial services and utilities.

In such contexts, data are sequences of values or events obtained through repeated measurements over time. Often these data arrive at a very high frequency so, the usual data mining tools are not suitable for discovering knowledge in useful times. For this reason it may be needed to define new knowledge discovery processes to apply to continuous, high-volume, open-ended data streams.

Usually, algorithms for data streams mining update, in incremental and on-line way, the knowledge about data by means of proper synopses. These provide suitable summaries which are substantially smaller than their base dataset and allow to discard the data just after they have been processed.

A. Balzanella (✉) · R. Verde
Second University of Naples, Naples, Italy
e-mail: antonio.balzanella@gmail.com; rosanna.verde@unina2.it

A. Giusti et al. (eds.), *Classification and Data Mining*, Studies in Classification, Data Analysis, and Knowledge Organization, DOI 10.1007/978-3-642-28894-4_13,

Due to the high frequency of data arrival and to the storage constraints imposed by the open ended nature of data streams, often, we have a trade off between accuracy and storage requirements. That is, we generally are willing to settle for approximate rather than exact answers.

Since data streams are usually generated by monitoring activities, an important issue is to discover and describe their evolution of over time. Understanding the evolutions in data, involves to be able to take proper decisions on the monitored process. Let's think, for instance, to surveillance systems, to financial markets, to geology.

In data stream framework, the evolution mining can be seen from two different points of view: concept drift detection and structural drift detection (Gama and Gaber, 2007; Ganguly et al., 2009).

The first, concerns to monitor the evolution of the distribution generating the examples of a single univariate or multivariate data stream. The second deals with multiple data streams, searching for changes in the proximity relations among the streams due to changes in single streams.

Concept drift detection has been widely dealt in literature (Aggarwal, 2007; Sebastiao and Gama, 2007), however, structural drift is a more recent challenge.

Structural drift mining looks at the data evolution from a systematic point of view where each data stream is seen as part of a set and a change in its data reflects some structural drift only if impacts on the proximities to the other streams. Thus, at the opposite, if all the observed streams evolve in the same way no structural drift is detected.

Structural drift is strongly related to clustering since changes in proximity relations among streams affect their clustering structure. Thus, it is possible to analyze the changes in the proximities, monitoring how the partitioning of the streams evolves over time.

Consistently with such intuition, an interesting proposal is Da Silva et al. (2009). It is a strategy focused on discovering changes in web usage over time. The basic idea is to split the incoming parallel data streams into non overlapping batches and then, to run a clustering algorithm on each batch as soon as it is recorded. The clustering algorithm provides local partitions of data which represent the proximity relations over each time period. Finally, a measure of the evolution in such proximity relations is obtained by computing the adjusted Rand Index and the F-measure (Maulik and Bandyopadhyay, 2002) between two local partitions.

In this paper, we introduce a new approach which aims both, at measuring, the evolution in a set of multiple streams and to keep track of such evolution through on-line discovered summaries. Unlike to the mentioned approach, in addition to a global change measure between two time periods, we allow to monitor how pairwise proximities evolve over time. Moreover, we allow to set, at query time, the size of the compared temporal intervals.

Basically, we propose a new strategy for the updating of the proximity relations among the streams through a suitable co-association matrix obtained by the clustering of non overlapping batches of data. Then we introduce two measures for evaluating the evolutions in data through the analysis of the changes in the co-association matrix.

The strategy is based on three steps which are performed with the arrival of new observations:

- Local clustering of non overlapping batches of data
- Updating of the co-association matrix to records the proximities among the streams
- Evolution detection by processing the co-association matrix

As we will show in the next sections, this processing schema allows to discover the evolution at several level of granularity and supports tilted windows for storing recent information at a finer detail and older information at a courser detail.

The paper is organized as follows. In Sects. 2 and 3 are shown the details of the proposed strategy. In Sect. 4, an application on simulated data shows the effectiveness the our proposal. Section 5 presents conclusions and perspectives.

2 Monitoring Proximity Relations Among Data Streams

In order to describe the details of our proposal, we need to introduce the main definitions used in the rest of the paper.

Let us note with $S = \{Y_1, \ldots, Y_i, \ldots, Y_n\}$ a set of n data streams $Y_i = [(y_1, t_1), \ldots, (y_j, t_j), \ldots, (y_\infty, t_\infty)]$ made by real valued ordered observations on a discrete time grid $T = \{t_1, \ldots, t_j, \ldots t_\infty\} \in \Re$.

The on-line processing of incoming parallel data streams is performed on batches of data detected through temporal windows.

Formally, a time window w_f with $f = 1, \ldots, \infty$ is an ordered subset of T having size s with $w_f \cap w_{f'} = \emptyset \;\; \forall f \neq f'$. Each time window w_f frames a subset Y_i^w of Y_i, called subsequence such that $Y_i^w = \{y_j, \ldots, y_{j+s-1}\}$.

According to the schema in Fig. 1, on each batch of data, a clustering algorithm is performed in order to provide a local partition of data and a set of suitable summaries. In particular we advice the use of the Dynamic Clustering Algorithm (DCA) proposed in Diday (1971) and De Carvalho et al. (2004).

The DCA looks for the best partitioning of data in K clusters and the representation of each cluster by means of a set of prototypes, optimizing a criterion of internal clusters homogeneity.

The algorithm performs a step of representation of the clusters and a step of allocation of the subsequences to the clusters according to the minimum dissimilarity to the prototypes.

In our case, DCA provides a local partitioning $P_w = C_1^w \cup \ldots \cup C_\kappa^w \cup \ldots \cup C_K^w$ into K clusters of the subsequences framed by w_f and the associated set of prototypes $B^w = (b_1^w, \ldots, b_\kappa^w, \ldots, b_K^w)$. The prototypes allow to summarize the behavior of the streams in time localized windows.

The output partition P_w of each window w_f is used to update a co-association matrix $A_f = [a_f(i,m)]$ (with $i, m = 1, \ldots, n$) which will record the proximities among the streams. For each couple of streams Y_i, Y_m allocated to the same cluster,

Fig. 1 Processing schema

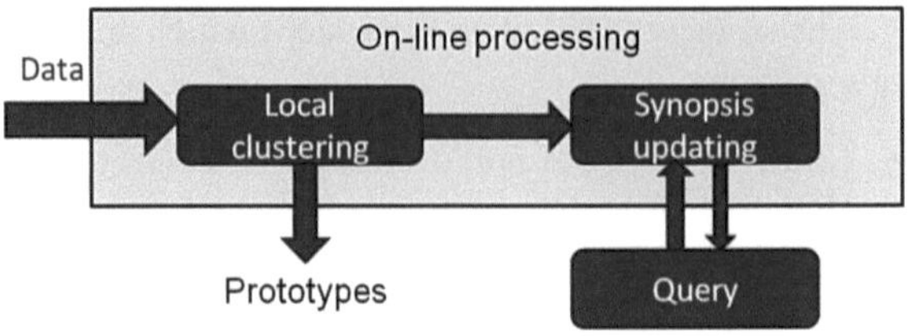

the matrix A_f is updated such that $a_f(i,m) = a_{f-1}(i,m)+1$, while for each couple of streams allocated to different clusters the set value is $a_f(i,m) = a_{f-1}(i,m)$, where a_{f-1} is the proximity corresponding to the previous time window w_{f-1}.

This involves that the following procedure has to be run on each window:

---

for each local cluster $C_\kappa^w \in P_w$ **do**
  Detect all the possible couples of subsequences Y_i^w, Y_m^w which are allocated to the cluster C_κ^w
  for each couple (i,m) **do**
    add 1 to the cells $a_f(i,m)$ and $a_f(m,i)$ of A_f
  end for
end for

For instance, let us assume to have five streams $(Y_1, Y_2, \ldots, Y_5)$ and a local partition $P_1 = (Y_1^w, Y_2^w)(Y_3^w, Y_4^w, Y_5^w)$, the updating of A consists in:

1. Adding 1 to the cells $a_f(1,2)$ and $a_f(2,1)$
2. Adding 1 to the cells $a_f(3,4)$ and $a_f(4,3)$
3. Adding 1 to the cells $a_f(3,5)$ and $a_f(5,3)$
4. Adding 1 to the cells $a_f(4,5)$ and $a_f(5,4)$

The co-association matrix allows to record the status of the proximity relations among the streams at a time stamp. Especially a high value in a cell highlights a strong relation between a couple of streams since it is the consequence of an high number of times in which the couple has been allocated to the same cluster. The opposite is still true since a low value indicates a weak proximity relation.

3 Evaluating the Evolution of Proximity Relations

Starting from the strategy introduced in the previous section for updating the co-association matrix, here we introduce the tools for measuring the evolutions in the proximities.

It is worth of noting that the co-association matrix A satisfies the additive property. Given two time stamps t_1, t_2 where the status of proximity relations is available through the co-association matrices A_{t_1} and A_{t_2}, it is possible to recover the updates in $t_2 - t_1$ by the difference $A_{t_2} - A_{t_1}$.

Consistently with this statement, we introduce the *Structural Change* (SD) measure for evaluating the evolution of the proximity relations between two time periods $\Delta t' = [t_1, t_2]$ and $\Delta t'' = [t_3, t_4]$:

$$SD_{\Delta t', \Delta t''} = \left\| \frac{A_{\Delta t'}}{b_{\Delta t'}} - \frac{A_{\Delta t''}}{b_{\Delta t''}} \right\|_2 \tag{1}$$

where: $\|\cdot\|_2$ is the Frobenius norm;
$A_{\Delta t'}$ and $A_{\Delta t''}$ are the co-association matrices computed for the time intervals $[t_1, t_2]$ and $[t_3, t_4]$;
$b_{\Delta t'}$ and $b_{\Delta t''}$ are the number of windows in $[t_1, t_2]$ and $[t_3, t_4]$.

The previous is a global change measure which provides the strength of the change between two user chosen time periods, however we can still explore the evolutions in pairwise proximities by means of the matrix PC:

$$PC_{\Delta t', \Delta t''} = ABS\left(\frac{A_{\Delta t'}}{b_{\Delta t'}} - \frac{A_{\Delta t''}}{b_{\Delta t''}} \right) \tag{2}$$

where ABS is the absolute value.

It returns a matrix storing, in each cell, the strength of the change in pairwise proximity relations.

The measure $SD_{\Delta t', \Delta t''}$ and the matrix $PC_{\Delta t', \Delta t''}$ introduced here, are based on the availability of a co-association matrix at each time stamp, however, due to the storage constraints of the data stream analysis framework, we need to select particular instants of time at which to store a snapshot of the co-association matrix.

We need to advice a strategy which provides an effective trade-of between the storage requirements and the ability to recall the proximities from different time horizons.

In stream data analysis, people are usually interested in recent changes at a fine scale but in long-term changes at a coarse scale. Naturally, we can register time at different levels of granularity. The most recent time is registered at the finest granularity; the more distant time is registered at a coarser granularity; and the level of coarseness depends on the application requirements and on how old the time point is (from the current time). Such a time dimension model is called a tilted time frame (Han and Kamber, 2006).

We introduce, for our aims, a logarithmic tilted time frame schema where the co-association matrix is stored in multiple granularities according to a logarithmic scale. Suppose that the most recent matrix A stores the proximities until the current time. The remaining slots for recording the proximity information, are, for example, the last quarter (15 min), the next two quarters (ago), 4 quarters, 8 quarters, 16 quarters, and so on, growing at an exponential rate.

According to the strategy proposed for updating the co-association matrix and in particular due to the additive property, the logarithmic time frame schema can be realized by the simple deletion of the non required matrices as shown in Fig. 2

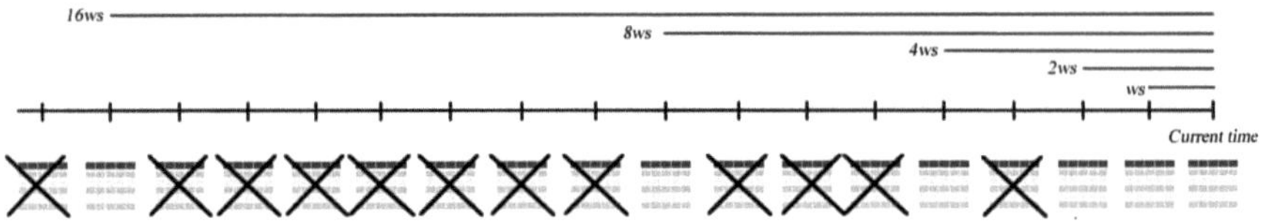

Fig. 2 Tilted time frames

4 Main Results

In order to evaluate the effectiveness of the proposed strategy in discovering changes in proximity relations in data, we have performed several tests on a simulated dataset. The choice of using a simulated dataset allows, in our context, to evaluate the results in a controlled environment where comparisons to concurrent methods can be still performed since the whole set of data is stored on some available media.

We generate four datasets, each one made by $n = 100$ time series having 6,000 observations. In all the datasets, the streams are generated according to two clusters however, at the time stamp $t = 3{,}000$ (in the middle of the series generation), there is a change in the proximities obtained modifying the values of the equations parameters used to generate the data.

The datasets differ in the strength of this proximity change in terms of number of time series which move from the first cluster to the second one and from the second to the first one. Especially, the percentages of change are respectively 50, 25, 15, 5%.

Taking the first dataset as example, this involves that the data partition is $P' = \{Y_1, \ldots, Y_{50}\}\{Y_{51}, \ldots, Y_{100}\}$ for $t \leq 3{,}000$ and $P'' = \{Y_1, \ldots, Y_{25}, Y_{76}, \ldots, Y_{100}\}$ $\{Y_{26}, \ldots, Y_{75}\}$ for $t > 3{,}000$ which corresponds to a shifting of one half of the series.

Our aim is to evaluate if the proposed strategy is able to discover the time point of the evolution, to measure its strength and to understand which streams have the strongest evolution.

In order to run our procedure, we need to set the size of each window s, the number of clusters K for the clustering of data batches and, finally, the two time intervals over which to compute the $SD_{\Delta t', \Delta t''}$ measure and $PC_{\Delta t', \Delta t''}$ matrix.

The choice of the windows size s impacts on the maximum detail at which the procedure is able to discover changes in the proximities. This is because the co-association matrix A is updated each time a new batch of data whose size is s arrives. For our tests we set $s = 50$.

The number K of local clusters has been set to 2 since 2 is the number of clusters existent in data. Since in real applications the true value can be unknown and it can variate over the flowing of data, we have still performed tests using $K = 3, \ldots, 6$ in order to evaluate if the procedure is still able to capture the proximity changes.

Finally,we need to set the two time intervals over which the tests are made. We have chosen to perform a dynamic monitoring such that the first time interval $\Delta t' = [t_1, t_2]$ is made by the most recent 200 observations corresponding to four windows of data, while the second time interval $\Delta t'' = [t_3, t_4]$ is made by the previous 200 observations not included in $\Delta t' = [t_1, t_2]$.

The main results for the $SD_{\Delta t',\Delta t''}$ measure for the tested datasets, are shown in the following figure (Fig. 3):

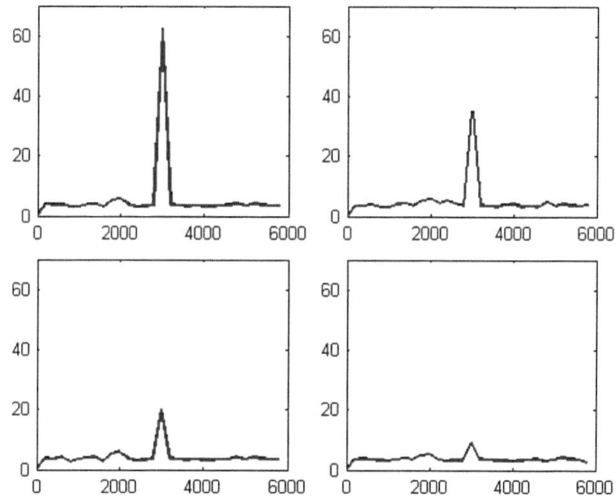

Fig. 3 $SD_{\Delta t',\Delta t''}$ measure over time for the four datasets

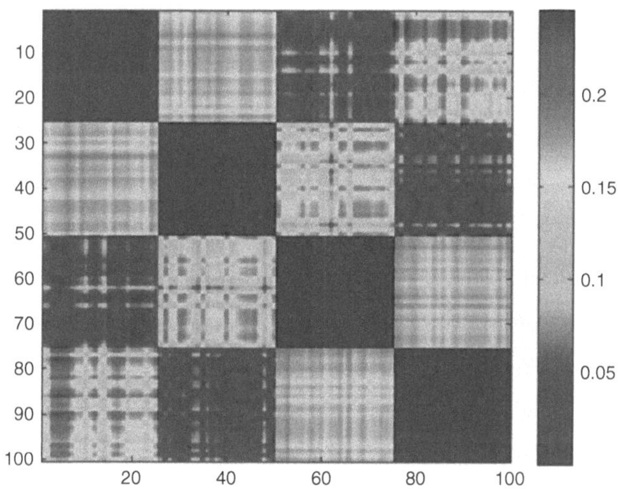

Fig. 4 $PC_{\Delta t',\Delta t''}$ matrix at $t = 3{,}000$ for the first dataset

It is possible to note that the peak in the plot is in correspondence of the middle of the monitoring activity such as expected. Moreover looking at the Fig. 4, which illustrates the values of the $PC_{\Delta t',\Delta t''}$ matrix for the first dataset, it is possible to discover which pairs of streams highlight strong changes in proximity relations at the time point $t = 3{,}000$

5 Conclusions and Perspectives

In this paper we have introduced a new strategy for discovering changes in proximity relations among multiple data streams. Such strategy provides both a measure for evaluating the strength of the change and a representation of the local behaviors by means of cluster prototypes. Future developments will be the evaluation of several distance measure to use in the local clustering and how these impact on the change detection. Moreover other real and simulated datasets will be tested.

References

Aggarwal, C. (2007). *Data streams models and algorithms* (Springer series: Advances in database systems, Vol. 31). New York: Springer.

De Carvalho, F., Lechevallier, Y., & Verde, R. (2004). Clustering methods in symbolic data analysis. In D. Banks, L. House, F. R. McMorris, P. Arabie, & E. Gaul (Eds.), *Classification, clustering, and data mining applications* (Studies in classification, data analysis, and knowledge organization, pp. 299–317). Berlin: Springer.

Da Silva, A., Lechevallier, Y., & de Carvalho, F. (2009). Vers la simulation et la détection des changements des données évolutives d'usage du Web. *EGC*, 453–454.

Diday, E. (1971). La methode des Nuees dynamiques. *Revue de Statistique Appliquee, 19*(2), 19–34.

Gama, J., & Gaber, M. M. (2007). *Learning from data streams: Processing techniques in sensor networks*. Berlin/New York: Springer.

Ganguly, A. R., Gama, J., Omitaomu, O. A., Gaber, M. M., & Vatsavai, R. R. (2009). *Knowledge discovery from sensor data*. Boca Raton: CRC.

Han, J., & Kamber, M. (2006). *Data mining: Concepts and techniques* (Series in data management systems). San Francisco: Kaufmann.

Maulik, U., & Bandyopadhyay, S. (2002). Performance evaluation of some clustering algorithms and validity indices. *IEEE Transactions on Pattern Analysis and Machine Intelligence, 24*(12), 1650–1654.

Sebastiao, R., & Gama, J. (2007). Change detection in learning histograms from data streams. In *Progress in artificial intelligence* (Lecture notes in computer science, Vol. 4874). Berlin/Heidelberg: Springer.

A Model-Based Approach for Qualitative Assessment in Opinion Mining

Maria Iannario and Domenico Piccolo

Abstract Data mining is an increasing area of interest where the collection of large amount of data is characterized by heterogeneous information with respect to origin and content; thus, a high degree of specialization is required for a correct analysis. In this paper, we limit ourselves to consider opinions that are expressed as ordered preferences and may be delivered as rating or ranking evaluations. Such situations are different and deserve careful considerations. In both cases, we discuss the framework of CUB models introduced to analyse the ordinal responses by which people express their opinions. Specifically, the approach may be inserted as a useful routine in data mining area for improving the study of essential features supported by empirical evidence.

1 Introduction

Data mining has been defined as "the non-trivial extraction of implicit, previously unknown, and potentially useful information from data" (Frawley et al., 1991). Specifically, it allows to perform the task of *knowledge discovery* in databases "which is the non-trivial process of identifying valid, novel, potentially useful, and ultimately understandable patterns in data" (Fayyad et al., 1996). A common characteristic of such environment is the large amount of data which are available in heterogeneous manner with respect to both origin and content. As a consequence, data mining theory and methods call for a high degree of specialization where sophisticated statistical analyses and efficient numerical algorithms are heavily involved (Nisbet et al., 2009).

M. Iannario (✉) · D. Piccolo
Department of Statistical Sciences, University of Naples Federico II, Naples, Italy
e-mail: maria.iannario@unina.it; domenico.piccolo@unina.it

A. Giusti et al. (eds.), *Classification and Data Mining*, Studies in Classification, Data Analysis, and Knowledge Organization, DOI 10.1007/978-3-642-28894-4_14,

In this context, we limit ourselves to consider opinions that are the result of explicit questions and are often related to liking, disliking or indifference positions with regard to a specific "object" (item, sentence, vignette, and so on). We will focus on ordinal data involving evaluation, comparison or perception: from a statistical point of view, we model opinions of a large number of subjects when they are expressed as preferences (via ratings and/or rankings) in a consistent way.

The paper is organized as follows: next Section motivates our approach and clarifies the relationship among opinions, evaluation/preferences and expressed scores. Section 3 introduces a class of models (denoted as CUB) able to specify and fit probability structures to observed patterns in preference data; then, a few generalizations are discussed in Sect. 4 and some empirical evidence are shown in Sect. 5. A multivariate model for ranks is proposed in Sect. 6 and checked on a sample of people expressing ordered opinions about some Italian newspapers. Some concluding remarks end the paper.

2 Opinions Expressed by Ordinal Data

Ordinal data measuring opinions and preferences may be available as ratings or rankings, and this information is structurally different. When we collect *ratings* we have numbers which convey the level of a "stimulus" as perceived by respondents, whereas *rankings* generate grades which represent the preferred position of an item in a given list. Formally, rating an "object" on a scale $[1, m]$ produces a unique integer, that is the realization of a discrete *univariate random variable* defined on the support $\{1, 2, \ldots, m\}$. Instead, ranking analysis of m "objects" produces a permutation of the first m integers, that is the realization of a *multivariate random variable* of $(m - 1)$ dimensions. As a consequence, rating/scoring needs a reference scale whereas ranking implies comparison with other items.

It is often difficult to summarize and visualize thousands of expressed preferences on several objects in classical approaches (McCullagh and Nelder, 1989), since standard plots and functions are not immediately related to the final responses. Moreover, some explorative measures (such as average, median, range, etc.) do not capture the relevant components of preferences and evaluations.

To analyse this kind of data, we introduced mixture models of discrete probability distributions based on the focal motivations that lead people to transform their own opinions into an expressed preference through a given ordinal sequence of categories (Piccolo, 2003; D'Elia and Piccolo, 2005; Iannario, 2008). The cornerstone of this approach is the idea that opinions towards an item (fact, person, concept, political party, etc.) derive from lived experience (empirical, emotional, educational, and so on) and logical connections. Opinions are the weighted result of subjective (internal) and objective (external) factors expressed with some fuzziness. This assessment (of qualitative nature) is generally performed in more than one step and may be modified by several circumstances concerning the experiment: thus, we move towards a stochastic approach. When we ask people to express an opinion (preference) on a rating scale, we are collecting the realization of a combination of personal *feeling* and intrinsic *uncertainty*.

In this regard, the awareness of the role of *uncertainty* in empirical information is significantly increasing in data mining approach and some statistical tools have been proposed for detecting specific style of responses: choices at extreme or midpoint values, misunderstanding of questions, fatigue (Chen, 2001), low involvement in the topics, willingness to joke and faking (Liu, 2010), and so on. On the other hand, *feeling* is a sentiment generated by several causes (and well fitted by a unimodal distribution) which are more related to the subjective background of the respondent.

These considerations motivate the nature and the shape of the proposed distribution as a weighted mixture of two discrete random variables generated by unobservable components. Finally, we relate those components to subjects' covariates (for instance, by a logistic link) for strengthening the interpretative content of this framework.

3 Mixture Models for Ordinal Data

Formally, for given m categories, we denote the Uniform and shifted Binomial random variable distributions defined on the support $\{1, 2, \ldots, m\}$ as U_r and $b_r(\xi)$, respectively. Then, we interpret opinions expressed by means of ratings $(r_1, r_2, \ldots, r_n)'$ as realizations of a discrete random variable R whose probability mass distribution:

$$P_r(R = r) = \pi b_r(\xi) + (1 - \pi) U_r, \qquad r = 1, 2, \ldots, m$$

is well defined on the parametric space $\Omega(\pi, \xi) = \{(\pi, \xi) : 0 < \pi \leq 1; 0 \leq \xi \leq 1\}$. Its identifiability has been proved for $m > 3$ (Iannario, 2010).

Model parameters (π, ξ) are related to uncertainty and feeling components, respectively. This random variable will be denoted as CUB model (the acronym may be read as a *C*ombination of *U*niform and shifted *B*inomial) and it turns out to be extremely parsimonious with respect to standard approaches; in fact, it adheres to latent variables paradigm without estimating cut-points.

A strong assumption of the proposed mixture is that it does not strictly require the existence of two subgroups, although this circumstance is admissible within the CUB modelling framework. We are conjecturing that any response is the result of the joint effect of two random variables with a varying weight (estimable from data). Thus, we are modelling the behaviour of each respondent.

Indeed, each respondent acts with a *propensity* to adhere to a thoughtful and to a completely uncertain opinion, measured by (π) and $(1 - \pi)$, respectively. As a consequence, in a rating survey, $(1 - \pi)$ is a *measure of uncertainty*. In addition, this parameter is strictly related to heterogeneity of data as shown by a formal relationship with Gini index (Iannario, 2012).

Then, $(1 - \xi)$ may be interpreted as a *measure of adhesion* toward a given opinion: usually, high values of the responses imply high consideration. Then, in rating studies, the quantity $(1 - \xi)$ increases with agreement towards the item; on the contrary, in ranking analyses ξ increases with preference (since a low rank implies a preferred item).

Since there is a one-to-one correspondence among CUB probability distributions and parameters, we may represent each CUB model as a point in the unit square, as successfully exploited in several fields (Iannario, 2008; Piccolo and D'Elia, 2008; Iannario and Piccolo, 2009, 2010; Corduas et al., 2010). In this manner, CUB model visualization becomes immediate and it adds value to experimental results based on ordinal data. For instance, it is immediate to see how the introduction of covariates and/or the analysis of subgroups modify the behaviour of respondents.

From an operational point of view, we may assess and summarize expressed preferences as a collection of points and test the possible effect of covariates, when space, time and circumstances are modified. In this regard, we generalize the standard assumptions by introducing subjects' and objects' covariates, multi-objects models and shelter choices (Iannario and Piccolo, 2012). Moreover, for a more accurate discussion of these and related inferential issues we refer to Piccolo (2006) whereas for computational purposes a program in **R** is freely available (Iannario and Piccolo, 2009).

4 A Multistage Ranking Model

It is possible to introduce a multistage model in the same framework since ranking may be considered as a stepwise procedure: people choose the best in the given list, then the second best, and so on, until the worst. It generalizes that originally proposed by Plackett (1975) who assessed a multivariate ranking distribution as function of the probability of the object as being set in the first position (Marden, 1995; Xu, 2000). Instead, we consider that the simultaneous probability of a given permutation vector is the product of conditioned probabilities of selecting a rank for an object given that it cannot be assigned to previously assigned ranks. In addition, we assume that marginal distributions of ranks are CUB random variables for saving interpretation and parsimony; indeed, this multivariate model requires $2(m - 1)$ parameters to define probabilities of any permutation vector.

For a given $m > 3$, the discrete distribution of the multivariate random variable $(R_1, R_2, \ldots, R_m)$ with support the set of all permutation vectors $(r_1, r_2, \ldots, r_{m-1}, r_m)$ of the first m integers is:

$$
\begin{aligned}
&P_r(R_1 = r_1, R_2 = r_2, \ldots, R_{m-1} = r_{m-1}, R_m = r_m) \\
&= P_r(R_1 = r_1)\ P_r(R_2 = r_2 \mid R_1 = r_1)\ P_r(R_3 = r_3 \mid R_1 = r_1,\ R_2 = r_2) \ldots \\
&= P_r(R_1 = r_1)\ \frac{P_r(R_2 = r_2)}{1 - P_r(R_2 = r_1)}\ \frac{P_r(R_3 = r_3)}{1 - P_r(R_3 = r_1) - P_r(R_3 = r_2)} \cdots
\end{aligned}
$$

Here, we are denoting the marginal CUB distribution of any object as: $Pr(R_s = r_s) = Pr(R_s = r_s \mid \pi_s, \xi_s)$, for any $s = 1, 2, \ldots, m$.

From an inferential point of view, given a multivariate random sample $\boldsymbol{R} = (r_{i,1}, r_{i,2}, \ldots, r_{i,m-1}, r_{i,m})'$, for $i = 1, 2, \ldots, n$, we will apply an optimization routine to search for a parameter vector $\boldsymbol{\theta} = (\pi_1, \xi_1, \pi_2, \xi_2, \ldots, \pi_{m-1}, \xi_{m-1})'$ which maximizes the log-likelikehood function defined as:

$$\ell(\boldsymbol{\theta}; \boldsymbol{R}) = \sum_{i=1}^{n} \log \Big[Pr(R_1 = r_1, R_2 = r_2, \ldots, R_{m-1} = r_{m-1}, R_m = r_m; \boldsymbol{\theta}) \Big].$$

In order to accelerate the convergence,[1] we let the univariate CUB model estimates as starting values for the multivariate parameters.

By exploiting asymptotic likelihood tests, this model allows to check some hypotheses of interest: homogeneous uncertainty in the responses ($H_0 : \pi_s = \pi, \forall s$), idiosyncratic random behaviour towards some opinion ($H_0 : \pi_t = 0$, for a given t), common feeling for subgroups of items ($H_0 : \xi_{s_1} = \xi_{s_2} = \cdots = \xi_{s_t}$, for a given collection $\{s_1, s_2, \ldots, s_t\}$ of t objects), and so on.

5 A Case Study on Political Opinions

According to some political studies, it is possible to investigate ideologies by asking to define their own Left/Right political self-placement. During the period 29 March-11 April 2010, a survey has been conducted and sample data consisting of several socio-economic covariates have been collected and then validated on 707 subjects. In addition, interviewees have been requested to rank their preferences with respect to $m = 7$ Italian newspapers ("Corriere della Sera", "Il Foglio", "Il Giornale", "Il Manifesto", "La Repubblica", "Libero", "L'Unità"). Respondents are asked to mark their overall political position over a 9-point Likert scale according to the following wording: 1 = "Extremely Left"; 2 = "Strong Left"; 3 = "Left", 4 = "Slightly Left"; 5 = "Moderate-Centre"; 6 = "Slightly Right"; 7 = "Right"; 8 = "Strong Right"; 9 = "Extremely Right". These observational data have been investigated to experiment the possibility of using the proposed modelling framework for detecting significant interrelationships among perceived political self-placement and individual characteristics.

First of all, we model the marginal rank for each newspaper and show in Fig. 1 (left panel) the parametric representation by estimated CUB model. Then, the ranking position given to newspapers has been related to the expressed political self-placement of each respondent and the coefficient of the link function (denoted

[1] Such optimizations have been pursued by the application module CML of *GAUSS* language (Aptech, 2002), aimed at constrained maximum likelihood estimation.

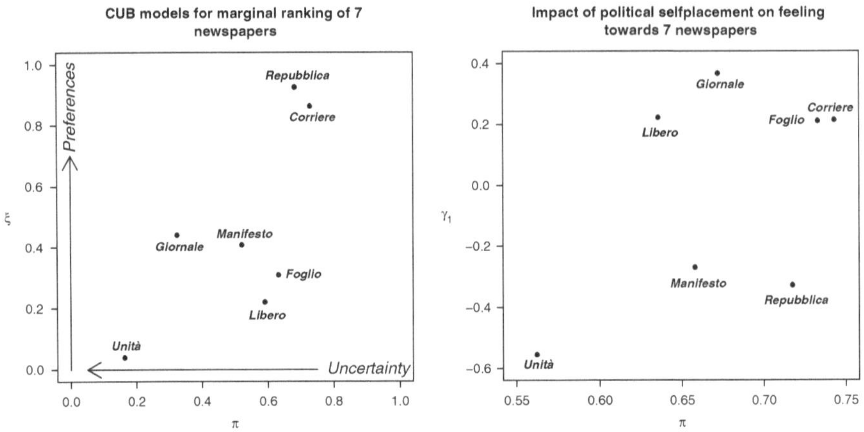

Fig. 1 Representation of estimated CUB models for preferences towards newspapers (*left*) and visualization of the impact of political self-placement on the feeling towards newspapers (*right*)

by γ_1) has been found always significant; empirical evidence shows that the value of this parameter is positively related to Right-wing and Moderate self-placement of respondents. Then, we plot these estimated γ_1 versus estimated π (right panel of Fig. 1) in order to investigate the relationship among uncertainty (generally intermediate) and self-placement for all the newspapers. Thus, we may observe that "Corriere della Sera" is in opposite direction with respect to "La Repubblica" whereas, if we consider only the expressed preferences (left panel of Fig. 1), they have been almost evenly rated. This result shows that both newspapers are the most preferred at all; Left-wing oriented respondents strongly prefer "La Repubblica" whereas "Corriere della Sera" has a wider spectrum of supporters.

This data set enhances that education and opinions towards some sentences are significant covariates to predict political self-placement as a comprehensive CUB model (here not reported) would confirm. Briefly, we analyse the predicted profiles of political self-placement conditioned by some level of education, as in Fig. 2 (left panel).

Finally, right panel of Fig. 2 shows how the parameters of the marginal univariate models of the ranks for the seven newspapers are substantially unchanged if we move to a multivariate setting. Indeed, numerical estimates of the feeling parameters are quite similar to the univariate one and thus multivariate parameters interpretation may be easily derived form the marginal random variables. Instead, uncertainty as measured by the π parameter estimates is somehow modified in small amounts.

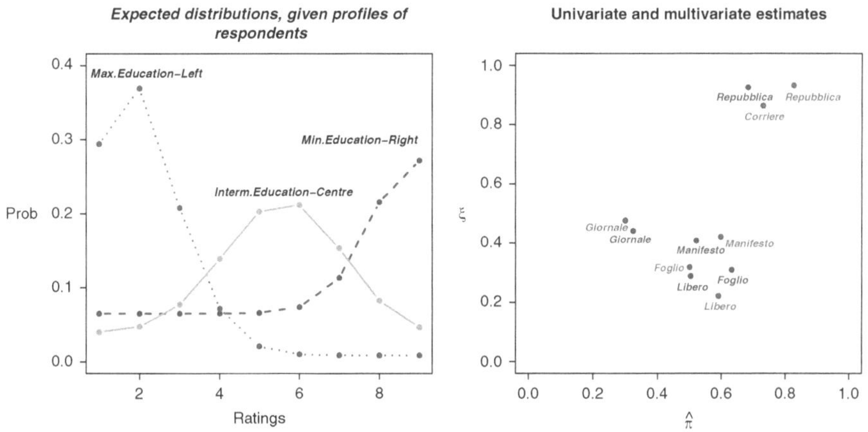

Fig. 2 Estimated profiles of political self-placement as function of education (*left*) and comparison of marginal and multivariate parameter estimates of models for ranking (*right*)

6 Concluding Remarks

Both previous statistical considerations and empirical evidence confirm the usefulness to model opinions expressed as ordinal data when either ratings or rankings about such opinions are available. Indeed, models add qualified and testable interpretation scattered throughout huge amount of heterogeneous opinion data. A benefit of this methodology is the possibility to easily detect and visualize patterns, similarities and anomalies with respect to space, time and circumstances.

Improvements are still necessary in order to produce an automatic classification and discrimination analysis in such contexts but submitted results seem encouraging enough for such objectives.

Acknowledgements The paper has been partly supported by a MIUR grant (code 2008WKHJPK-PRIN2008) for the project: "Modelli per variabili latenti basati su dati ordinali: metodi statistici ed evidenze empiriche" within the Research Unit of University of Naples Federico II (PUC number E61J10000020001).

References

Aptech Systems, Inc. (2002). *Constrained maximum likelihood estimation for GAUSS, version 2.0.3*. Mapley Valley, WA.

Chen, Z. (2001). *Data mining and uncertain reasoning: An integrated approach*. New York: Wiley.

Corduas, M., Iannario, M., & Piccolo, D. (2010). A class of statistical models for evaluating services and performances. In M. Bini, et al. (Eds.), *Statistical methods for the evaluation of educational services and quality of products* (Contribution to Statistics, pp. 99–117). New York: Springer.

D'Elia, A., & Piccolo, D. (2005). A mixture model for preference data analysis. *Computational Statistics & Data Analysis, 49*, 917–934.
Fayyad, U., Piatelsky-Shapiro, G., Smyth, P., & Uthurusamy, R. (1996). *Advances in knowledge discovery and data mining*. Cambridge: AAAI/MIT.
Frawley, W., Piatelsky-Shapiro, G., & Matheus, C. (1991). Knowledge discovery in databases: An overview. In G. Piatetsky-Shapiro & W. Frawley (Eds.), *Knowledge discovery in databases* (pp. 1–30). Menlo Park: AAAI/MIT. Reprinted also in: *AI Magazin*, Fall (1992).
Iannario, M. (2008). A class of models for ordinal variables with covariates effects. *Quaderni di Statistica, 10*, 53–72.
Iannario, M. (2010). On the identifiability of a mixture model for ordinal data. *Metron, LXVIII*, 87–94.
Iannario, M. (2012). Preliminary estimators for a mixture model of ordinal data, Advances in Data Analysis and Applications, 6, DOI 10.1007/s11634-012-0111-5.
Iannario, M., & Piccolo, D. (2009). A program in R for CUB models inference, Version 2.0, Available at http://www.dipstat.unina.it/CUBmodels1/.
Iannario, M., & Piccolo, D. (2010). Statistical modelling of subjective survival probabilities. *GENUS, LXVI*, 17–42.
Iannario, M., & Piccolo, D. (2012). CUB models: Statistical methods and empirical evidence. In R. Kenett & S. Salini (Eds.), *Modern analysis of customer satisfaction surveys*: with applications using R (pp. 231–258) Chichester: Wiley.
Liu, B. (2010). Sentiment analysis and subjectivity. In N. Indurkhya & F. J. Damerau (Eds.), *Handbook of natural language processing* (pp. 627–666). London: Chapman and Hall.
Marden, J. I. (1995). *Analyzing and modeling rank data*. London: Chapman and Hall.
McCullagh, P., & Nelder, J. A. (1989). *Generalized linear models* (2nd ed.). London: Chapman and Hall.
Nisbet, R., Elder, J., & Miner, G. (2009). *Handbook of statistical analysis and data mining applications*. Amsterdam: Elsevier/Academic.
Piccolo, D. (2003). On the moments of a mixture of uniform and shifted binomial random variables. *Quaderni di Statistica, 5*, 85–104.
Piccolo, D. (2006). Observed information matrix for MUB models. *Quaderni di Statistica, 8*, 33–78.
Piccolo, D., & D'Elia, A. (2008). A new approach for modelling consumers' preferences. *Food Quality and Preference, 19*, 247–259.
Plackett, R. L. (1975). The analysis of permutations. *Applied Statistics, 24*, 193–202.
Xu, L. (2000). A multistage ranking model. *Psychometrika, 65*, 217–231.

An Evaluation Measure for Learning from Imbalanced Data Based on Asymmetric Beta Distribution

Nguyen Thai-Nghe, Zeno Gantner, and Lars Schmidt-Thieme

Abstract Hand (Mach Learn 77:103–123, 2009) has shown that the AUC has a serious deficiency since it implicitly uses different misclassification cost distributions for different classifiers. Thus, using the AUC can be compared to using different metrics to evaluate different classifiers. To overcome this incoherence, the H measure was proposed, which uses a *symmetric* Beta distribution to replace the implicit cost weight distributions in the AUC. When learning from imbalanced data, *misclassifying a minority class example is much more serious than misclassifying a majority class example.* To take different misclassification costs into account, we propose using an *asymmetric* distribution (B42) instead of a symmetric one. Experimental results on 36 imbalanced datasets using SVMs and logistic regression show that the asymmetric B42 could be a good choice for evaluating in imbalanced data environments since it puts more weight on the minority class.

1 Introduction

Class imbalance is a phenomenon in which the class distribution is far from the uniform distribution (in this paper, we consider the problem of binary classification). It appears in many machine learning applications such as fraud detection, intrusion detection, and so on (Chawla et al., 2004; He and Garcia, 2009). Most classifiers are designed to maximize the accuracy of their models. Thus, when learning from imbalanced data, they are usually overwhelmed by the majority class examples. This is the main cause for the performance degradation of such classifiers, and is also considered as one of ten challenging problems in data mining research (Yang and Wu, 2006). For example, in fraud credit card detection, suppose that the data

N. Thai-Nghe (✉) · Z. Gantner · L. Schmidt-Thieme
University of Hildesheim, Hildesheim, Germany
e-mail: nguyen@ismll.uni-hildesheim.de; gantner@ismll.uni-hildesheim.de; schmidt-thieme@ismll.uni-hildesheim.de

A. Giusti et al. (eds.), *Classification and Data Mining*, Studies in Classification, Data Analysis, and Knowledge Organization, DOI 10.1007/978-3-642-28894-4_15,

set has 999 legitimate transactions (majority class) and only 1 fraudulent transaction (minority class – the one we would like to detect). To maximize the accuracy, in this case, the classifiers optimized for accuracy will classify all transactions as belonging to the majority class to get 99.9% accuracy. However, this result has no meaning because the fraudulent transaction is misclassified.

Obviously, to evaluate the classifiers in this case, the accuracy metric becomes useless, and the area under the ROC curve (AUC) is commonly used instead (Hanley and McNeil, 1982; Bradley, 1997). The AUC has been widely used to evaluate the performance of classifiers. However, Hand (2009) has shown that using the AUC is equivalent to averaging the misclassification loss over cost ratio distributions which depend on the score distributions. Since the score distributions depend on the classifier itself, employing the AUC as an evaluation measure actually means measuring different classifiers using different metrics. To overcome this incoherence, the "H measure" was proposed, which uses a *symmetric* Beta distribution to replace the implicit cost weight distributions in the AUC. When learning from imbalanced data, misclassifying a minority class example (e.g., a fraud credit card transaction) is much more serious than misclassifying a majority example. Thus, we propose using an *asymmetric* Beta distribution such as $beta(x; 4, 2)$ (called **B42**) instead of the symmetric one as in the H measure.

Furthermore, as investigated in He and Garcia (2009), there are two open problems for the future research in this area: The need for a *standardized evaluation* protocol and the need for *uniform benchmarks* as well as *large data sets* (Jamain and Hand, 2009). The contributions of this work are (1) to propose an evaluation metric for learning from imbalanced data, (2) to introduce large benchmark data sets for systematic studies on imbalanced data.

2 The H Measure: A Replacement for the AUC

To overcome the incoherence of the AUC, the "H measure" was proposed, which is determined by

$$H = 1 - \frac{\int Q(T(c); b, c) u_{\alpha,\beta}(c) dc}{\pi_0 \int_0^{\pi_1} c u_{\alpha,\beta}(c) dc + \pi_1 \int_{\pi_1}^{1} (1-c) u_{\alpha,\beta}(c) dc}. \tag{1}$$

where π_0 and π_1 are prior probabilities; c_0 and c_1 are the misclassification costs for class 0 (majority) and class 1 (minority); $b = c_0 + c_1$ and $c = c_1/(c_0 + c_1)$; $f_0(s)$ and $f_1(s)$ are the probability density functions; and $F_0(s)$ and $F_1(s)$ are the cumulative distribution functions for class 0 and class 1, respectively.

$$Q(t; b, c) \triangleq \{c\pi_1(1 - F_1(t)) + (1 - c)\pi_0 F_0(t)\}b$$

is the loss for an arbitrary choice of threshold t and

$$u_{\alpha,\beta}(c) = beta(c;\alpha,\beta) = \frac{c^{\alpha-1}(1-c)^{\beta-1}}{B(1;\alpha,\beta)}$$

is a symmetric Beta distribution. Please refer to Hand (2009, 2006) for details.

3 B42: A New Evaluation Measure for Learning from Imbalanced Data

The new metric is based on the Beta distributions which are a popular model for random variables (Degroot and Schervish, 2002) with values in the interval [0,1]. The Beta function, also known as Euler's Beta integral (Degroot and Schervish, 2002), is defined as

$$B(1;\alpha,\beta) = \int_0^1 c^{\alpha-1}(1-c)^{\beta-1}dc.$$

It can also be defined by using the Gamma function $B(1;\alpha,\beta) = \frac{\Gamma(\alpha)\,\Gamma(\beta)}{\Gamma(\alpha+\beta)}$.

A generalization of the Beta function is the incomplete Beta function:

$$B(x;\alpha,\beta) = \int_0^x c^{\alpha-1}(1-c)^{\beta-1}dc.$$

The probability density function of the Beta distribution has its mode at $\frac{\alpha-1}{\alpha+\beta-2}$ and is determined by

$$f(x;\alpha,\beta) = \frac{1}{\mathrm{B}(1;\alpha,\beta)}x^{\alpha-1}(1-x)^{\beta-1} = \frac{\Gamma(\alpha+\beta)}{\Gamma(\alpha)\,\Gamma(\beta)}x^{\alpha-1}(1-x)^{\beta-1}$$

As discussed in Hand (2009), the alternative cost distribution, which can replace the implicit cost weight distribution in the AUC, should be a non-uniform one. Thus, an asymmetric Beta distribution would be a good choice for this replacement. As we can see in Fig. 1, for two balanced classes, a symmetric Beta distribution acts as a cost weight distribution, which places most probabilities at 0.5, is used in the H.

However, when learning from imbalanced data sets, misclassifying a minority class example (e.g., in terrorist detection system, misclassifying a terrorist who can carry a bomb on a flight) is much more serious than misclassifying a majority class example (e.g., misclassifying a normal passenger as a terrorist (Thai-Nghe et al., 2010)). Thus, the misclassification cost c_1 (false negative cost) of the minority is much higher than the misclassification cost c_0 (false positive cost) of the majority, therefore, the cost ratio $c = c_1/(c_0 + c_1)$ should be higher than 0.5. For the aforementioned reason, we use the *asymmetric* Beta distribution B42 as a cost weight distribution. The B42 *places higher weight on minority class examples* and is a unimodal distribution with mode at 0.75.

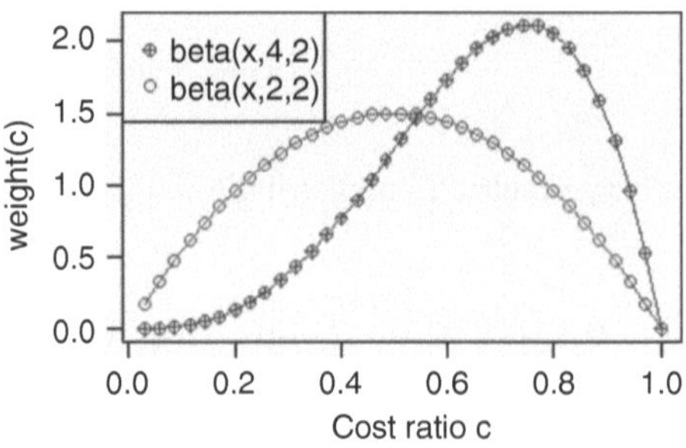

Fig. 1 Symmetric and asymmetric beta distributions

Please note that one can choose some other values for α (e.g., *beta(x,6,2)*, *beta(x,8,2)*...). In those cases, the absolute values of the metrics can be higher, but the relative values are not significantly different. Thus, we decide to use *beta(x,4,2)*.

4 Empirical Evaluation

We compare two classifiers – ℓ_2-regularized logistic regression (ℓ_2-LR) and ℓ_2-loss SVMs (ℓ_2-SVM) – wrt. the AUC, H, and B42 on 36 data sets using five-fold cross-validation. To test for significance, we perform paired t-tests with significance level 0.05. We use the LIBLINEAR software (Fan et al., 2008) with some small modifications to get posterior probability outputs. We perform hyperparameter search as described in (Thai-Nghe et al., 2010) to determine the best hyperparameters for all methods, e.g. the ratio between C^+ and C^-, since our previous results shown that this solution was helpful.

4.1 Data Sets

We have experimented on both small and large data sets collected from the UCI repository (archive.ics.uci.edu/ml) and the Netflix Prize (www.netflixprize.com). We group them into three groups as in Table 1. Nominal attributes are converted to binary numeric attributes. For multi-class data sets, many of them (e.g., RCV1, News20, etc.) were already transformed to binary-class data sets as in the LIBSVM data set library (www.csie.ntu.edu.tw/$\sim$cjlin/libsvmtools/datasets). The remaining multi-class datasets are converted to binary-class using one-versus-the-rest. We encoded the class which has the smallest number of examples as the minority (positive) class, and the rest as the majority (negative) one.

The Netflix (nf) data set originally has 100,480,507 ratings from 480,189 customers for 17,770 movies. To create a binary matrix, in which rows represent users/customers and columns represent items/movies, we assign 1 for each observed rating, and 0 otherwise. We then sort the columns based on their class distributions

Table 1 Comparison of ℓ_2-SVM (base) and ℓ_2-LR using three metrics: B42, AUC, and H

			B42			AUC			H		
Data set	%Minor.	Size	ℓ_2-SVM	ℓ_2-LR		ℓ_2-SVM	ℓ_2-LR		ℓ_2-SVM	ℓ_2-LR	
r2l ◇	0.02	743.0 MB	.413 ± .371	.519 ± .356		.963 ± .082	.980 ± .044		.345 ± .290	.444 ± .278	
nf-005p ♣	0.05	2.6 GB	.006 ± .002	.005 ± .003		.523 ± .033	.617 ± .038	○	.002 ± .001	.003 ± .002	
nf-05p ♡	0.50	2.6 GB	.005 ± .001	.022 ± .005		.628 ± .008	.767 ± .013	○	.002 ± .001	.010 ± .001	
probe ◇	0.83	743.0 MB	.358 ± .364	.543 ± .283		.726 ± .119	.818 ± .122		.324 ± .270	.467 ± .196	
nf-1p ♣	1.00	2.6 GB	.010 ± .001	.039 ± .002		.670 ± .007	.784 ± .004	○	.005 ± .001	.019 ± .001	
appetency ♡	1.78	1.6 GB	.012 ± .003	.026 ± .007		.735 ± .020	.775 ± .014	○	.007 ± .003	.013 ± .005	
ann	2.30	436.0 KB	.615 ± .093	.659 ± .068		.929 ± .025	.984 ± .010	○	.536 ± .105	.591 ± .057	
allhyper	2.70	270.0 KB	.459 ± .125	.328 ± .105	•	.862 ± .084	.886 ± .030		.254 ± .226	.227 ± .116	
w1a	2.97	3.4 MB	.169 ± .183	.214 ± .106	○	.810 ± .108	.853 ± .034		.108 ± .124	.160 ± .133	○
allrep	3.29	275.0 KB	.441 ± .064	.385 ± .058	•	.970 ± .006	.967 ± .008		.343 ± .072	.264 ± .042	•
anneal	4.45	80.0 KB	.635 ± .155	.411 ± .116	•	.957 ± .028	.911 ± .042		.573 ± .181	.392 ± .297	
allbp	4.75	200.0 KB	.374 ± .169	.324 ± .082		.886 ± .135	.859 ± .112		.280 ± .132	.207 ± .081	
hypothyroid	4.77	281.0 KB	.134 ± .179	.343 ± .139	○	.834 ± .103	.843 ± .056		.292 ± .208	.266 ± .143	
nf-5p ♣	5.00	2.6 GB	.112 ± .005	.126 ± .005		.766 ± .036	.804 ± .004		.061 ± .003	.068 ± .003	
sick	6.10	205.0 KB	.625 ± .071	.596 ± .065	•	.929 ± .035	.941 ± .023		.535 ± .080	.517 ± .070	
churn ♡	7.34	1.6 GB	.011 ± .003	.023 ± .005		.605 ± .017	.648 ± .018	○	.005 ± .002	.011 ± .002	
abalone	9.36	259.0 KB	.206 ± .054	.205 ± .054		.847 ± .024	.845 ± .024		.125 ± .043	.122 ± .041	
ijcnn	9.70	7.6 MB	.313 ± .158	.300 ± .150		.861 ± .059	.858 ± .058		.227 ± .144	.214 ± .135	
nf-10p ♣	10.00	2.6 GB	.194 ± .008	.226 ± .010	○	.756 ± .005	.817 ± .006	○	.118 ± .006	.137 ± .007	
nf-20p ♣	20.00	2.6 GB	.223 ± .004	.237 ± .003		.752 ± .003	.772 ± .003		.149 ± .003	.157 ± .003	
hepatitis	20.64	23.0 KB	.422 ± .195	.484 ± .147		.645 ± .368	.736 ± .257		.344 ± .234	.417 ± .341	
transfusion	23.80	24.0 KB	.060 ± .077	.399 ± .253	○	.562 ± .177	.761 ± .178		.132 ± .142	.372 ± .255	
a9a	23.93	3.4 MB	.266 ± .033	.270 ± .009		.792 ± .006	.794 ± .005		.176 ± .006	.178 ± .007	
a2a	24.08	2.3 MB	.293 ± .011	.318 ± .011	○	.792 ± .009	.793 ± .005		.178 ± .013	.180 ± .007	

(Continued)

Table 1 (Continued)

Data set	%Minor.	Size	B42			AUC			H		
			ℓ_2-SVM	ℓ_2-LR		ℓ_2-SVM	ℓ_2-LR		ℓ_2-SVM	ℓ_2-LR	
real-sim	30.75	88.2 MB	.474 ± .278	.768 ± .200	∘	.812 ± .218	.959 ± .057		.455 ± .263	.735 ± .231	∘
url	33.05	2.2 GB	.075 ± .021	.095 ± .034		.546 ± .025	.565 ± .045		.072 ± .020	.086 ± .032	
cod-rna	33.30	25.4 MB	.166 ± .115	.108 ± .076		.586 ± .226	.640 ± .094		.155 ± .106	.079 ± .056	
pima	34.89	41.0 KB	.216 ± .144	.169 ± .085		.587 ± .197	.621 ± .037		.186 ± .124	.158 ± .077	
diabetes	34.90	68.0 KB	.396 ± .052	.398 ± .049		.671 ± .055	.695 ± .052		.144 ± .059	.165 ± .072	
heartdisease	36.00	22.0 KB	.024 ± .034	.108 ± .090	∘	.317 ± .177	.550 ± .105	∘	.023 ± .033	.092 ± .068	
breastcancer	37.99	60.0 KB	.087 ± .096	.138 ± .144		.404 ± .158	.488 ± .176	∘	.099 ± .112	.150 ± .154	
nf-47p ♣	47.00	2.6 GB	.006 ± .001	.007 ± .000		.463 ± .003	.462 ± .002		.007 ± .001	.008 ± .001	
rcv1	47.54	1.2 GB	.006 ± .004	.086 ± .022	∘	.533 ± .015	.559 ± .025		.006 ± .004	.063 ± .017	∘
splice	48.30	699.0 KB	.106 ± .024	.111 ± .027		.584 ± .030	.589 ± .031		.084 ± .020	.086 ± .021	
covtype	48.76	70.0 MB	.087 ± .101	.103 ± .112		.533 ± .106	.543 ± .108		.072 ± .084	.084 ± .090	
news20	49.99	136.7 MB	.099 ± .074	.088 ± .046		.490 ± .251	.486 ± .241		.101 ± .169	.091 ± .146	
Average			.225	.256		.703	.749		.181	.202	

◇: KDD Cup 1999 data set; ♡: KDD Cup 2009 data set; ♣: Netflix data set.

∘, • statistically significant improvement or degradation (level = 0.05)

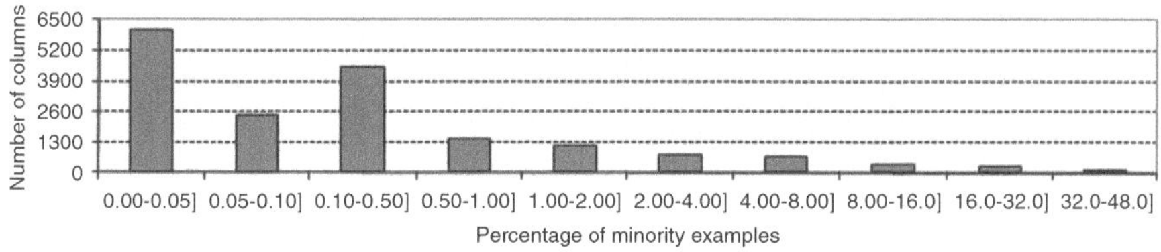

Fig. 2 Distribution of columns and %minority examples on Netflix data set

Table 2 Win/tie/lose results aggregated from Table 1 to three groups; ℓ_2-SVM (base) versus ℓ_2-LR

Groups (12 data sets/group)	%Minority	B42	AUC	H
Group 1 (highly imbalanced group)	0.02–5	**1/8/3**	5/7/0	1/10/1
Group 2	5–30	4/7/1	2/10/0	0/12/0
Group 3 (nearly balanced group)	30–49	**3/9/0**	2/10/0	2/10/0

as in Fig. 2. To create a data set, we choose one column (movie) to be the target, whereas the other columns represent the input features. This way, we can generate 17,770 different data sets. For example, the data set "nf-05p" means that we choose a target column which has 0.5 % minority.

Please note that the last five data sets are not imbalanced. We use them to see how the results are affected when learning from "nearly balanced" to "highly imbalanced" class distributions.

4.2 B42 Versus AUC and H

Table 1 presents the detailed results of three metrics: B42, AUC, and H. The AUC evaluates ℓ_2-LR outperforming ℓ_2-SVM (at least equal) on three groups, while B42 shows that when the imbalance ratio increases, ℓ_2-LR shifts from win (3/9/0) to lose (1/8/3) results, as illustrated in Table 2. For example, 1/8/3 means that the ℓ_2-LR wins one time, ties eight times, and loses three times, compared to the ℓ_2-SVM.

Tables 3 and 4 summarize the agreed/disagreed results of B42 vs. AUC and B42 vs. H on 36 data sets when comparing ℓ_2-LR with ℓ_2-SVM (base). The bold number in the diagonal (e.g. 10 and 7) means that B42 evaluates ℓ_2-LR significantly outperforming/degrading ℓ_2-SVM ten times, but that AUC disagrees on those results, while the reverse is seven times the case. These agreed/disagreed results could be because the B42 places more weight on the minority examples, thus, it has more statistically significant improvements or degradations compared to the AUC and the H. However, a deeper analysis needs to be done here. The results are presented in the next paragraph.

Let us analyze more details for the specific data set "nf-05p" in Fig. 3, which displays an example of cost weight distribution implicitly used in the AUC (for "nf-05p") and explicitly used in B42 and H.

Table 3 The B42 disagrees with the AUC 17 times out of 36 data sets

	Significant difference	No significant difference
Significant difference	2	**7**
No significant difference	**10**	17

Table 4 The B42 disagrees with the H 8 times out of 36 data sets

	Significant difference	No significant difference
Significant difference	4	**0**
No significant difference	**8**	24

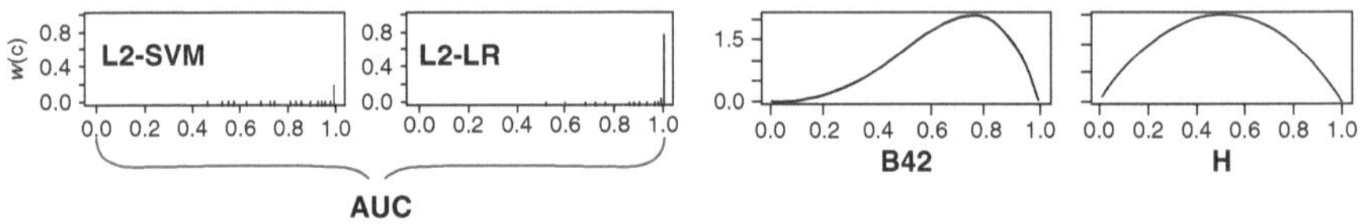

Fig. 3 Cost weight distribution of the AUC (on nf-05p data set), of B42, and of H

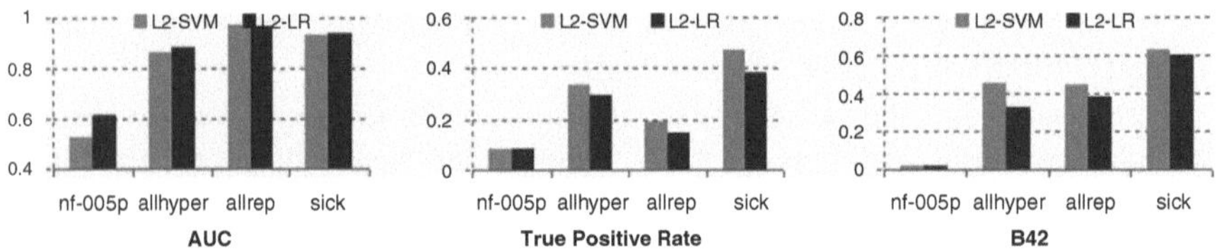

Fig. 4 Typical results of the AUC, the true positive rate, and the B42

Clearly, the AUC places *different cost weight distributions for* ℓ_2*-LR (higher at 1.0) and* ℓ_2*-SVM on the same "nf-05p" data set.* This means that the AUC uses different metrics to evaluate different classifiers (Hand, 2009), while B42 and H use the same distribution for all data sets and classifiers. This is the reason why the result of ℓ_2-LR significantly outperforms ℓ_2-SVM regarding the AUC while it only ties regarding B42. The same situation happens with other data sets e.g., "nf-005p", "nf-1p" and "ann".

Furthermore, Fig. 4 shows four typical results of the AUC, the true positive rate, and the B42. We can see that the AUC evaluates the ℓ_2-LR outperforming the ℓ_2-SVM, however, the true positive rate and the B42 show the reversed results.

The B42 is consistent with the true positive rate while the AUC is not. Thus, if we would like to take the minority class into account then the B42 is a better choice. In addition, the empirical results also show that B42 is not only suitable for evaluating on imbalanced data but also for evaluating on balanced data sets (in group 3 in Table 2, its results are also consistent with other metrics, e.g. the H measure).

5 Conclusion

We propose a new evaluation measure (B42) which bases on the asymmetric Beta distribution to evaluate the classifiers when learning from imbalanced data sets, instead of using the AUC, which has known shortcomings, and the H measure, which fixes the AUC's deficiencies, but is more suitable for balanced class distribution. The experimental results show that the B42 can take the minority class into account when evaluating. We will study how to directly optimize the B42 and H measures as well as study how to apply B42 for multi-class problem.

Acknowledgements The first author was funded by the "Teaching and Research Innovation Grant" Project of Can Tho University-Vietnam.

References

Bradley, A. P. (1997). The use of the area under the ROC curve in the evaluation of machine learning algorithms. *Pattern Recognition, 7*(30), 1145–1159.

Chawla, N. V., Japkowicz, N., & Kotcz, A. (2004). Editorial: Special issue on learning from imbalanced data sets. *SIGKDD Explorations, 6*(1), 1–6.

Degroot, M. H., & Schervish, M. J. (2002). *Probability and Statistics* (3rd ed.). Boston: Addison-Wesley.

Fan, R. E., Chang K. W., Hsieh C. J., Wang X. R., & Lin C. J. (2008). Liblinear: A library for large linear classification. *Journal of Machine Learning Research, 9*, 1871–1874.

Hand, D. J. (2006). Classifier technology and the illusion of progress. *Statistical Science, 21*(1), 1–14.

Hand, D. J. (2009). Measuring classifier performance: A coherent alternative to the area under the ROC curve. *Machine Learning, 77*(1), 103–123.

Hanley, J., & McNeil, B. (1982). The meaning and use of the area under receiver operating characteristics (ROC) curve. *Radiology, 143*(1), 29–36.

He, H., & Garcia, E. A. (2009). Learning from imbalanced data. *IEEE Transactions on Knowledge and Data Engineering, 21*(9), 1263–1284.

Jamain, A., & Hand, D.J. (2009). Where are the large and difficult datasets? *Advances in Data Analysis and Classification, 3*(1), 25–38.

Thai-Nghe, N., Gantner, Z., & Schmidt-Thieme, L. (2010). Cost-sensitive learning methods for imbalanced data. In *Proceeding of IEEE IJCNN10, IEEE CS, Barcelona* (pp. 1–8).

Yang, Q., & Wu, X. (2006). 10 challenging problems in data mining research. *International Journal of Information Technology and Decision Making, 5*(4), 597–604.

Outlier Detection for Geostatistical Functional Data: An Application to Sensor Data

Elvira Romano and Jorge Mateu

Abstract In this paper we propose an outlier detection method for geostatistical functional data. Our approach generalizes the functional proposal of Febrero et al. (Comput 5 Stat 22(3):411–427, 2007; Environmetrics 19(4):331–345, 2008) in the spatial framework. It is based on the concept of the kernelized functional modal depth that we have opportunely defined extending the functional modal depth. As an illustration, the methodology is applied to sensor data corresponding to long-term daily climatic time series from meteorological stations.

1 Introduction

In the last years many scientific contexts, where complete functions presenting spatial dependence are the object of the analysis, have given rise to the development of a new branch of Functional Data Analysis (Ramsay and Silverman, 2005): *Spatial Functional Data Analysis* (*SFDA*).

A common feature of this analysis is to describe and analyze the underlying mechanism that generates specific structures in space and time. This is performed by studying mathematical characteristics of temporal functions and their spatial relations in a continuous setting.

Recent contributions in this framework deal with different topics. Most of them are extension of methods for spatial data to spatially correlated functional data. A good review can be found in (Delicado et al., 2010).

E. Romano (✉)
Dipartimento di Studi Europei e Mediterranei, Seconda Università degli Studi di Napoli, Napoli, Italy
e-mail: elvira.romano@unina2.it

J. Mateu
Departamento de Matematicas, Universitat Jaume I, Castellon de la Plana, Spain
e-mail: mateu@mat.uji.es

A. Giusti et al. (eds.), *Classification and Data Mining*, Studies in Classification, Data Analysis, and Knowledge Organization, DOI 10.1007/978-3-642-28894-4_16,

They range from methods to model correlated variables in space and time to methods that perform spatial prediction of functional data, and passing through explorative methods. To give an outline of these methods, we find for the first group: regression models (Yamanishi and Tanaka, 2003), hierarchical models for spatially correlated functional responses, functional multiple regression and functional analysis of variance for spatially correlated variables (Baladandayuthapani et al., 2008). Methods for the second purpose are: ordinary kriging for function-valued data (Giraldo et al., 2011), pointwise functional kriging predictors (Giraldo et al., 2009b), functional kriging (total model) (Giraldo et al., 2010), a cokriging method for spatial functional data (Nerini and Monestiez, 2008; Nerini et al., 2010). In the explorative context, methods for clustering spatially dependent functional data have been proposed (Giraldo et al., 2009a; Romano et al., 2010).

In this paper we consider the problem of the outlier detection for geostatistical functional data. This type of problem can be considered a natural extension of the spatio-temporal outlier detection. It aims to find spatial outliers also, however instead of just looking for a single snapshot in time, it searches for outliers curves.

In general, *an outlier is an observation that deviates so much from other observations as to arouse suspicion that it was generated by a different mechanism* (Hawkins, 1980).

Their identification plays an important role in the analysis since they adversely lead to model misspecification, biased parameter estimation and incorrect results. Recently, in the functional context this problem has led to the development of some methods. These can be distinguished as follows:

- Robust methods which aim at avoiding unsatisfactory results coming from the problem inherent to the violation of theoretical assumptions together with the presence of certain amount of outlying observations. This is the case of the method using successive likelihood ratio tests and smoothed bootstrapping for identifying outliers (Febrero et al., 2007, 2008).
- Graphical methods for visualizing and identifying functional outliers. These are functional versions of the bagplot and boxplot, which make use of the first two robust principal component scores, apply the bivariate bagplot and map the features of the bagplot into the functional space (Hyndman and Shang, 2010). These are also be considered a strong analog to the classical boxplot by using Modified Band Depth to order functional data, visualizing functional data directly in the functional space (Sun and Genton, 2011).

All of these are generally limited to the assumption of independence between curves. The only existing approach, taking into account the problem of outliers detection for spatially correlated temporal data is based on an adjusted functional boxplot (Sun and Genton, 2012). In particular a constant factor in the functional boxplot is selected to control the probability of correctly detecting no outliers.

Here we propose an outlier detection strategy for geostatistical functional data quite different from this last one. It is an extension of the strategy proposed by Febrero et al. (2007) to the spatial context. Thus a depth measure considering the spatial correlation in the data is proposed.

The main idea consists in using the spatial variability among curves in order to detect the real anomalies in the spatial functional data sets. Specifically we will consider as outlier a curve which does not belong to the spatial variability structure.

The rest of the paper is organized as follows. Section 2 provides a short description of geostatistical functional data. In Sect. 3 the proposed method is presented. A real data application is analyzed in Sect. 4. The paper ends with some discussion and open questions.

2 Geostatistical Functional Data

In environmental, geochemical, geophysical monitoring research, data are characterized by spatial and temporal variability. In all these cases, due to the development of real-time monitoring instruments, data are available as functions of the time spatially located. Typically these data are processed by space-time geostatistical approaches. However, since data flow continuously and at short inter arrival time, traditional methods cannot support the computational costs. In addition, the observations come from a process intrinsically smooth and are affected by error. This motivates the introduction of functional data analysis techniques within geostatistical framework.

Geostatistical functional data concern data providing information about curves (varying over a continuum) located in a fixed subset D of $\mathbb{R}^d$ (d-dimensional Euclidean Space) with positive volume. Let $\{\chi_s : s \in D \subset R^d, t \in T \subset \mathbb{R}\}$ be a stationary isotropic functional random process. We assume to observe a realization of this random process observed at n locations, $\chi_{s_1}(t), \ldots, \chi_{s_i}(t), \ldots, \chi_{s_n}(t)$ for $s_i \in D$ $t \in T$.

The observed data for a fixed site s_i, follows the model:

$$\chi_{s_i}(t_j) = \mu_{s_i}(t_j) + \epsilon_{s_i}(t_j), \quad j = 1, \ldots, M \quad i = 1, \ldots, n \tag{1}$$

where $\epsilon_{s_i}(t_j)$ are residuals with independent zero mean and $\mu_{s_i}(\cdot)$ is the mean function which summarizes the main structure of χ_{s_i}.

We assume that the process is a stationary isotropic functional random process, that formally means that the expected value $\mathbb{E}(\chi_s(\cdot))$ and the variance $\mathbb{V}(\chi_s(\cdot))$ do not depend on the spatial location. In particular we assume that:

- $\mathbb{E}(\chi_s(t)) = m(t), \forall\, t \in T,\ s \in D$.
- $\mathbb{V}(\chi_s(t)) = \sigma^2(t), \forall\, t \in T,\ s \in D$.
- $\mathbb{C}ov(\chi_{s_i}(t), \chi_{s_j}(t)) = \mathbb{C}(h,t)$ where $h = \left\| s_i - s_j \right\|$ $\forall\, s_i, s_j \in D$.
- $\frac{1}{2}\mathbb{V}(\chi_{s_i}(t) - \chi_{s_j}(t)) = \gamma(h,t) = \gamma_{s_i s_j}(t)$ where $h = \left\| s_i - s_j \right\|$ $\forall\, s_i, s_j \in D$.

The function $\gamma(h,t)$ is called semivariogram of $\chi_s(t)$ and can be expressed as:

$$\gamma(h,t) = \gamma_{s_i s_j}(t) = \frac{1}{2}\mathbb{V}(\chi_{s_i}(t) - \chi_{s_j}(t)) = \frac{1}{2}\mathbb{E}\left[\chi_{s_i}(t) - \chi_{s_j}(t)\right]^2 \tag{2}$$

which by using Fubini's theorem, becomes $\gamma(h) = \int_T \gamma_{s_i s_j}(t)dt$ for $\left\| s_i - s_j \right\| = h$.

This function, called trace-variogram function can be estimated by the classical methods of the moments by means of Giraldo et al. (2011):

$$\hat{\gamma}(h) = \frac{1}{|2N(h)|} \sum_{i,j \in N(h)} \int_T \left(\chi_{s_i}(t) - \chi_{s_j}(t)\right)^2 dt \tag{3}$$

where $N(h) = (s_i; s_j) : \|s_i - s_j\| = h$ for regular spaced data and $|N(h)|$ is the number of distinct elements in $N(h)$. When data are irregularly spaced the $N(h)$ becomes $N(h) = \{(s_i; s_j) : \|s_i - s_j\| \in (h - \epsilon, h + \epsilon)\}$ with $\epsilon \geq 0$ being a small value.

In computing the empirical semivariogram, we consider that the functions are expanded in terms of some basis functions by:

$$\chi_{s_i}(t) = \sum_{l=1}^{Z} a_{il} B_l(t) = \mathbf{a}_i{}^T \mathbf{B}(t),\ i = 1, \ldots, n \tag{4}$$

Thus, the empirical trace-variogram function can be expressed by

$$\gamma(h) = \frac{1}{2\,|N(h)|} \sum_{i,j \in N(h)} \left[\left(\mathbf{a}_i - \mathbf{a}_j\right)^T \mathbf{W} \left(\mathbf{a}_i - \mathbf{a}_j\right)\right] \forall i, j \mid \|s_i - s_j\| = h \tag{5}$$

where $\mathbf{a}_i, \mathbf{a}_j$ are vectors of the basis coefficients for the χ_{s_i} and χ_{s_j} curves, and $W = \int_T \mathbf{B}(t) \mathbf{B}(t)^T dt$ is the Gram matrix that is the identity matrix for any orthonormal basis while for other basis as B-Spline basis function, W is computed by numerical integration.

The empirical trace-variogram cannot be computed at every lag distance h and due to variation in the estimation it is not ensured that it is a valid variogram. As in applied geostatistics the empirical trace-variograms are approximated (by ordinary least squares (ols) or weighted least squares (wls) by model function ensuring validity (Cressie, 1993).

3 Outlier Detection for Geostatistical Functional Data

Intuitively a geostatistical functional data is an *outlier* if it is in some way *significantly different* from its *neighbors*. Thus the problem becomes to define the concept of neighborhood and of significantly different. According to these observations, we pay attention to local differences among spatial neighborhood by considering the spatial correlation structure among curves. In this sense, outliers are curves spatially located (geostatistical functional data) which are inconsistent with their neighbourhoods. Given $\chi_{s_1}(t), \ldots, \chi_{s_i}(t), \ldots, \chi_{s_n}(t)$ belonging to a separable Hilbert Space, with $s_i \in D$, $t \in T$. A georeferenced curve is an outlier if it is far in terms of the spatial functional variability, from most and not necessarily all the other observations, since it may be an isolated outliers or a group of outliers throughout D.

In order to quantify the outlygness of a curve we need of an instrument to compare and measure the *behavior* of a georeferenced curve with the others.

Thus we extend the notion of a depth function for functional data by considering the spatial component. Depth measurement for functional data is a concept defined in order to measure the centrality of a curve with respect to a set of curves. It provides a center-outward ordering of a sample of curves in the Hilbert Space from the center (Febrero et al., 2008).

Several depth functions have been proposed in the functional framework (Cuevas et al., 2006, 2007). Among these, we consider and extend the functional modal depth to the spatial functional context.

The modal depth is a functional depth based on the concept of mode. It is defined as the curve most dense to the other curves in the sample. We define the modal spatial functional depth as the curve most *spatially* dense to the other curves. Specifically we replace a spatial functional distance among the curves to a distance among curves in order to consider their spatial dependence. Then, we define the spatial functional modal depth as the curve that attains the maximum value of the following expression:

$$SMD(\chi_{s_i}) = \sum_{j=1}^{n} K\left(\frac{\|\chi_{s_i} - \chi_{s_j}\|_w}{b}\right) \tag{6}$$

where $K : R^+ \rightarrow R^+$ is a kernel function, $\|\cdot\|_w$ is the distance among the georeferenced curves weighted by the spatial variability among the sites, expressed by:

$$d_w(\chi_{s_i}, \chi_{s_j}) = d(\chi_{s_i}, \chi_{s_j})\gamma_{ij}(h) \tag{7}$$

where $d(\chi_{s_i}, \chi_{s_j}) = \sqrt{\int_T (\chi_{s_i}(t) - \chi_{s_j}(t))^2 dt}$ is the distance between the curves without considering the spatial component, and $\gamma_{ij}(h)$ corresponds to the variogram calculated for the distance between sites s_i and s_j.

In the definition of this kernelized spatial functional depth function, the kernel function and the bandwidth b have to be chosen. We chose as in the functional context the Gaussian kernel function and, define a bandwidth parameter by considering the spatial location of the georeferenced curves. The bandwidth parameter, in the functional context, is chosen as the 15th percentile of the empirical distribution of the L^2, L^∞ distance among curves. We propose to choice it following the same procedure, but by evaluating the empirical distribution of $d_w(\chi_{s_i}, \chi_{s_j})$, s_i, $s_j = 1, \ldots, n$, that is the spatial weighted distance among the georeferenced curves at different lag h. Based on these notions, coherently with the functional framework an outlier for a spatial functional sample will have considerably less depth.

Thus, we can generalize the definition of functional outlier to the spatial functional framework as follows:

Definition 1. A georeferenced curve χ_{s_i} is an outlier if $SMD(\chi_{s_i}) < \alpha$ with the cutoff α selected in such a way that the percentage of correct observations poorly identified as outliers was approximately 1 %. That is, such that:

$$Pr(SMD(\chi_{s_i}) < \alpha) = 0.01 \tag{8}$$

Since we have not the theoretical distribution as reference, we need to consider the empirical distribution of *SMD* in order to estimate the value of α. We use the method based on bootstrap of the curves in the original set with a probability proportional to depth (Febrero et al., 2008), in order to obtain a robust estimate of the percentile since the sample could be contaminated by outliers. The procedure to detect outliers is performed at the same way of the functional context as follows:

- Computing the spatial functional depths $SMD(\chi_{s_1}), \ldots, SMD(\chi_{s_n})$.
- Selecting the set of curves such that $SMD(\chi_{s_i}) < \alpha$ for a given cutoff α and remove them from the dataset since these are considered outliers.
- Finally come back to the first step with the new dataset without the outliers found in the second step and repeat this last step until no more are found.

In the following section we underline the main characteristics of the proposed strategy through its application to sensor data.

4 An Application to Sensor Data

The data, available at http://eca.knmi.nl/, refers to long-term daily resolution climatic time series from meteorological stations throughout Europe and the Mediterranean provided from over 40 countries. Most series cover at least the period 1946 up to now. We focus on temperature data recorded from 1, January, 2000 to 28, February, 2010 and perform our test only on 500 stations available over the whole time period. Such type of data may contain a percentage of data outliers which are considerably dissimilar to the rest of the data based on some measurement. Outliers may be noisy observations or alternatively they may indicate abnormal behaviors that are very important and may lead to significant discoveries. Usually the method used to discovery such kind of anomalies are spatio-temporal data, that are not time consuming when there are many time stamp to consider.

The first step before applying the outlier detection strategy is reconstructing a functional form of the time series coming from the signal sources of interest. The functional observations spatially located were smoothed using B-splines basis Since a large number of basis functions Z could cause over fitting, and a too-small Z may lose important aspects of the functions Ramsay and Silverman (2005), the number of basis functions selected by cross-validation. Then in order to evaluate from an

Table 1 Outliers detected by the proposed procedure

	α	SMD	N outliers
Trimming	10.50	10.42	20
Weighting	11.04	10.42	20

explorative point of view, how the spatial dimension influences the results of the analysis, we apply the outlier detection strategy with the modal depth function for functional data and our proposal which considers the spatial component.

The aim in this case is only to monitor how much of the presence of the functional dependence influences the spatial functional variability structure and as consequence, the stability of the analysis.

From the results, we obtain that seven curves identified as functional outliers are a subset of the set of 20 georeferenced curves identified as spatial functional outliers. Looking at the causes of these results we can observe that these data correspond to curves located in the ocean and curves subject to measurement errors.

A further test was performed in order to evaluate the bootstrap procedures used for detecting the cutoff α. Especially we evaluated two different smoothed procedures: Trimming and Weighting (Febrero et al., 2007, 2008). The results in the Table 1 highlight that although the value is different for the two procedures, in both the cases the number of outliers is 20 confirming a small possibility of detecting false outliers. As further result, we note that the introduced spatial weighted norm leads to detect spatial functional outliers different from functional outliers.

5 Conclusion and Perspectives

In this article we have considered the outlier detection problem for spatially correlated functional data. After defining the concept of spatial functional outlier, we have extended a functional data analysis method to the spatial framework by defining the spatial functional kernelized depth function for georeferenced curves. The performance of the method is shown by an application on a real data set. We have extended only one possible concept of functional modal depth for functional data, however it will be interesting to consider other kind of measures. Moreover it will be interesting to compare our method with the approach proposed by Sun and Genton (2012).

References

Baladandayuthapani, V., Mallick, B., Hong, M., Lupton, J., Turner, N., & Caroll, R. (2008). Bayesian hierarchical spatially correlated functional data analysis with application to colon carcinoginesis. *Biometrics, 64*, 64–73.

Cressie, N. (1993). *Statistics for spatial data*. New York: Wiley.

Cuevas, A., Febrero, M., & Fraiman, R. (2006). On the use of bootstrap for estimating functions with functional data. *Computational Statistics and Data Analysis, 51*, 1063–1074.

Cuevas, A., Febrero, M., & Fraiman, R. (2007). Robust estimation and classification for functional data via projection-based depth notions. *Computational Statistics, 22*, 481–496. doi:10.1007/s00180-007-0053-0.

Delicado, P., Giraldo, R., Comas, C., & Mateu, J. (2010). Statistics for spatial functional data: Some recent contributions. *Environmetrics, 21*, 224–239.

Febrero, M., Galeano, P., & Gonzalez-Manteiga, W. (2007). Functional analysis of NOx levels: Location and scale estimation and outlier detection. *Computational Statistics, 22*(3), 411–427.

Febrero, M., Galeano, P., & Gonzalez-Manteiga, W. (2008). Outlier detection in functional data by depth measures, with application to identify abnormal NOx levels. *Environmetrics, 19*(4), 331–345.

Giraldo, R., Delicado, P., & Mateu, J. (2010). Continuous time-varying kriging for spatial prediction of functional data: an environmental application. *Journal of Agricultural, Biological, and Environmental Statistics (JABES), 15*, 66–82.

Giraldo, R., Delicado, P., & Mateu, J. (2011). Ordinary kriging for function-valued spatial data. *Environmental and Ecological Statistics, 18*, 411–426. doi:10.1007/s10651-010-0143-y

Giraldo, R., Delicado, P., & Mateu, J. (2012). Hierarchical clustering of spatially correlated functional data. Statistica Neerlandica. doi:10.1111/j.1467-9574.2012.00522.x

Giraldo, R., & Mateu, J. (2012). Kriging for functional data. In Encyclopedia of Environmetrics, (2nd ed.). Forthcoming

Hawkins, D. M. (1980). *Identification of outliers*. London: Chapman and Hall.

Hyndman, R. J., & Shang, H. L. (2010). Rainbow plots, bagplots, and boxplots for functional data. *Journal of Computational and Graphical Statistics, 19*, 29–45.

Nerini, D., & Monestiez, P. (2008). *A cokriging method for spatial functional data With applications in oceanology*. Long summary sent to "The First International Workshop on Functional and Operational Statistics", Toulouse.

Nerini, D., Monestiez, P., & Manté, C. (2010). A cokriging method for spatial functional. *Journal of Multivariate Analysis, 101*, 409–418.

Ramsay, J. E., & Silverman, B. W. (2005). *Functional data analysis* (2nd ed.). Springer: New York.

Romano, E., Balzanella, A., & Verde, R. (2010). A regionalization method for spatial functional data based on variogram models: An application on environmental data. In *Proceedings of the 45th Scientific Meeting of the Italian Statistical Society*, Padova.

Sun, Y., & Genton, M. G. (2011). Functional boxplots. *Journal of Computational and Graphical Statistics, 20*, 316–334.

Sun, Y., & Genton, M. G. (2012). Adjusted functional boxplots for spatio-temporal data visualization and outlier detection. *Environmetrics, 23*, 54–64.

Yamanishi, Y., & Tanaka, Y. (2003). Geographically weighted functional multiple regression analysis: A numerical investigation. *Journal of Japanese Society of Computational Statistics, 15*, 307–317.

Graphical Models for Eliciting Structural Information

Federico M. Stefanini

Abstract The structure of a Bayesian Network is a priori plausible if the directed acyclic graph has one or more plausible structural features. Expert beliefs about the structure of a Bayesian Network may be substantial but limited both to a subset of nodes or to a set of network features indirectly related to network edges. Complex elicitation tasks involving dozens of reference features may be cognitively too difficult for the expert, unless limited subsets of features may be considered at one time. In this paper chain graph models on descriptors of structural features are proposed as a tool to elicit the degree of belief associated to the structure of a Bayesian Network. An algorithm and a parameterization are developed to support the elicitation.

1 Introduction

Bayesian Networks (BN) are increasingly exploited to represent probabilistic and causal relationships in biology, for example in the so called 'omics' fields (Wilkinson, 2007). Learning the structure of a BN is still challenging for the combinatorial explosion of candidate structures with the increase of the number of nodes.

Bayesian structural learning of BNs depends on the joint distribution function of data, parameters and structure, or the marginal distribution after integrating out model parameters. Widely adopted elicitation techniques define the initial distribution on the space of Directed Acyclic Graphs (DAGs) given a fixed set of nodes by placing restrictions like: a total ordering of nodes, the presence of sharp order constraints on some nodes, the marginal independence of unknowns

F.M. Stefanini (✉)
Department of Statistics "G.Parenti", University of Florence, Florence, Italy
e-mail: stefanini@ds.unifi.it

A. Giusti et al. (eds.), *Classification and Data Mining*, Studies in Classification, Data Analysis, and Knowledge Organization, DOI 10.1007/978-3-642-28894-4_17,

or the existence of a prior network which is a good summary of expert prior beliefs. All these (and others) restrictions aim at making the structural learning task feasible even in medium-to-large networks. In complex problem domains like system biology, the elicitation should be based on structural features (Stefanini, 2008) as preeminent building blocks.

Here recent proposals from the literature (Stefanini, 2009a,b) are reconsidered to provide a general definition of structural features based on propositional logic. The circumstances under which DAGs, undirected graphs and chain graphs are suited for the elicitation are established and a parameterization is developed to provide a bridge between elicited odds and the degree of belief about plausible structural features.

2 Methods

An approach to the elicitation of a prior distribution on the space of DAGs is developed by means of graphical models which define the joint plausibility of several structural features. Background on graphs and Markov properties is provided, for example, by Whittaker (1990), Studeny and Bouckaert (1998) and Cowell et al. (1999).

A graph $\mathcal{G}$ is a pair (V, E) where $V = \{v_1, v_2, \ldots, v_K\}$ is a finite set of nodes and $E \subset V \times V$ is the set of edges. In a Directed Acyclic Graph (DAG) relevant subsets of V are: children $ch(v_i)$ of node v_i, parents $pa(v_i)$ of node v_i, ancestors $an(v_i)$, descents $de(v_i)$ (Cowell et al., 1999). A graph without directed edges is called undirected graph (UG). An UG with $E = V \times V$ is said to be complete. A clique C is a maximal complete subgraph of an UG. A moralized DAG is an undirected graph obtained by joining pairs of nodes sharing a children (if not yet connected) with an undirected edge and by removing the orientations of all edges. A chain graph (CG) is a DAG of chain components $(\tau_1, \tau_2, \ldots)$ where nodes within each chain component may be linked only by undirected edges, and only directed edges may link nodes located in different chain components: the arrow $v_i \rightarrow v_j$ may be present only if v_i belongs to a chain component preceding the chain component in which v_j is located. A moralized chain component is an undirected graph on $\tau_i \cup pa(\tau_i)$ where the subgraph on $pa(\tau_i)$ is complete and where the direction of arrows to τ_i is removed.

A random vector of observables $X_V = X_{v_1,\ldots,v_K} = (X_{v_1}, \ldots, X_{v_K})$ is indexed by a fixed set of nodes V. The sample space is indicated as Ω_{X_V}. A sub-vector defined by indexes $A \subset V$ is X_A. The random vector of descriptors (see next sections) $R = (R_1, R_2, \ldots, R_{n_f})$ is indexed by a set of integers $V_{\mathcal{R}} = \{1, 2, \ldots, n_f\}$.

Hereafter the set of DAGs defined on the fixed set of nodes V is indexed by the variable Z defined on a subset of integers Ω_Z, that is the set of DAGs is one-to-one with Ω_Z. Therefore $p(z \mid \xi)$ indicates the initial distribution to be elicited given the context information ξ.

2.1 Structural Reference Features

Bayesian networks are popular tools to describe causal relationships (Pearl, 2009) and probabilistic conditional independence (Dawid, 2008).

In the causal semantic, an arrow $v_i \rightarrow v_j$ indicates that X_{v_i} is a direct cause of X_{v_j} with respect to the considered variables X_V. In principle the intervention on variable x_{v_i} may determine a change of x_{v_j}. The intervention on a subset of variables $D \subset V$ indicates the external setting of variables in X_D to prescribed values, therefore the system or process is perturbed, not merely observed. The granularity of a causal DAG depends on the variables included in the model, thus if X_{v_i} is a direct cause of X_{v_j} in V then it may became a causal ancestor after enlarging the original set of variables (and nodes). A key property of a casual DAG $\mathscr{G}$ is the stability under external intervention, that is, changes forced on a variable x_v do not change the relationships described by $\mathscr{G}$.

In the probabilistic semantic the conditional independence of a subvector X_A from a subvector X_B given the subvector X_S, with A, B, S disjoint subsets of V, may be derived by d-separation (Pearl, 2009) or equivalently by checking separation in the moral graph of the smallest ancestral set containing $A \cup B \cup S$ (Cowell et al., 1999, Proposition 5.13).

Structural features of a DAG are causal or probabilistic statements referred to a fixed set of nodes V. A candidate structure z is plausible for an expert if it has one or more elicited structural features.

Definition 1 (Structural Feature in a Reference Set). A structural feature (SF) $\mathscr{R}_j(z, w)$ in a reference set $\mathscr{R}$ for the set of DAGs on V is a predicate describing a plausible probabilistic or causal characteristic of the unknown Directed Acyclic Graph $z \in \Omega_Z$. Argument w is in the partition $\mathscr{W}$ of a given numeric domain Ω_W of variable W. An Atomic Structural Feature (ASF) $\mathscr{R}_j(z)$ does not depend on any auxiliary variable W.

A reference set $\mathscr{R}$ is a collection $\mathscr{R} = \{\mathscr{R}_j : j \in J\}$ of SFs indexed in a set J, with $n_f = |\mathscr{R}|$. An ASF $\mathscr{R}_j(z)$ takes the value true or false if applied to a structure z, and $\mathscr{Z}_j = \{z : \mathscr{R}_j(z)\}$ is the equivalence class of DAGs where $\mathscr{R}_j(z)$ is true. Examples of SFs include: $\mathscr{R}_1 =$ 'The maximum number of arrows reaching a node in V is 3', $\mathscr{R}_2 =$ 'Variable X_4 is an immediate cause of variable X_2', $\mathscr{R}_3 =$ 'Variables X_1, X_8 are conditionally independent from X_6 given X_5', $\mathscr{R}_4 =$ 'Node v_2 is a hub node of at least out-degree w'. Features from $\mathscr{R}_1$ to $\mathscr{R}_3$ are atomic, but $\mathscr{R}_4$ is not, because it depends on the auxiliary variable W 'number of arrows leaving a node'; for example the numeric domain could be $\Omega_W = \{4, 5, 6, 7, 8, 9\}$ and $\mathscr{W} = (\{4, 5\}, \{6\}, \{7, 8, 9\})$.

SFs are defined so that the presence of a feature in a candidate structure z increases its plausibility. Moreover, for all $z \in \Omega_Z$, the proposition $\mathscr{R}(z, w)$ is true for one element $w \in \mathscr{W}_i$ or none. A simple representation of the configurations that features in a reference set may take is based on descriptors.

Definition 2 (Descriptors). A descriptor R_i for the SF $\mathscr{R}_i$ is a map

$$R_i : \mathscr{W}_i \times \{false, true\} \rightarrow \{0, 1, \ldots, \mid \mathscr{W}_i \mid\}$$

so that $(w, false) \mapsto 0 \ \ \forall w \in \mathscr{W}_i$ and $(w, true) \mapsto h_w \ \ \forall w \in \mathscr{W}_i$, that is a different integer is associated to each w if *true*. The vector $R = (R_1, R_2, \ldots, R_{n_f})$ defined on the cartesian product $\Omega_R = \bigotimes_{i=1}^{n_f} \Omega_{R_i} = \bigotimes_{i=1}^{n_f} \{0, 1, \ldots, \mid \mathscr{W}_i \mid\}$ is called vector of descriptors. The descriptor of an ASF is defined by *false* $\mapsto 0$ and *true* $\mapsto 1$.

A descriptor takes value 0 to indicate the configuration 'other', that is an unspecified set not in $\mathscr{W}$, or an integer h_w representing the configuration $w \in \mathscr{W}$. The jth configuration of descriptors in R is indicated as $r_j = (r_{1,j}, r_{2,j}, \ldots, r_{n_f,j}) \in \Omega_R$ while a generic configuration is indicated $r \in \Omega_R$. The jth configuration of a sub-vector of R defined by indexes in A is $r_{A,j} \in \Omega_R$, for example $r_{\{1,3\},j} = (r_{1,j}, r_{3,j})$.

Proposition 1 (Partition of DAGs). *The vector R induces the partition* $\mathscr{Z} = \{\mathscr{Z}_r : \forall r \in \Omega_R\}$ *of the fixed set of DAGs* Ω_Z *on* V*, with* $\mathscr{Z}_r$ *the collection of all DAGs showing configuration* r.

The prior distribution $p(z \mid \xi)$ on the space of DAGs on V may be elicited by 'extending the argument' (Stefanini, 2008):

Proposition 2 (Elicitation through descriptors). *The prior distribution* $p(z \mid \xi)$ *is elicited as:*

$$p(z \mid \xi) = \frac{1}{n_{r^{[z]}}} P[R = r^{[z]} \mid \xi] \tag{1}$$

with $n_{r^{[z]}}$ *the cardinality of the equivalence class* $\mathscr{Z}_r$ *in which* z *is located, and* $r^{[z]}$ *the configuration of SFs in DAG* z.

2.2 The Degree of Belief

The elicitation of subjective beliefs based on Eq. (1) may take several forms. The cardinality of equivalence classes in $\mathscr{Z}$ was estimated in (Stefanini, 2008) by Markov chain simulation: $\hat{n}_{r^{[z]}} = (N_r + 1)N_V / N_T$, where N_T is the total number of DAGs uniformly sampled from the space of all DAGs on V, N_V is the size of such space and $N_r \leq N_T$ is number of sampled DAGs showing configuration r. As regards vector R in Eq. (1), in Stefanini (2008) the elicitation of $\mid \Omega_R \mid -1$ parameters has been considered for ASFs, and a reduction of elicitation effort was described under the condition that just the total number of features possessed by a DAG is relevant. Further saving in parameters and elicitation effort may be obtained by exploiting conditional independence relationships believed to hold among SFs, for example through the elicitation of a graphical model on the vector of descriptors R.

Definition 3 (Order relation on descriptors). The ordered partition $\mathcal{O}$ of descriptors is defined by the expert to indicate disjoints subsets of SFs to be jointly considered during the elicitation.

For the sake of brevity, here an outline involving both the order relation and the chain graph is provided. The first question to be posed is 'Which is the subset of descriptors that you would jointly elicit at first, and which C.I. relations would you consider?'. Answers define the first group of nodes and their relations, that is an undirected graph which is the first chain component. The following two steps may be iterated up to the consideration of all descriptors in R: (a) 'Define a subset of descriptors not yet selected to be jointly considered in the elicitation and provide C.I. relations among them', thus the chain component τ_i is defined; (b) 'Is there any dependence on descriptors belonging to previous subsets to be taken into account?', so that arrows representing conditioning information are introduced from some nodes in τ_{i-k} to nodes in $\tau_i, k > 0$. The above questions are prototypes for actual questions which should be formulated in a language as close as possible to the problem domain of interest. The need of control questions is just mentioned here.

In Stefanini (2009b), the special case of a strict order has been introduced thus the elicitation takes the form of odds values for the descriptor R_i conditioned to the configuration of descriptors which precede it in $\mathcal{O}$, and if C.I. relations among descriptors are elicited then they are exploited by defining a DAG of reference features. The second special case has been discussed in Stefanini (2009a) as a tool to revise elicited beliefs and it involves ASFs: the order relation is degenerate thus just one set of unordered descriptors is obtained, and in this case an undirected graphical model describes the C.I. relations among SFs.

Here the general case in which the joint probability distribution of descriptors $p(r_1, r_2, \ldots, r_{n_f} \mid \mathcal{O}, \xi)$ is assumed to be Markov with respect to the elicited CG is introduced. The resulting CG model may support the elicitation if model parameters are cognitively suited for the elicitation, that is interpretable and easy to assess for the expert, at least after some training.

The parameterization introduced here for CG models generalizes that one described in Stefanini (2009a) for UG models of ASFs. For a generic chain component, $p(r_{\tau_i} \mid r_{pa(\tau_i)}, \xi) = D_t^{-1} \prod_{C \in \mathcal{C}} \phi_C(r_C)$, with $\mathcal{C}$ a collection of graph cliques in the moralized chain component τ_i; ϕ_C are non negative and not unique functions called clique potentials; $D_t^{-1}, t \in \Omega_{pa(\tau_i)}$ are normalization constants, one for each value of the conditioning sub-vector $R_{pa(\tau_i)}$; in the first chain component τ_1 just one constant is needed (empty conditioning). Here the log-linear expansion (Whittaker, 1990) of the distribution $p(r_{\tau_i \cup pa(\tau_i)} \mid \xi)$ defined on the moralized chain component τ_i is considered with usual first-cell (treatment) constraint to zero $ln\left(p(r_{\tau_i \cup pa(\tau_i)} \mid \xi)\right) = \sum_{a \subseteq \mathcal{A}} u_a(r_a)$ with $\mathcal{A} = \{\emptyset\} \cup \tau_i \cup pa(\tau_i)$; the model is graphical if the parameters take arbitrary values but with the constraint that if a pair of coordinates is not linked by an edge in the UG then all u-terms containing the selected coordinates are identically null, a constraint indicated as $\mathcal{A}_{\mathcal{G}}$.

Algorithm 1: Elicitation through UGs for chain components

```
Input: A chain graph.
for each chain component τ_i ∈ (τ_1, τ_2, ...) do
    if τ_i ≠ τ_1 then
        Augment the UG of current component with a complete graph on parents
    end if
    Find cliques (model generators)
    for each configuration of conditioning variables (empty for τ_1) do
        for each model generator containing at least one node in τ_i do
            for each subset in the current generator, in a simple-to-complex order, containing
            at least one node in τ_i do
                Elicit current parameter {ψ_a} if not yet elicited;
            end for
        end for
        Calculate the normalization constant
    end for
end for
```

After exponentiating, the odds of a configuration for R_{τ_i} against the 'no-feature' configuration conditional on a configuration $R_{pa(\tau_i)}$ of parents is:

$$\frac{p(r_{\tau_i \cup pa(\tau_i)} \mid \xi)}{\psi(\emptyset, r_{pa(\tau_i)})} = \prod_{\emptyset \neq a \subseteq \mathscr{A}_{\mathscr{G}}} \psi(r_{\tau_i,a}, r_{pa(\tau_i),a}) \tag{2}$$

where $\psi(\emptyset, r_{pa(\tau_i),a})$ is the baseline of no features for τ_i given a conditioning configuration $r_{pa(\tau_i),a}$; $\psi(r_a, r_{pa(\tau_i),a})$ are exponentiated u-terms in which coordinate projection functions are exploited to separate the sub-configuration of a chain component from that of its parents. After eliciting the above odds the model parameters are calculated in a straightforward way, with a normalization constant for each configuration of parents. The revision of elicited values may be performed by inspection of implied conditional odds and of marginal distributions on reduced reference sets (Stefanini, 2008, 2009a).

Algorithm 1 summarizes the steps to elicit the degree of belief. From an operational standpoint, an UG model is defined for each chain component and its parents, but such model is normalized for each configuration taken by parents. Odds values are defined for each configuration of parents while the elicitation of terms in the complete graph made just by parents is skipped because those terms are absorbed into the normalization constants.

2.3 A Simple Case Study

A recently published case study (Stefanini et al., 2009) deals with classical biomarkers in breast cancer studies without exploiting structural prior information. Here a set of plausible features is defined by a trained expert, who also defines the order $\mathscr{O}$.

Table 1 A simple case study

Variable	Label	Reference features	Type
Age	AGE	$\mathscr{R}_1$ ='ER regulates many proteins'	Causal
Oestrogen receptor	ER	$\mathscr{R}_2$ ='AGE regulates ER'	Causal
Progesterone receptor	PR	$\mathscr{R}_3$ ='AGE regulates PR'	Causal
Ki/47 protein	NEU	$\mathscr{R}_4$ ='NEU depends on up to 3 variables'	Probabilistic
Protein P53	P53	$\mathscr{R}_5$ ='ER regulates NEU'	Causal
Proliferation index	PROLN	$\mathscr{R}_6$ ='ER is informative for PROLN'	Probabilistic
		$\mathscr{R}_7$ ='PR is informative for PROLN'	Probabilistic

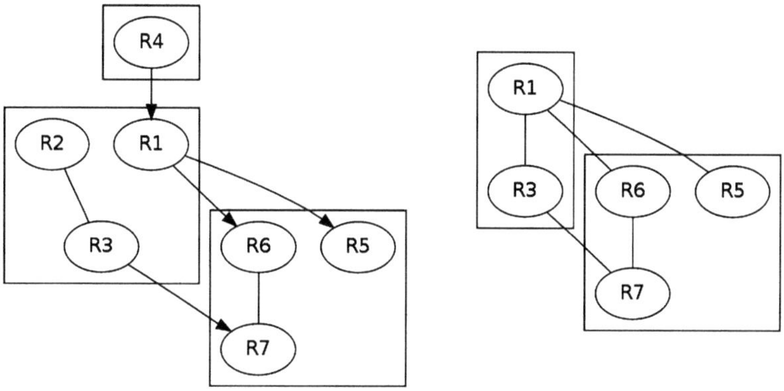

Fig. 1 A chain graph of features (*left*) and the working UG for last chain component (*right*)

In Table 1, first column left, the list of variables is shown, together with a label (second column), the elicited reference features (third column) and the kind of SF (forth column from left). The elicited order is $\mathscr{O} = (\{R_4\}, \{R_1, R_2, R_3\}, \{R_5, R_6, R_7\})$, thus the first chain component is $\tau_1 = \{4\}$, while subsequent chain components are $\tau_2 = \{1, 2, 3\}$ and $\tau_3 = \{5, 6, 7\}$. In Fig. 1, the elicited CG of all descriptors is shown on the left, while on the right the last chain component is shown after moralization into a working UG. The illustration here is limited to a sketch of the quantities to be elicited for the last component of the above CG model involving five binary descriptors, without reaching the numerical assessment of $p(z \mid \xi)$ and of the posterior distribution given a set of collected data. Indexes of two descriptors, R_1 and R_3, are parents of indexes of R_5, R_6, R_7 thus 4 conditioning configurations are possible, namely R_1, R_3 equal to: $(0,0)$, $(1,0)$, $(0,1)$, $(1,1)$. The example involves ASFs, thus a simplification of notation exploiting binary descriptors is possible and the graphical multiplicative model for the moralized chain component is $p(r_5, r_6, r_7, r_1, r_3 \mid \xi) = \psi_\emptyset \cdot \psi_1^{r_1} \cdot \psi_3^{r_3} \cdot \psi_{1,3}^{r_1 r_3} \cdot \psi_{1,5}^{r_1 r_5} \psi_{1,6}^{r_1 r_6} \psi_{3,7}^{r_3 r_7} \psi_5^{r_5} \cdot \psi_6^{r_6} \cdot \psi_7^{r_7} \cdot \psi_{6,7}^{r_6 r_7}$, with all $\Omega_{R_i} = \{0, 1\}$. After absorbing parameters involving only conditioning variables into the normalization constant, the odds values to be elicited take the following form: $\frac{p(r_5, r_6, r_7, r_1, r_3 \mid \xi)}{\psi(\emptyset, r_1, r_3)} = \psi_{1,5}^{r_1 r_5} \psi_{1,6}^{r_1 r_6} \psi_{3,7}^{r_3 r_7} \psi_5^{r_5} \cdot \psi_6^{r_6} \cdot \psi_7^{r_7} \cdot \psi_{6,7}^{r_6 r_7}$ and values of model parameters follow by straightforward algebraic manipulation. Similar expressions may be obtained for chain components τ_1 and τ_2.

3 Discussion

The elicitation of prior beliefs about the structure of a Bayesian network may benefit from flexibility and parsimony of graphical models with some advantages over standard practice. First, hard constraint on structures are weakened to soft constraints, e.g. the belief 'enzyme *A* regulates protein *B*' may be elicited with its inherent uncertainty. Second, the typically huge size of the space of structures is reduced to a smaller but equally informative space of structural features. Third, the cardinality of equivalence classes is properly taken into account.

A general reference set of SFs may be logically incompatible with DAGs. A logical analysis of SFs should be always performed to associate a null prior probability value to configurations of SFs not consistent with DAGs. The auxiliary Monte Carlo simulation performed to estimate the cardinality of equivalence classes suggests critical configurations of SFs that deserve further inspection: a null number of DAGs in large samples may be a sampling zero or a structural zero.

Further work is needed to make the approach suited to daily use, for example by developing a software targeted towards experts in specific problem domains to support the elicitation without the help of a statistician. Human cognitive peculiarities and shortcut strategies involved in this scheme should be characterized to develop corrections for potential causes of bias. Similarly, the effect of the way propositions are formulated is open to investigation.

References

Cowell, R. G., Dawid, P. A., Lauritzen, S., Spiegelhalter, D. J. (1999). *Probabilistic networks and expert systems*. New-York: Springer.

Dawid, A. P. (2008). Beware of the DAG. *Journal of Machine Learning Research, NIPS 2008 Workshop on Causality and Conference Proceedings, 6*, 59–86.

Pearl, J. (2009). Causal inference in statistics: An overview. *Statistics Surveys, 3*, 96–146.

Stefanini, F. M. (2008). Eliciting expert beliefs on the structure of a Bayesian Network. In *PGM2008: Probabilistic Graphical Models 2008*, Hirtshals.

Stefanini, F. M. (2009a). The revision of elicited beliefs on the structure of a Bayesian Network. In *S.Co. 2009, Book of short papers*, Milano.

Stefanini, F. M. (2009b). Prior beliefs about the structure of a probabilistic network. In *SIS2009, Book of short papers*, Pescara.

Stefanini, F. M., Corradini, D., Biganzoli, E. (2009). Conditional independence relations among biological markers may improve clinical decision as in the case of triple negative breast cancers. *BMC Bioinformatics, 10*(Suppl 12), S13. doi: 10.1186/1471-2105-10-S12–S13.

Studeny, M., Bouckaert, R. R. (1998). On chain graph models for description of conditional independence structures. *Annals of Statistics, 26*(4), 1434–1495.

Whittaker, J. (1990). *Graphical models in applied multivariate statistics*. New-York: Wiley.

Wilkinson, D. J. (2007). Bayesian methods in bioinformatics and computational systems biology. *Briefings in Bioinformatics, 8*(2), 109–116.

Adaptive Spectral Clustering in Molecular Simulation

Marcus Weber

Abstract In this chapter, PCCA+ is described as a special spectral clustering algorithm which is applicable for molecular simulation data. From a mathematical point of view, only PCCA+ is able to correctly identify the physical timescales of molecular motion. In order to decrease the statistical error of this timescales analysis, an adaptive clustering algorithm is necessary.

1 Introduction

Molecules are made of atoms. The *conformational state* x of an N-atoms molecule can, therefore, be described by $3N$ cartesian coordinates in a three dimensional space. The different conformational states are not equally probable. Assuming a canonical ensemble (i.e. observing molecules at constant temperature), there is a probability density function $\pi(x), x \in R^{3N}$. This function can be approximated (up to an unknown scaling factor) by computational methods on the basis of molecular modelling. This probability density function π is (never exactly but) approximately zero for almost all $3N$-dimesional vectors x. Only a "few" conformational states x are physically meaningful. This meaningful set of states is not a convex set in R^{3N}. It is a rather complicated network. There are some "islands" (clusters) in R^{3N} connected by "narrow canyons". A very important task in computational molecular design is to identify these clusters of conformational states and to characterize

M. Weber (✉)
Zuse Institute Berlin (ZIB), Takustrasse 7, D-14195 Berlin, Germany
e-mail: weber@zib.de

A. Giusti et al. (eds.), *Classification and Data Mining*, Studies in Classification, Data Analysis, and Knowledge Organization, DOI 10.1007/978-3-642-28894-4_18,

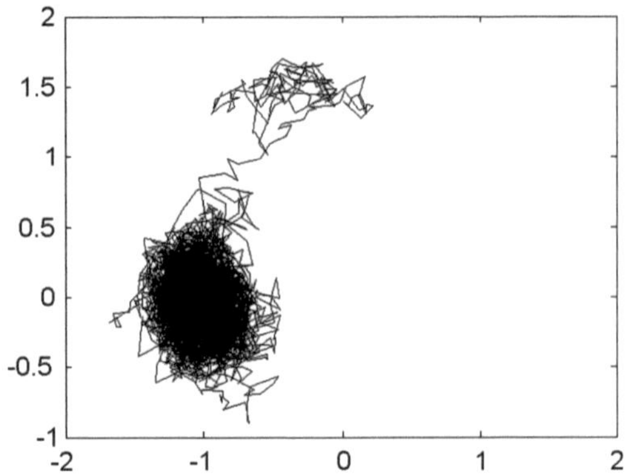

Fig. 1 MCMC simulation of a 3-clusters probability density function (10,000 steps). Starting in the upper central cluster, the trajectory crosses a barrier to the left cluster and is trapped. The third cluster is not found

the network of these clusters (Bujotzek and Weber, 2009). Usually, a trajectory-based solution to this problem is applied, which is not intended in this chapter. The *first step* of such an approach is given by generating a sample of conformational states $\{x_1, \ldots, x_n\} \subset R^{3N}$ which is distributed according to the density function π. This is by far not trivial due to the high-dimensional space R^{3N} and due to the complicated (and unknown) network of clusters. In Fig. 1, an example for the difficulties of such a trajectory-based approach is shown. The sampling points in this figure are generated by a Markov chain Monte-Carlo method MCMC (either by solving equations of motion or by stochastic methods). The density function π is symmetric with regard to the y-axis. Three clusters of points should be found, but only two clusters were identified. The reason is, that the MCMC method hardly finds the "narrow canyons" between the clusters and is trapped inside the clusters. In a *second step* of the trajectory-based approach, a clustering algorithm is applied to the generated set of samples. In this second step, transitions between the clusters are counted in order to characterize the transition network. Unfortunately, due to the rare transitions in MCMC, this information is based on bad statistics. Obviously, this two-step approach has a lot of disadvantages with regard to statistical errors and reproducability of the results (Röblitz, 2008; Weber et al., 2007).

In fact, molecular dynamics simulations or MCMC methods are not very suitable to generate enough statistics for the aimed cluster analysis. These sampling methods mainly collect data from the "boring" part of the conformational space (the basins of attraction, the clusters). They only rarely sample transitions between the clusters, which is the most important information to estimate transition rates. Obviously, we are looking for something that can be seen as "orthogonal" to conventional molecular simulation methods.

The characterization of the network of conformational clusters is a very hard task. Recently it turned out that only adaptive PCCA+, a special spectral clustering algorithm, is able to characterize the network of clusters in a physically interpretable and reproducable way and can also be seen as an "orthogonal" approach to molecular simulation (Kube and Weber, 2007; Weber, 2009).

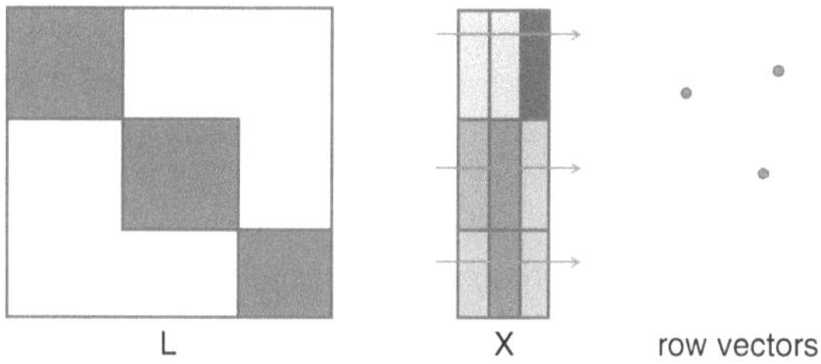

Fig. 2 A block-diagonal Laplacian L has leading eigenvectors X with blockwise constant entries. For the case of k blocks, the row vectors of X are the vertices of a k-simplex

2 Spectral Clustering and PCCA+

The idea of spectral clustering is to make use of eigenvector (spectral) data from an $(m \times m)$-graph Laplacian L in order to find $k \ll m$ clusters in a set of m objects. Usually, the spectral clustering algorithm starts with defining a nonnegative and symmetric $(m \times m)$-similarity matrix K. The similarity measure for an application to molecular simulation data will be given in Sect. 3. In many practical cases, the similarity measure is based on an euclidean distance between the objects. In order to yield L, a diagonal matrix D is determined such that the row-sums of $L := K - D$ are zero. In the case of a huge set $m \gg 1$ of objects, the similarity matrix K is intended to be sparse for the efficient computation of the eigenvectors and eigenvalues of L. L is symmetric and negative-semidefinite. The constant vector $e \in R^m, e_i = 1, i = 1, \ldots, m$, is an eigenvector corresponding to the leading eigenvalue $\lambda_1 = 0$. From the Theorem of Frobenius and Perron, we can assume that the eigenvalue λ_1 is algebraically and geometrically simple, if the matrix L is not decomposable into block diagonal form.

In order to understand spectral clustering, assume that the Laplacian L (after proper ordering of the row and column indexes) has block diagonal form with k blocks on its diagonal. Thus, there are k disconnected clusters of objects (K has the same block diagonal structure as L). Every block is an indecomposable graph Laplacian. Thus, the leading eigenvalue $\lambda_1 = \ldots = \lambda_k = 0$ is k-fold (Deuflhard and Weber, 2005). The corresponding space of eigenvectors is spanned by an $(m \times k)$-matrix X with blockwise constant elements, see Fig. 2. Thus, there is a $(k \times k)$-matrix $\mathscr{A}$, such that $\chi := X\mathscr{A} \in R^{m \times k}$ is a matrix with $\{0, 1\}$-entries. The k columns of this matrix χ can be interpreted as the indicator vectors of the k clusters of objects, i.e. $\chi_{ij} = 1$, if object i corresponds to cluster j and $\chi_{ij} = 0$ otherwise. Furthermore, the row-sums of χ are 1, so that each object is uniquely assigned to its cluster. This transformation $\mathscr{A}$ can be visualized as follows. The matrix X has blockwise constant elements. Thus, if we think of the rows of X as m points in k-dimensional space, there are only k different points. Furthermore, the constant vector e linearly depends on the columns of X, such that these k points lie on a plane in k-dimensional space, i.e. they form a k-simplex. The matrix $\mathscr{A}$ linearly maps this simplex to the standard k-simplex, spanned by the k unit vectors.

In practical applications of spectral clustering, the matrix L does not have the assumed disconnected block diagonal structure. In molecular simulation, e.g. the term "narrow canyons between the islands" is an image for an almost disconnected block diagonal structure of L. Thus, the simplex structure of the row vectors of X is perturbed. Instead of defining only k different points in the k-dimensional space, the rows of X "spread" into k clusters of points (Weber, 2006).

Usually, after computation of the row vectors of X, spectral clustering aims at finding the k clusters of points by standard clustering methods, e.g., by k-means. Spectral clustering algorithms can be distiguished by the different similarity measures for the construction of K, by the different formulation of the eigenvalue problem (sometimes a certain scaling of L is used), and by the different types of algorithms used to identify the k cluster of points in the set of the row vectors of X. For an excellent overview see von Luxburg (2007).

PCCA+ is a spectral clustering algorithm (Deuflhard and Weber, 2005; Weber, 2006). It is used in the context of molecular simulation. In Sect. 3 it will be shown, that the definition of the "Laplacian" is motivated physically in this case. Thus, there is no free choice for defining the similarity measure or for solving the eigenvalue problem. Like usual spectral clustering algorithms, PCCA+ takes the row vectors of X into account. However, instead of clustering the row vectors of X into k clusters, PCCA+ aims at computing a transformation matrix $\mathscr{A}$ such that the columns of $\chi = X\mathscr{A}$ can be interpreted as membership vectors of the k clusters. Thus, the result of PCCA+ is not given by "crisp" sets but by grades of membership between 0 and 1.

$\mathscr{A}$ is determined in such a way, that the elements of χ are nonnegative and the row-sums are 1 (partition of unity). Like in the crisp case, the transformation $\mathscr{A}$ maps the row vectors of X into the simplex spanned by the k unit vectors. There are different variants of the PCCA+ algorithm. These variants can be distinguished by the way the transformation matrix $\mathscr{A}$ is defined (Röblitz, 2008; Weber, 2006, 2009). In all cases, the entries of $\mathscr{A}$ are the solution of an optimization problem. In molecular simulation, every transformation $\mathscr{A}$ is allowed which maps X into the standard simplex, but for the reason of interpretation of the clustering results one aims at optimized transformations. E.g., the matrix $\mathscr{A}$ (and the number of clusters) can be optimized in such a way that χ is as crisp as possible (Weber et al., 2006; Röblitz, 2008).

3 Application of PCCA+ in Molecular Simulation

After having introduced the PCCA+ algorithm, the set of objects to be clustered is defined. We will not use a set $\{x_1, \ldots, x_n\} \subset R^{3N}$ of sampling points as an input for the clustering in molecular simulation due the discussed disadvantages of MCMC-simulation. The derivation of a graph Laplacian for the application of PCCA+ in molecular simulation is a little bit more complicated (Weber, 2009). Instead of clustering points x_i in R^{3N}, the conformational state space is decomposed into m subsets. These subsets are the objects of the clustering. In many cases,

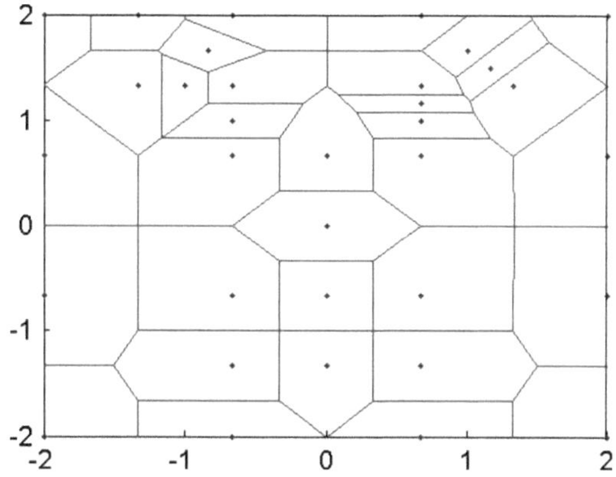

Fig. 3 A decomposition of the conformational space into Voronoi cells. One step of an adaptive discretization for the example problem of Fig. 1 is shown

Voronoi cells are suitable for the decomposition of high-dimensional spaces, see Fig. 3. Assume, we have defined m Voronoi cells. The off-diagonal elements k_{ij} of the symmetric $(m \times m)$-similarity matrix K are defined as the statistical weights of the intersecting surfaces between Voronoi cell V_i and Voronoi cell V_j. This statistical weight can be computed by the following surface integral:

$$k_{ij} = \int_{\delta V_i V_j} \pi(x)\, dS,$$

with $\delta V_i V_j$ being the intersecting surface between the Voronoi cells, and dS denoting a suitable surface measure. The matrix K has a zero diagonal, $k_{ii} = 0$. This is a definition of a sparse matrix K. I.e., $k_{ij} \neq 0$ if there exists an intersecting surface between V_i and V_j. Two neighboring Voronoi cells are the less similar the less molecular states are "observed" in their intersecting surface. For a theoretical derivation of K and an algorithmic realization of a numerical quadrature see Weber (2009).

On the basis of K one can define the symmetric matrix $L := K - D$ with vanishing row-sums (see Sect. 2). In contrast to the described spectral clustering approach, a rescaling of the rows of L is done, i.e. a positive diagonal matrix R is multiplied, such that $Q := R^{-1}L$ is the "Laplacian" of PCCA+. Q shares the important property of vanishing row-sums with the matrix L, and Q is also negative semi-definite. Thus, it can be used for spectral clustering via PCCA+. The rescaling of L is needed in order to yield a physical meaningful matrix Q, see Sect. 4. The diagonal elements of R are the statistical weights of the Voronoi cells, i.e., if V_i is a Voronoi cell, then $r_i := \int_{V_i} \pi(x)\, dx$ is the corresponding diagonal element of R. These statistical weights can be computed by direct free energy estimation algorithms (Klimm et al., 2011; Weber and Andrae, 2010). In contrast to many other spectral clustering algorithms, the "Laplacian" Q is not symmetric, but it can be symmetrized in the following way. Instead of computing the spectral properties of Q, the eigenvalue problem is solved for the symmectric matrix $Q_{sym} := R^{1/2}\, Q\, R^{-1/2}$. The matrix Q_{sym} has the same eigenvalues as Q, but the $(m \times k)$-eigenvector matrix X_{sym} for the leading k eigenvectors of Q_{sym} has to be transformed, such that $X = R^{-1/2} X_{sym}$ are the leading k eigenvectors of Q.

Although, the construction of Q seems to be very difficult because of the numerical computation of surface and volume integrals, it does not differ too much from the algorithmic approach of an MCMC method. For the numerical computation of the high-dimensional integrals Monte-Carlo quadrature (and a corresponding sampling) is needed. The difference between the convetional MCMC and the presented approach is mainly based on a different distribution of the sampling points. Whereas MCMC mainly samples from the "boring" parts of the conformational space, the sampling of the "Laplacian" Q mainly concentrates on the intersecting surfaces. As we will see in the next Section, these surfaces are mainly located in the transition region between the clusters. Thus, the Q-sampling is an "orthogonal" approach to conventional MCMC sampling.

4 Adaptive Decomposition

What is the physical meaning of Q? Exactly speaking, the matrix Q is a Galerkin discretization of the infinitesimal generator of the probability flux in R^{3N}. Instead of Q, it could be easier to understand a different matrix, namely $P(\tau)$. In molecular simulation there are two important operators, one is the infinitesimal generator $\mathcal{Q}$ and the other one is a lagtime-dependent Markov operator $\mathcal{P}(\tau)$. The relation between these two operators, $\mathcal{P}(\tau) = \exp(\tau \mathcal{Q})$, means that $\mathcal{Q}$ is the infinitesimal generator of $\mathcal{P}(\tau)$. So, what is the physical meaning of the operator $\mathcal{P}(\tau)$?

Given a (normalized) probability density function $g : R^{3N} \to R_+$ of conformational states in R^{3N}. This function can be written as $g = f \circ \pi$, where $f : R^{3N} \to R_+$ denotes the ratio between g and the stationary density π. If $f \circ \pi$ is the probability density of conformational states at time $t_0 = 0$, then $(\mathcal{P}(\tau)\; f) \circ \pi$ is the probalility density of conformational states at time $t_1 = \tau$. Thus $\mathcal{P}(\tau)$ denotes the "movement" of the probability density in conformational state space between two time steps t_1 and $t_2 = t_1 + \tau$. By discretizing the conformational state space into m Voronoi cells, this Markov operator $\mathcal{P}(\tau)$ $(m \times m)$-transition matrix $P(\tau)$. The Markov operator $\mathcal{P}$ is projected (Galerkin (Galerkin discretization) to a finite dimensional matrix P. Although $\mathcal{P}(\tau) = \exp(\tau \mathcal{Q})$, this equation does not hold in the finite dimensional case, $P(\tau) \neq \exp(\tau\, Q)$. The identity $P(\tau) = \exp(\tau\, Q)$ only is true, if the Galerkin discretization of the operator $\mathcal{Q}$ is based on a linear combination(!) χ of the leading eigenvectors of $\mathcal{Q}$ (Kube and Weber, 2007; Weber and Kube, 2008; Weber, 2009). This is the reason, why only PCCA+ is a good clustering approach for the application in molecular simulation, because PCCA+ is the only spectral clustering approach for which the result χ is the linear combination of eigenvectors of Q. The matrices $Q_c := (\chi^\top R\, \chi)^{-1}(\chi^\top, R\, Q\, \chi)$ and $P_c(\tau) := (\chi^\top R\, \chi)^{-1}(\chi^\top R\, P(\tau)\, \chi)$ have the desired property $P_c(\tau) = \exp(\tau Q_c)$, if the

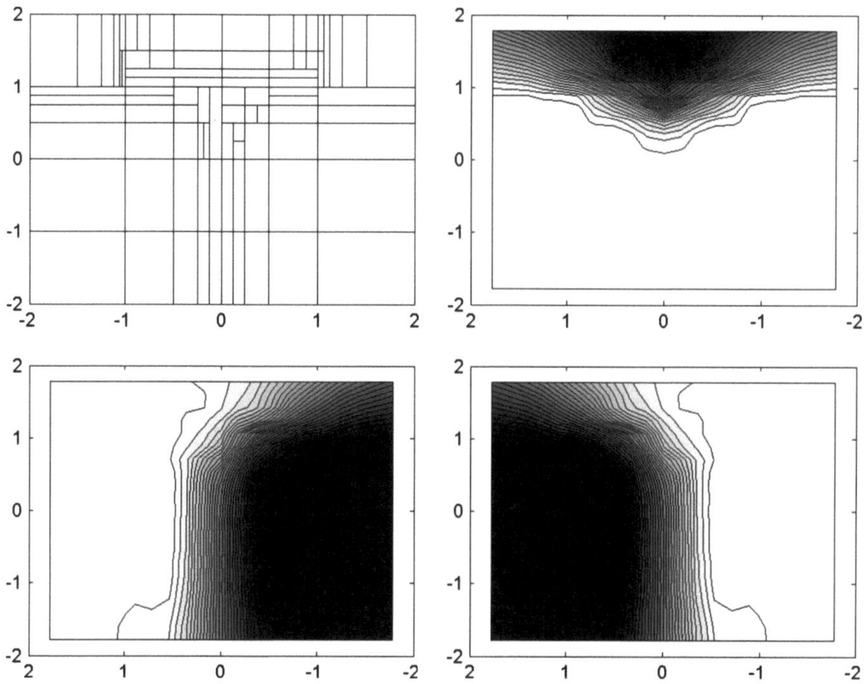

Fig. 4 An adaptive spectral clustering and the adaptive box discretization. The same potential density function as in Fig. 1 is used. For the visualization of the clusters, the Voronoi cells yield grayscale values between 0 and 1 according to their grade of membership to the three different clusters (columns of χ). This clustering has been generated with about 5,000 Monte-Carlo quadrature points. In contrast to Fig. 1, all three clusters have been identified

eigenvectors X based on a Voronoi discretization of R^{3N} are a good approximation of the eigenfunctions of $\mathcal{Q}$. Thus, the entries of Q_c are physically interpretable as transition rates between the clusters. In order to find a good discretization of R^{3N}, one has to refine neighboring Voronoi sets V_i and V_j hierachically and adaptively (Haack et al., 2010), if the difference between the row i and row j of χ is high. Thus, this kind of hierachical refinement mainly takes place in transition regions between the clusters, see Fig. 4. In contrast to the bad statistics in Fig. 1, the adaptive spectral clustering approach finds three clusters and can identify the transition network with the correct symmetry:

$$Q_c = \begin{pmatrix} -0.012366 & 0.006183 & 0.006183 \\ 0.000030 & -0.000047 & 0.000017 \\ 0.000030 & 0.000017 & -0.000047 \end{pmatrix}.$$

5 Summary

We have shown that Q-sampling and PCCA+ are the key methods to compute physical meaningful data efficiently from molecular simulation. The sampling approach is "orthogonal" to conventional MCMC methods. PCCA+ provides clusters χ, that can be used for a Galerkin discretization of the infinitesimal generator of the molecular process and preserves the time-scales of the system.

References

Bujotzek, A., & Weber, M. (2009). Efficient simulation of ligand-receptor binding processes using the conformation dynamics approach. *Journal of Bioinformatics and Computational Biology, 7*(5), 811–831.

Deuflhard, P., & Weber, M. (2005). Robust perron cluster analysis in conformation dynamics. *Linear Algebra and its Applications, 398c*, 161–184.

Haack, F., Röblitz, S., Scharkoi, O., Schmidt, B., & Weber, M. (2010). Adaptive spectral clustering for conformation analysis. *ICNAAM 2010: International Conference of Numerical Analysis and Applied Mathematics, AIP Conference Proceedings, 1281*, 1585–1588.

Klimm, M., Bujotzek, A., & Weber, M. (2011). Direct reweighting strategies in conformation dynamics. *MATCH, 65*, 333–346.

Kube, S., & Weber, M. (2007). A coarse-graining method for the identification of transition rates between molecular conformations. *The Journal of Chemical Physics, 126*(2), 024103.

Röblitz, S. (2008). *Statistical error estimation and grid-free hierarchical refinement in conformation dynamics*. Doctoral Thesis, FU Berlin.

von Luxburg, U. (2007). A Tutorial on spectral clustering. *Statistics and Computing, 17*(4), 395 416.

Weber, M. (2006). *Meshless methods in conformation dynamics*. München: Verlag Dr. Hut. ISBN 3-89963-307-5.

Weber, M. (2009). A Subspace Approach to Molecular Markov State Models via an Infinitesimal Generator, *ZIB Report 09–27*, Zuse-Institut Berlin.

Weber, M., & Andrae, K. (2010). A simple method for the estimation of entropy differences. *MATCH Communications in Mathematical and in Computer Chemistry, 63*(2), 319–332.

Weber, M., & Kube, S. (2008) Preserving the markov property of reduced reversible markov chains. *Numerical Analysis and Applied Mathematics, International Conference on Numerical Analysis and Applied Mathematics 2008, AIP Conference Proceedings, 1048*, 593–596.

Weber, M., Kube, S., Walter L., & Deuflhard, P. (2007). Stable computation of probability densities for metastable dynamical systems. *SIAM Journal of Multiscale Modeling and Simulation, 6*(2), 396–416.

Weber, M., Rungsarityotin, W., & Schliep, A. (2006). An indicator for the number of clusters using a linear map to simplex structure. In: *From Data and Information Analysis to Knowledge Engineering, 29th Annual Conference of the German Classification Society 2005, March 9–11, Studies in Classification, Data Analysis, and Knowledge* (pp. 103–110). Heidelberg: Springer.

Part III
Applications

Spatial Data Mining for Clustering: An Application to the Florentine Metropolitan Area Using RedCap

Federico Benassi, Chiara Bocci, and Alessandra Petrucci

Abstract The paper presents an original application of the recently proposed RedCap method of spatial clustering and regionalization on the Florentine Metropolitan Area (FMA). Demographic indicators are used as the input of a spatial clustering and regionalization model in order to classify the FMA's municipalities into a number of demographically homogeneous as well as spatially contiguous zones. In the context of a gradual decentralization of governance activities we believe the FMA is a representative case of study and that the individuation of new spatial areas built considering both the demographic characteristics of the resident population and the spatial dimension of the territory where this population insists could become a useful tool for local governance.

1 Introduction

For several years spatial data mining has been considered as the multi-dimensional equivalent of temporal data-mining (Roddick and Spiliopoulou, 1999). Today, however, there is a consensus among researchers to consider spatial data mining as an independent approach to data analysis and measuring phenomena as confirmed by recent studies (Angayarkkani and Radhakrishnan, 2009; Behnisch and Ultsch, 2010; Jin and Guo, 2009).

One of the most important assumptions of classical statistical analysis is that the data samples are independently generated; on the contrary, the spatial approach removes this assumption and theorizes that the spatial location of the samples is an item that cannot be ignored (Tobler, 1970). Thus, it follows that data mining is connected to the concept of patterns while spatial data mining is connected to

F. Benassi · C. Bocci (✉) · A. Petrucci
Department of Statistics "G. Parenti", University of Florence, Florence, Italy
e-mail: benassi@ds.unifi.it; bocci@ds.unifi.it; alessandra.petrucci@unifi.it

A. Giusti et al. (eds.), *Classification and Data Mining*, Studies in Classification, Data Analysis, and Knowledge Organization, DOI 10.1007/978-3-642-28894-4_19,

the concept of spatial patterns. Obviously, these theoretical differences between classic and spatial data mining have important repercussions in operational terms. To apply a spatial (data mining) approach implies that the dimension of large databases become larger as spatially referenced objects also carry information concerning their representation in space by geometrical and topological properties. This implies: (a) more powerful techniques to manipulate the data and extract knowledge, (b) a new kind of cartographic knowledge to represent the results obtained and make them readable to a non-technical stakeholders (policy makers, local administrators etc.) and (c) a more flexible software to encourage users to interact with the data (Koperski et al., 1996).

The paper is structured as follows. In Sect. 2 we briefly describe the regionalization process and RedCap's major features. Data description and results presentation are discussed in Sect. 3. Finally, we conclude with some final remarks.

2 Regionalization Method

According to Guo (2008) we define regionalization as a process that divides a large set of spatial objects into a number of spatially contiguous regions while optimizing an objective function, normally a homogeneity (or heterogeneity) measure of the identified regions. Therefore regionalization is a special kind of spatial clustering where the condition of spatial contiguity among spatial objects plays a priority role.

RedCap is a method of spatial clustering and regionalization elaborated by Guo (2008). It is essentially based on a group of six methods for regionalization given by the combination of three agglomerative clustering methods (Single Linkage clustering, SLK; Average Linkage clustering, AVG; Complete Linkage clustering, CLK) and two different spatial constraining strategies (First-Order constraining and Full-Order constraining). Guo (2008) shows that among the six methods, the Complete Linkage clustering with Full Order constraining strategy (CLK-Full Order) achieve the best performance. We refer to the work of Guo et al. (2005) and Guo (2008) for technical and computational details about these six methods of regionalization.

RedCap consists of two fundamental steps. In the first step, based on the iterative algorithms of the Self Organizing Map (SOM) and developed in Guo et al. (2005), the method finds spatial clusters without imposing any spatial constraining strategy. The results of this first step are visualized by the SOM unified distance matrix and by the Parallel Coordinate Plot (PCP), where we can observe the profile of the clusters and their level of similarity. In the second step, based on a contiguity matrix and a set of constrained strategies, the method completes the regionalization process. The results of these two steps are then related and visualized on an interactive map.

3 Data and Results

The Florentine Metropolitan Area (FMA) was created by the deliberation of the regional council of Tuscany n.130 on 13/2/2000. This area is composed of three provinces (Firenze, Pistoia and Prato) and is divided into 73 municipalities. Due to its recent definition only few studies on its population structure and dynamic are available in the literature (Petrucci et al., 2008; Vignoli et al., 2007). In particular, there are no studies that consider directly the spatial dimension in the analysis. The FMA is very heterogeneous in terms of demographic structures and dynamics, settlements models, geo-morphological structures and economic specialization. In addition the FMA is strongly affected by many phenomena of mobility: residential migrations, daily migrations (commuting), international migrations. Due to this strong heterogeneity and in the context of an increasing decentralization of the governance activities we believe that the FMA represents an interesting case of study.

We select as input variables five demographic indexes (computed for each municipality) plus the spatial attributes of each municipality. The five demographic indexes, computed by using data on the resident population produced by the Italian National Institute of Statistics (Istat), are: Youth dependency index (IDG), Aging index (IV), Elderly dependency index (IDA), Population in active age substitution index (IS), Population in active age replacement index (IR).

Firstly, an explorative analysis based on the results of the SOM algorithms is carried out. Applying a visual approach we build groups of similar clusters through the results of clustering process visualized on the unified distance matrix without taking into account the condition of spatial contiguity among them.

Starting from 73 municipalities we identify 16 clusters that are defined by the node hexagons on the SOM. Territorial units with the same color belong to the same cluster and clusters with similar colors present a low level of dissimilarity.

From the results of the explorative analysis, we classify the 16 clusters in three main groups. The first group, that we define "young", is composed by 5 clusters and 29 municipalities (Fig. 1). This group has a relatively young age structure and a high level of inner homogeneity as the colors of the node hexagons and the PCP show. The five clusters have a low level of the IV, IDA and IS indexes, a high/medium level of IDG and, finally, a low/medium level of IR index (Fig. 1b).

The second group, that we define "old", is composed by 4 clusters and 15 municipalities. This second groups of clusters is characterized by a population with a very old age structure as the PCP shows clearly (Fig. 2c). In fact, the level of IDG is low while the level of the others indexes (IV, IDA, IS, IR) is medium/high in one case and very high in the others three cases. This group of clusters presents a great level of inner homogeneity as the colors of node hexagons in SOM and the profiles of the PCP show (Fig. 2).

The third group of clusters, that we define "medium", is composed by 6 clusters and 26 municipalities. As we can see in the Fig. 3 this third group is composed by clusters that present a medium level of all indexes. The inner homogeneity of this

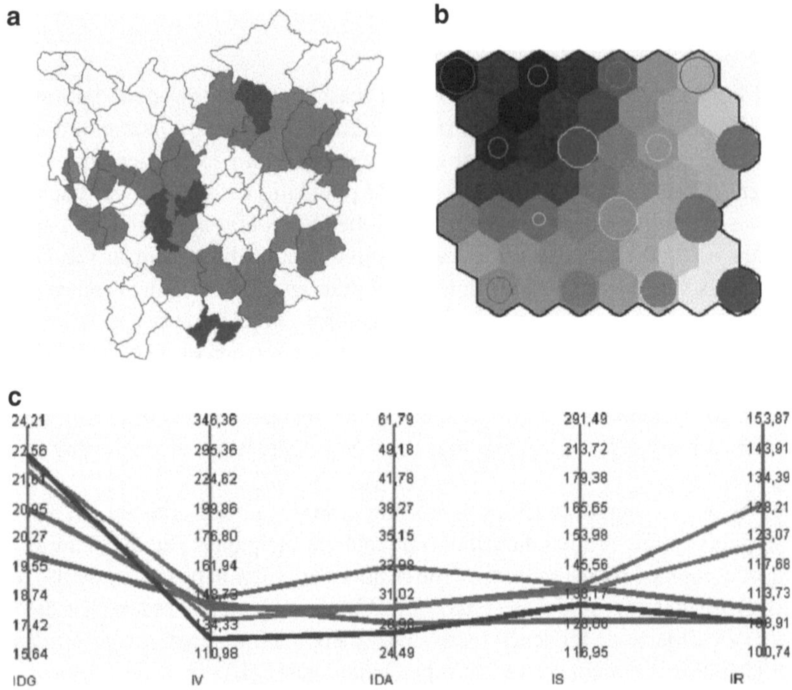

Fig. 1 Group 1 – "Young". (**a**) Multivariate mapping. (**b**) Clustering with SOM. (**c**) Multivariate visualization of clusters (parallel coordinate plot)

group is relatively low as we can see by the colors of the node hexagons of the SOM (Fig. 3b) and in the profile of the clusters represented in the PCP (Fig. 3c). Actually, looking at the indexes values, this group could be divided in two subgroups: "medium high" and "medium low".

After the explorative analysis we apply the Complete Linkage Clustering method together with a Full Order constraining strategy (CLK-Full Order) in order to obtain n areas that minimize the inner heterogeneity of the demographic structure of the population under the condition of spatial contiguity. We identify six areas (Fig. 4). Areas A and B are very similar to each other and both are characterized by a high level of inner homogeneity. In fact, by a visual analysis we can clearly see that the municipalities that belong to these areas have basically the same colors (Fig. 4a, 4b). Therefore in areas A and B the resident population has a very old structure as shown in the PCP of the old group of clusters (Fig. 2c). The similarity between areas A and B is not only in terms of demographic structure: these areas are in fact mountainous areas characterized by rural settlements and both are localized on the boundaries of the FMA. The fact that these areas present a very old demographic structure is probably connected to their recent demographic history. It is known that these areas have been interested by a sustained depopulation process – common to the majority of the Italian mountainous areas – caused by a strong internal rural-urban migration

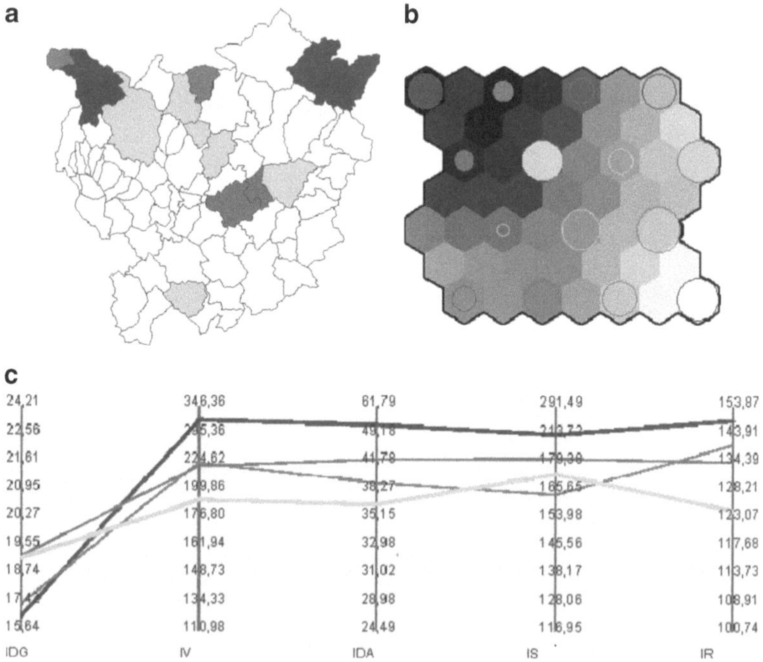

Fig. 2 Group 2 – "Old". (**a**) Multivariate mapping. (**b**) Clustering with SOM. (**c**) Multivariate visualization of clusters (parallel coordinate plot)

flows. The internal migrants were mainly young people and young couples in search of a more modern and dynamic social environment (typically urban) which offered higher schooling and employment opportunities. For similar reasons, these areas have not become destination areas for international migrants (especially for international labor migrants) that usually are attracted by the dynamic and highly informal labour market typical of FMA's urban and peri-urban areas. These migration patterns combine with the aging process that involve the whole Italian population, have determined the extremely old structure of the population in these two areas.

Area C presents a medium level of inner homogeneity as indicated by the different shades of colors of the municipalities that belong to this area (Fig. 4c). The age structure of the population of this area is medium-old: some municipalities have an old age structure while others have a relatively younger age structure. Area C is composed by two important urban centers, Firenze and Pistoia, and by a peri-urban area around these two cities. Therefore, we can say that area C presents a urban settlement and a medium old demographic structure but with an internal spatial structure that can be divided in two sub-structures: a more properly urban area – the core of FMA – with a relatively old demographic structure and a peri-urban area with a relatively younger structure. These results, especially with regards to Firenze and Pistoia, are confirmed by the evidences of Petrucci et al. (2008). From a theoretical point of view these results can be explained by the following

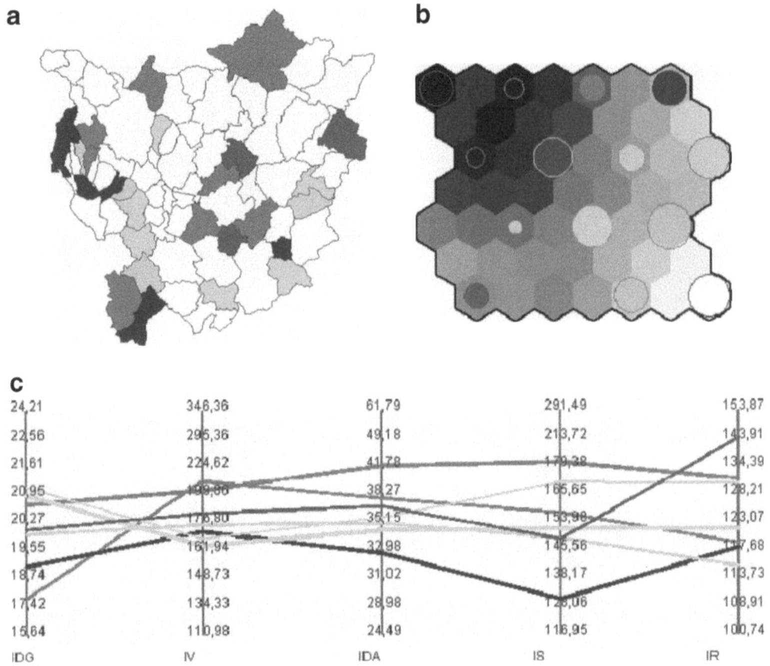

Fig. 3 Group 3 – "Medium". (**a**) Multivariate mapping. (**b**) Clustering with SOM. (**c**) Multivariate visualization of clusters (parallel coordinate plot)

considerations: (a) in recent years Firenze and Pistoia (like the majority of medium and large size cities of Italy) were involved in the suburbanization process. Typically this process involves mainly people with a relatively old age structure that decide to move to less urbanized areas with a higher quality of life (Benassi et al., 2009). In the case of Pistoia and, especially, Firenze this process probably involves also young people and young couples that decide to leave their parental home and, owing to the extremely asymmetric structure of the housing market in these cities, they probably migrated to less central areas; (b) the high cost of life and housing in Firenze and Pistoia affects also the residential choice of international migrants; (c) the urban way of life is typically associated with relatively lower Total Fertility Rates.

According to the core-ring model for the study of urban development (Van den Berg et al., 1982) areas D and F (Fig. 4d, 4f) can be defined as a ring around the proper urban area. Area D presents a medium/old demographic structure and a medium inner homogeneity as the different colors of the municipalities of this area underline. This area is characterized by a semi-urban and rural settlements. Some specific sub-areas of area D (the Chianti area for example) are a well-known destination of many retirees from Northern European countries – particularly from the UK and Germany – and from the United States (Benassi and Porciani, 2010). Due to this spatial proximity to the core of the metropolitan system this area is also

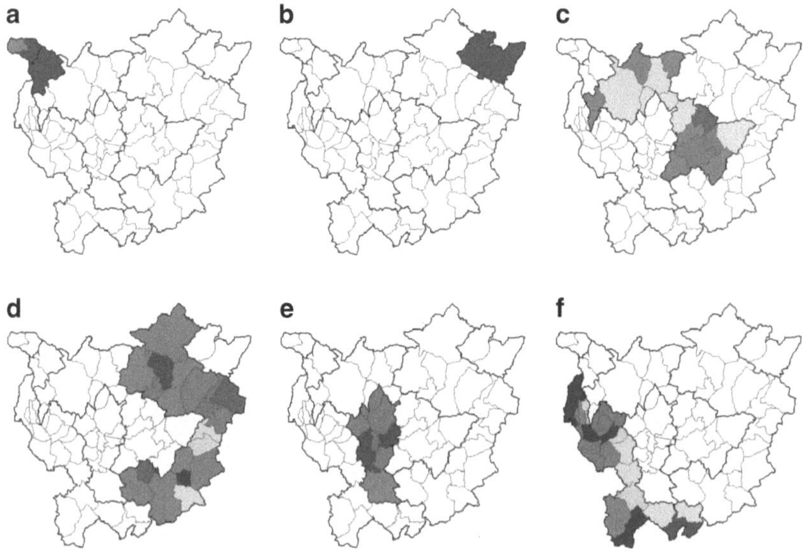

Fig. 4 Regionalization results. (a)–(f) Areas A through F

probably the destination area for internal migrants involved in the suburbanization process of area C. Area F is the second part of the ring around the core of the FMA system. The inner homogeneity of this area is relatively low but presents a younger demographic structure compared to the area D. This is probably due to the fact that this area is not a retirement destination for foreingers but, on the contrary, is certainly a destination area for international labor migrants and also for young people and young couples. Area A is in fact more influenced by the dynamics of area E, that represents the alter ego of area C. Like area C, area E is in fact a urban area characterized by urban settlements but, differently from area C, it presents a very young demographic structure. The reason of this dual situation is that area E is strongly involved in international migration movements, particularly from China. On the other hand the suburbanization process mainly driven by young people and young couples probably makes this area a destination for people migrating from area C.

4 Concluding Remarks

The RedCap method has some advantages: it is very ductile, user-friendly, free, allows to interact directly with the data, and takes into account directly the spatial dimension. The spatial analysis of the demographic structure of the resident population of the FMA produces results that clearly show how the spatial attributes influence the demographic structure of the population. The FMA is a complex demographic spatial system where the following items coexist: (a) mountainous areas with a very old demographic structure (areas A and B); (b) dual core metropolitan areas composed by a relatively young area (E) and a relatively old area (C); (c) two ring areas that are basically the spatial extension of the FMA core (areas D and F).

Starting from this empirical evidence we want to underline that ignoring the spatial dimension can lead to misleading inference. The use of appropriate methods for the detection of spatial clusters can improve the measurement and interpretation of urban socio-economic phenomena and provide a useful information to local authorities and policy makers for regional and urban planning.

References

Angayarkkani, K., & Radhakrishnan, N. (2009). Efficient forest fire detection system: A spatial data mining and image processing based approach. *International Journal of Computer Science and Network Security, 9*(3), 100–107.

Behnisch, M., & Ultsch, A. (2010). Are there cluster of communities with the same dynamic behavior? In *Classification as a tool for research* (Part. 3, pp. 445–453). Berlin: Springer

Benassi, F., & Porciani, L. (2010). The dual demographic profile of migration in Tuscany. In T. Salzmann, B. Edmonston, & J. Raymer (Eds.), *Demographic aspects of migration* (pp. 209–226). Berlin: VS-VERLAG Springer.

Benassi, F., Bottai, M., & Giuliani, G. (2009). Migrazioni e processi di urbanizzazione in Italia. Spunti interpretativi in unottica biografica. In M. J. Macchi (Ed.), *Geografie del popolamento: Metodi, casi e teorie* (pp. 71–78). Siena: Edizioni dell'Università di Siena.

Guo, D. (2008). Regionalization with dynamically constrained agglomerative clustering and partitioning. *International Journal of Geographical Information Sciences, 22*(7), 801–823.

Guo, D., Gahegan, M., MacEachren, A. M., & Zhou, B. (2005). Multivariate analysis and geovisualization with an integrated geographic knowledge discovery approach. *Cartography and Geographic Information Science, 32*(2), 113–132.

Jin, H., & Guo, D. (2009). Understanding climate change patterns with multivariate geovisualization. In *Proceedings of the International Conference on Data Mining Workshops* (pp. 217–222). Los Alamitos: IEEE. doi: 10.1109/ICDMW.2009.109.

Koperski, K., Adhikany, J., & Han, J. (1996). Knowledge discovery in spatial database: Progress and challenges. In *Proceedings of the Workshop on Research Issues on Data Mining and Knowledge Discovery,* Montreal. pp. 55–70.

Petrucci, A., Salvati, N., Salvini, S., & Vignoli D. (2008). Invecchiamento e mobilità nell'area metropolitana fiorentina. *Rivista di Economia e Statistica del Territorio, 2*, 81–103.

Roddick, J. F., & Spiliopoulou, M. (1999). A bibliography of temporal, spatio and spatio-temporal data mining research. *SIGKDD Explorations, 1*(1), 34–38.

Tobler, W. (1970). A computer movie simulating urban growth in the Detroit region. *Economic Geography, 46*(2), 234–240.

Van den Berg, L., Drewett, R., Klaassen, L. H., Rossi, A., & Vijverberg, C. H. T. (1982). *Urban Europe: A study of growth and decline*. Oxford: Pergamon.

Vignoli, D., Dugheri, G., Ferro, I., Salvini, S., & Secondi, L. (2007). *L'area fiorentina: quanti siamo e quanti saremo*. Serie La statistica per la città, Ufficio Statistica del Comune di Firenze.

Misspecification Resistant Model Selection Using Information Complexity with Applications

Hamparsum Bozdogan, J. Andrew Howe, Suman Katragadda, and Caterina Liberati

Abstract In this paper, we address two issues that have long plagued researchers in statistical modeling and data mining. The first is well-known as the "curse of dimensionality". Very large datasets are becoming more and more frequent, as mankind is now measuring everything he can as frequently as he can. Statistical analysis techniques developed even 50 years ago can founder in all this data. The second issue we address is that of model misspecification – specifically that of an incorrect assumed functional form. These issues are addressed in the context of multivariate regression modeling. To drive dimension reduction and model selection, we use the newly developed form of Bozdogan's *ICOMP*, introduced in Bozdogan and Howe (Misspecification resistant multivariate regression models using the genetic algorithm and information complexity as the fitness function, Technical report 1, (2012)), that penalizes models with a complexity measure of the "sandwich" model covariance matrix. This information criterion is used by the genetic algorithm as the objective function in a two-step hybrid dimension reduction process. First, we use probabilistic principle components analysis to independently reduce the number of response and predictor variables. Then, we use the genetic

H. Bozdogan (✉)
Department of Statistics, Operations, and Management Science, University of Tennessee, Knoxville, TN, USA
e-mail: bozdogan@utk.edu

J.A. Howe
Business Analytics Trans Atlantic Petroleum Istanbul, Turkey
e-mail: ahowe42@gmail.com

S. Katragadda
Advanced Analytics Express-Scripts Company St. Louis, MO, USA
e-mail: katragaddasuman11@gmail.com

C. Liberati
Economics Department, Universita' degli Studi Milano-Bicocca, Milan, Italy
e-mail: caterina.liberati@unimib.it

A. Giusti et al. (eds.), *Classification and Data Mining*, Studies in Classification, Data Analysis, and Knowledge Organization, DOI 10.1007/978-3-642-28894-4_20,

algorithm with the multivariate Gaussian regression model to identify the best subset regression model. We apply these methods to identify a substantially reduced multivariate regression relationship for a dataset regarding Italian high school students. From 29 response variables, we get 4, and from 46 regressors, we get 1.

1 Introduction

In this paper, we address two issues that have long plagued researchers in statistical modeling and data mining. The first is well known as the "curse of dimensionality." Very large datasets are becoming more and more frequent, as mankind is now measuring everything he can as frequently as he can. Statistical analysis techniques developed even 50 years ago can founder in all this data. The second issue we address is that of model misspecification – specifically that of an incorrect assumed functional form. These issues are addressed in the context of multivariate regression modeling, in which we present a novel hybrid dimension reduction technique. We apply these methods to identify a substantially reduced multivariate regression relationship for a data set regarding Italian high school students. From 29 response variables, we get 4, and from 46 regressors, we get 1.

2 Multivariate Regression Modeling with *ICOMP*

2.1 Multivariate Gaussian Regression

In the usual multivariate regression (MVR) problem, we have a matrix of responses $Y \in \mathbb{R}^{n \times p}$; n observations of p measurements on some physical process. The researcher also has k variables that have some theoretical relationship to Y: $X \in \mathbb{R}^{n \times q}$, of course, we usually include a constant term as an intercept for the hyperplane generated by the relationship, so $q = k + 1$. The predictive relationship between X and Y has both a deterministic and a stochastic component, such that the model is

$$Y = XB + E, \tag{1}$$

in which $B \in \mathbb{R}^{q \times p}$ is a matrix of coefficients relating each column of X to each column of Y, and $E \in \mathbb{R}^{n \times p}$ is a matrix of error terms. The usual assumption in multivariate regression is that the error terms are uncorrelated, homoskedastic Gaussian white noise:

$$Y \sim N_p(XB, \Sigma \otimes I_n), \text{ where } \mathrm{E}(Y) = XB, \text{ and } Cov(Y) = \Sigma \otimes I_n. \tag{2}$$

Under the assumption of Gaussianity, the log likelihood of the multivariate regression model is given by

$$\log L(\theta \mid Y) = -\frac{np}{2}\log(2\pi) - \frac{n}{2}\log|\Sigma| - \frac{1}{2}\mathrm{tr}[(Y - XB)' \, \Sigma^{-1} \, (Y - XB)']. \tag{3}$$

The estimated model covariance matrix, (*inverse Fisher information matrix*) can be derived using the results of Magnus and Neudecker (1988, p. 321), and is given by

$$\widehat{Cov}(vec(\hat{B}), \text{vech}(\hat{\Sigma})) \equiv \hat{\mathscr{F}}^{-1} = \begin{bmatrix} \hat{\Sigma} \otimes (X'X)^{-1} & \mathbf{0} \\ \mathbf{0}' & \frac{2}{n} D_p^{+}(\hat{\Sigma} \otimes \hat{\Sigma}) D_p^{+'} \end{bmatrix}. \quad (4)$$

The IFIM provides the asymptotic variance of the ML estimators when the model is correctly specified. Its *trace* and *determinant* provide scalar measures of the asymptotic variance, and they play a key role in the construction of information complexity. It is also very useful, as it provides standard errors for the regression coefficients on the diagonals.

In most statistical modeling problems, we almost always fit a wrong model to the observed data. This can introduce bias into the model due to model misspecification. The most common causes of model misspecification include: multicollinearity, autocorrelation, heteroskedasticity, and incorrect functional form. This final type is the type of misspecification we address. The common answer in the literature to nonnormality has been the utilization of *Box-Cox transformations* of Box and Cox (1964), which does not seem to work consistently well, especially in the context of multivariate regression. Of course, when performing regression analysis, it is not usually the case that all variables in X have significant predictive power over Y. Choosing an optimal subset model has long been a vexing problem, and there are many approaches to this problem. We follow Bozdogan and Howe (2012) and use the genetic algorithm to select a subset MVR model.

2.2 *Robust Misspecification-Resistant Information Complexity Criteria*

Acknowledging the fact that any statistical model is merely an approximate representation of the true data generating process, information criteria attempt to guide model selection according to the *principle of parsimony*. This principle of parsimony requires that as model complexity increases, the fit of the model must increase at least as much; otherwise, the additional complexity is not worth the cost. Virtually all information criteria penalize a poorly fitting model with negative twice the maximized log likelihood, as an asymptotic estimate of the KL information. The difference, then, is in the penalty for model complexity. In order to protect the researcher against model misspecification, Bozdogan and Howe (2012) generalized *ICOMP* to the case of a misspecified MVR model and introduce $ICOMP_{MISP}$, which can drive effective model selection **even when the Gaussian assumption is invalid**. Here we show their results without derivations or proofs.

If we note θ_g^* as the value of the parameters vector which minimizes the *Kullback-Liebler* distance (Kullback and Leibler, 1951) for some specified functional model $f(\theta_g^*)$ to the true functional model $g(\theta)$, and we use $\mathscr{R}$ to indicate the outer-product form of the Fisher information matrix, we have

Theorem 1. *Based on an iid sample, $y_1, \ldots, y_n$, and assuming regularity conditions of the log likelihood function hold, we have*

$$\hat{\theta} \sim N(\theta_g^*, \mathscr{F}^{-1}\mathscr{R}\mathscr{F}^{-1}), \text{ or } \sqrt{n}(\hat{\theta} - \theta_g^*) \sim N(0, \mathscr{F}^{-1}\mathscr{R}\mathscr{F}^{-1}). \tag{5}$$

Note that this tells us explicitly

$$Cov(\theta_g^*)_{Misspec} = \mathscr{F}^{-1}\mathscr{R}\mathscr{F}^{-1}, \tag{6}$$

which is called the sandwich *or* robust *covariance matrix, since it is a correct variance matrix whether or not the assumed or fitted model is correct.*

Of course, in practice the true model and parameters are unknown, so we estimate this with

$$\widehat{Cov}(\hat{\theta}) = \hat{\mathscr{F}}^{-1}\hat{\mathscr{R}}\hat{\mathscr{F}}^{-1}. \tag{7}$$

If the model is correct, we must have $\hat{\mathscr{F}}^{-1}\hat{\mathscr{R}} = I$, so

$$\widehat{Cov}(\hat{\theta}) = \hat{\mathscr{F}}^{-1}\hat{\mathscr{R}}\hat{\mathscr{F}}^{-1} = I\hat{\mathscr{F}}^{-1} = \hat{\mathscr{F}}^{-1}.$$

Thus, in the case of a correctly specified model, $\widehat{Cov}(\hat{\theta}) = \hat{\mathscr{F}}^{-1}$.

For multivariate regression, we have already seen the inner-product form of estimated IFIM in (4). The outer-product form $\mathscr{R}$ is derived in Magnus (2007), and we show the result in (8).

$$\hat{\mathscr{R}} = \begin{bmatrix} \hat{\Sigma}^{-1} \otimes X'X & \frac{1}{2}(\hat{\Sigma}^{-1/2} \otimes X')\hat{\Gamma}_1 D_p^{+\prime}\Lambda \\ \frac{1}{2}\Delta D_p^{+}\hat{\Gamma}_1'(\hat{\Sigma}^{-\frac{1}{2}} \otimes X) & \frac{1}{4}\Delta D_p^{+}\hat{\Gamma}_2^{*} D_p^{+\prime}\Delta \end{bmatrix}. \tag{8}$$

This matrix takes into consideration the actual sample skewness and kurtosis of the data. There is an issue of matrix stability to be addressed with the sandwich covariance matrix, however. Numerical issues with estimating the sandwich covariance matrix prevent it from approximating the FIM when the model is correctly specified. We employ the *Empirical Bayes covariance regularization* procedure

$$\widehat{Cov}(\hat{\theta}) = \widehat{Cov}(\hat{\theta}) + \frac{p-1}{(n)\,tr(\widehat{Cov}(\hat{\theta}))} I_p, \tag{9}$$

to ensure $\widehat{Cov}(\hat{\theta})$ is of full rank. Thus, the misspecification-resistant form of *ICOMP* for multivariate regression is computed as in (10). When the model is correctly specified, we expect $\widehat{Cov}(\hat{\theta}) = \hat{\mathscr{F}}^{-1}$, we get $ICOMP(\hat{\mathscr{F}}^{-1})$ in (11).

$$ICOMP(\widehat{Cov}(\hat{\theta}))_{MISP} = np\log 2\pi + n\log|\hat{\Sigma}| + np + 2C_1(\widehat{Cov}(\hat{\theta})) \tag{10}$$

$$ICOMP(\hat{\mathscr{F}}^{-1}) = np\log 2\pi + n\log|\hat{\Sigma}| + np + 2C_1(\hat{\mathscr{F}}^{-1}) \tag{11}$$

In both, C_1 is the first order maximal entropic complexity of Bozdogan (1988): a generalization of the model covariance complexity of Van Emden (1971), given by

$$C_1(\widehat{Cov}(\hat{\theta})) = \frac{s}{2}\log\frac{\mathrm{tr}(\widehat{Cov}(\hat{\theta}))}{s} - \frac{1}{2}\log|\widehat{Cov}(\hat{\theta})|,\ s = rank(\widehat{Cov}(\hat{\theta})). \quad (12)$$

3 Dimension Reduction with the Genetic Algorithm and Probabilistic Principle Components Analysis

3.1 Genetic Algorithm

The genetic algorithm (GA) is a search algorithm that borrows concepts from biological evolution. Unlike most search algorithms, the GA simulates a large population of potential solutions, encoded as binary strings. These solutions are allowed to interact over time; random mutations and natural selection allow the population to improve, eventually iterating to an optimal solution. The GA was popularized by Holland (1975), and it is a widely recognized as popular stochastic search and optimization algorithm. Today, there are many problems in science, economics, and research and development that are solved using the GA. We refer the reader to existing books and articles regarding details of the algorithm. Some excellent books are Goldberg (1989), Haupt and Haupt (2004) and Vose (1999). Articles specifically combining the GA with subset regression models would include Bozdogan (2004) in which the GA was implemented for multiple regression subset selection under the normality assumption. Also, Bozdogan and Howe (2012) extended this work to the case of misspecified multivariate regression.

3.2 Probabilistic Principle Components Analysis

In this paper, we employ Probabilistic Principle Component Analysis (PPCA) as a first step to independently reduce the dimensionality of the independent and dependent matrices. PPCA was developed in the late 1990s and popularized by Tipping and Bishop (1997). Here, we show some results from Tipping and Bishop (1997) and Bozdogan and Howe (2009) that are relevant to this research. Let $x \in \mathbb{R}^{1\times p}$ be a random vector; assume x can be expressed as a linear combination of *latent variables* and stochastic noise:

$$x = \Lambda f + \mu + \varepsilon, \quad (13)$$

where $f \in \mathbb{R}^{m\times 1}$ holds the latent variables, $\Lambda \in \mathbb{R}^{p\times m}$ is the loading matrix, and $\mu \in \mathbb{R}^{1\times p}$ defines the mean of x. Maximizing the PPCA likelihood function, we get the model covariance matrix in (14)

$$\widehat{Cov}(X) = \hat{\Sigma} = U_p \hat{L} U_p', \tag{14}$$

where U_p contains all the eigenvectors of $\hat{\Sigma}$. $\hat{L}$ is almost a $(p \times p)$ matrix with eigenvalues of $\hat{\Sigma}$ on the diagonals. Positions corresponding to variables not included in the given subset are replaced with the mean of the left-out eigenvalues. Using this, the inverse Fisher information matrix is given in (15):

$$\hat{\mathcal{F}}^{-1} = \begin{bmatrix} \widehat{Cov}(X) & \mathbf{0} \\ \mathbf{0}' & \frac{2}{n} D_p^{+} \widehat{Cov}(X) \otimes \widehat{Cov}(X) D_p^{+'} \end{bmatrix}. \tag{15}$$

The heavy-penalty form of *ICOMP* we use here is

$$ICOMP_{PEU}(\hat{\mathcal{F}}^{-1}) = -2\log L(\hat{\Lambda}, \mu, \hat{\sigma}^2 \mid x) + 2(\frac{nm}{n-m-2}) + \log(n) C_1(\hat{\mathcal{F}}^{-1}), \tag{16}$$

where m is the number of variables included from the original dataset. As with the MVR model, we can use the GA to reduce the dimensionality of a data set, with *ICOMP* as the objective function.

4 Numerical Results

Our dataset is a random sample of 1,400 students from the ALMALAUREA database. ALMALAUREA was started as a service for addressing the faculty choice of high school students based on interests, skills, and job expectations. All variables have been normalized to vary between -1 and 1. As response variables, we have $Mat_1, Mat_2, \ldots, Mat_{29}$: students judgements about different subjects (math, physics, chemistry, engineering, statistics...). Our regressor matrix is divided into two "sets." Answers regarding what the students think are important for ideal future work – collaboration, time flexibility, ...– are measured in variables $Nz_1, Nz_2, \ldots, Nz_{14}$. Variables $Np_1, Np_2, \ldots, Np_{32}$ measure students personal abilities (concentration, time management, curiosity, ...). The predictor variables are numbered from 1 to 14 for Nz, and 15 through 46 for Np.

For modeling this data, we first used the GA to identify optimal subset MVR models, driven by both *ICOMP*$(\hat{\mathcal{F}}^{-1})$ and *ICOMP*$(\widehat{Cov}(\hat{\theta}))_{MISP}$. If the Gaussian regression model was correctly specified, we would expect the criteria to select very similar models with similar scores. Results shown in the first third of Table 1 do not bear this out. While the substantially lower *ICOMP*$(\widehat{Cov}(\hat{\theta}))_{MISP}$ score indicates it has selected a better model, we have not been able to reduce the dimensionality at all. Mardia's tests for multivariate normal skewness and kurtosis (Mardia, 1974), reject the null hypothesis of normality, with results shown in Table 2, confirming the misspecification identified by *ICOMP*. Secondly, we used PPCA as a preliminary step to reduce the dimensionality of the matrix of responses.

Table 1 *ICOMP* Scores & Subsets of Predictors

Criteria	Score	Best set of predictors
No preliminary dimension reduction		
$ICOMP(\hat{\mathscr{F}}^{-1})$	64004	$\{1, 2, 4, 5, 6, 9, 12, 13, 16, 17, 19–21, 25–27, \ldots$ 29–31, 34, 35, 38, 41, 42$\}$
$ICOMP(\widehat{Cov}(\hat{\theta}))_{MISP}$	59701	{1–46}
Preliminary dimension reduction of only dependent variables matrix		
$ICOMP(\hat{\mathscr{F}}^{-1})$	9693	{1–46}
$ICOMP(\widehat{Cov}(\hat{\theta}))_{MISP}$	9483	{1–46}
Preliminary dimension reduction of both responses and regressors		
$ICOMP(\hat{\mathscr{F}}^{-1})$	10825	45
$ICOMP(\widehat{Cov}(\hat{\theta}))_{MISP}$	10963	45

Table 2 Normality test results for first identified model

Skewness		Kurtosis	
β_1	0	β_2	899
$\hat{\beta}_1$	32.55	$\hat{\beta}_2$	986.84
χ^{2*}	7594.92	Z^*	38.75
95 % Region	[0, 4652.09]	95 % Region	[−1.96, 1.96]
p-value	0.00000	p-value	0.00000
Conclusion	$\epsilon \nsim N(\mu, \Sigma)$	Conclusion	$\epsilon \nsim N(\mu, \Sigma)$

Using $ICOMP_{PEU}(\hat{\mathscr{F}}^{-1})$, the GA selected a model with only 4 dependent variables: $Mat_{26} - Mat_{29}$. We then attempted to identify a subset MVR model using just these responses. The *ICOMP* scores indicate that the Gaussian regression model is misspecified, with $ICOMP(\widehat{Cov}(\hat{\theta}))_{MISP} < ICOMP(\hat{\mathscr{F}}^{-1})$, though both criteria selected the fully saturated model. These results are shown in the middle third of Table 1. Mardia's expected and sample kurtosis values of 24 and 22.6 were very close; the test statistic for skewness, however, was 214 – much higher than the critical value of 31. Once again, we verify the misspecification identified by *ICOMP*.

Finally, we also used PPCA to select a subset of only 4 of the 46 independent variables. Those selected were $Np_{29} - Np_{32}$. We then ran two sets of the GA a third time, using both *ICOMP* versions, with results displayed in the bottom third of Table 1. Note how close the *ICOMP* scores are (relative to the other pairs), and that both criteria selected the same substantially reduced subset MVR model, using only a single predictor for the four responses. Thus, we have gone from an overly complex misspecified multivariate regression model, to a model that is both (very nearly) correctly-specified and parsimonious model.

While our end result would suggest the misspecification-resistant *ICOMP* was not needed, recall the first MVR subset model identified. If we had only used $ICOMP(\hat{\mathscr{F}}^{-1})$, we would have had less motivation to use PPCA to reduce the dimensionality of the model. We would have settled upon an MVR model with 32 responses and 24 regressors.

5 Concluding Remarks

In this research, we have applied a novel hybrid dimension reduction technique for multivariate regression. While independently reducing the number of dimensions in both the matrix of responses and regressors using PPCA and the GA, we used a new misspecification-resistant form of *ICOMP*. These methods allowed us to identify a nearly correctly-specified simple regression relationship with 4 of 29 dependent and 1 of 46 independent variables, rather than a misspecified overly complex relationship.

References

Box, G., & Cox, D. (1964). An analysis of transformations. *Journal of the Royal Statistical Society, Series B (Methodological), 26*, 211–246.

Bozdogan, H. (1988). Icomp: A new model-selection criteria. In H. Bock (Ed.), *Classification and related methods of data analysis* (pp. 599–608). Amsterdam: Elsevier.

Bozdogan, H. (2004). Intelligent statistical data mining with information complexity and genetic algorithms. In H. Bozdogan (Ed.), *Statistical data mining and knowledge discovery* (pp. 15–56). Boca Raton: Chapman and Hall/CRC.

Bozdogan, H., & Howe, J. (2009). *The curse of dimensionality in large-scale experiments using a novel hybridized dimension reduction approach*. The University of Tennessee. (Tech. Rep. 1).

Bozdogan, H., & Howe, J. (2012). *Misspecification resistant multivariate regression models using the genetic algorithm and information complexity as the fitness function. European Journal of Pure and Applied Mathematics, 5*(2), 211–249.

Goldberg, D. (1989). *Genetic algorithms in search, optimization and machine learning*. Boston: Addison-Wesley.

Haupt, R., & Haupt, S. (2004). *Practical genetic algorithms*. Hoboken: Wiley.

Holland, J. (1975). *Adaptation in natural and artificial systems: An introductory analysis with applications to biology, control, and artificial intelligence*. Ann Arbor: The University of Michigan Press.

Kullback, A., & Leibler, R. (1951). On information and sufficiency. *Annals of Mathematical Statistics, 22*, 79–86.

Magnus, J. (2007). The asymptotic variance of the pseudo maximum likelihood estimator. *Econometric Theory, 23*, 1022–1032.

Magnus, J., & Neudecker, H. (1988). *Matrix differential calculus with applications in statistis and econometrics*. New York: Wiley.

Mardia, K. (1974). Applications of some measures of multivariate skewness and kurtosis in testing normality and robustness studies. *Sankhya, B36*, 115–128.

Tipping, M., & Bishop, C. (1997). *Probabilistic principal component analysis* (Tech. Rep. NCRG/97/010). Neural Computing Research Group, Aston University.

Van Emden, M. (1971). An analysis of complexity. In *Mathematical centre tracts* (Vol. 35). Amsterdam: Mathematisch Centrum.

Vose, M. (1999). *The simple genetic algorithm: Foundations and theory*. Cambridge: MIT.

A Clusterwise Regression Method for the Prediction of the Disposal Income in Municipalities

Paolo Chirico

Abstract The paper illustrates a *clusterwise regression* procedure applied to the prediction of per capita disposal income (*PCDI*) in Italian municipalities. The municipal prediction is derived from the provincial *PCDI* taking into account the discrepancy between municipality and province in some indicators like per capita taxable income, per capita bank deposits, employment rate, etc. The relation between *PCDI* and indicators is shaped by a regression model. A single regression model doesn't fit very well all territorial units, but different regression models do it in groups of them. The aim of clusterwise regression is just that: detecting clusters where the correspondent regression models explain the data better than an overall regression model does. The application of the procedure to a real case shows that a significative reduction of the regression standard error can be achieved.

1 Introduction

The present work originates from a study of Unioncamere Piemonte (2009) about the prediction of the per capita disposal income (*PCDI*) in the Piedmont municipalities. More specifically Unioncamere Piemonte intended to predict the *PCDI* of the Piedmont municipalities by means of a regression model using some municipal indicators like "per capita taxable income", "per capita bank deposits", etc. Formally:

$$y_{ij} = \mathbf{x}'_{ij}\boldsymbol{\beta} + \varepsilon_{ij} \qquad (1)$$

where y_{ij} is the *PCDI* of the ith municipality in the jth province; $\mathbf{x}'_{ij}$ is the vector of regressors; $\boldsymbol{\beta}$ is the vector of the correspondent coefficients; ε_{ij} is the residual regression error.

P. Chirico (✉)
Department of Economics, University of Turin, Italy
e-mail: paolo.chirico@unito.it

A. Giusti et al. (eds.), *Classification and Data Mining*, Studies in Classification, Data Analysis, and Knowledge Organization, DOI 10.1007/978-3-642-28894-4_21,

Unioncamere knew the indicators for every Piedmont municipality but didn't know the *PCDIs*, even for a sample of municipalities, so that the model parameters couldn't be estimated on municipal data. On the other hand, all data were known at provincial level (the provincial *PCDIs* were provided by an external research institute). Therefore, the model parameters were estimated using the model (1) at provincial level; the *Ordinary Least Squares* estimation method was adopted considering all provinces on the same level of importance.

This paper proposes an evolution of that model in order to:

- Formalize better the regression errors and have municipal predictions consistent with the provincial *PCDI* (Sect. 2);
- Reduce the prediction errors by means of a *clusterwise regression* procedure (Sect. 3).

2 The Basic Model

Let's assume that the municipal *PCDIs* can be explained by some municipal indicators with a linear regression model like (1). The regression error ε_{ij} can be viewed as:

$$\begin{aligned}\varepsilon_{ij} &= y_{ij} - \mathbf{x}'_{ij}\boldsymbol{\beta} = \left[\sum_h y_{hij}\right]/n_{ij} - \mathbf{x}'_{ij}\boldsymbol{\beta} \\ &= \sum_h \left[y_{hij} - \mathbf{x}'_{ij}\boldsymbol{\beta}\right]/n_{ij} = \sum_h \varepsilon_{hij}/n_{ij} \end{aligned} \tag{2}$$

where ε_{hij} is the difference between the disposal income of the generic hth resident and the expected *PCDI* in its municipality; n_{ij} is the municipal population.

According to its definition, ε_{hij} is a random error and includes all individual factors determining the individual disposal income. At first, every ε_{hij} is assumed *independent* of every other error and regressor, and *identically distributed* with $E(\varepsilon_{hij}) = 0$ and $Var(\varepsilon_{hij}) = \sigma^2$. Such statements are clearly hard, but, at the moment, let's view them as a way to formalize better the features of ε_{ij}. Since $\varepsilon_{ij} = \sum_h \varepsilon_{hij}/n_{ij}$ and generally $n_{ij} > 1{,}000$, ε_{ij} can be assumed Gaussian. Now the model (1) can be better specified as:

$$y_{ij} = \mathbf{x}'_{ij}\boldsymbol{\beta} + \varepsilon_{ij} \tag{3}$$

with $\varepsilon_{ij} \sim N(0, \sigma^2/n_{ij})$.

As the provincial *PCDI* is $y_j = \sum_{hj} y_{hij}/n_j$, then:

$$y_j = \mathbf{x}'_j\boldsymbol{\beta} + \varepsilon_j \tag{4}$$

with $\varepsilon_j \sim N(0, \sigma^2/n_j)$.

In our case the *PCDIs* of the municipalities are unknown, even for a sample of municipalities, so that the model (3) is not useful for the parameters estimation. Nevertheless the *PCDIs* of the provinces are known so that the model parameters can be estimated through provincial data (model 4). Since the provincial regression errors have different variances, each of them equal to σ^2/n_j, the *Weighted Least Squares* (*WLS*) estimation method should be used:

$$\widehat{\boldsymbol{\beta}} = (\mathbf{X}'\mathbf{N}\mathbf{X})^{-1}\mathbf{X}'\mathbf{N}\mathbf{y} \tag{5}$$

where $\mathbf{X}$ is the data matrix of provincial regressors; $\mathbf{y}$ is the vector of provincial *PCDIs*; $\mathbf{N}$ is the diagonal matrix of provincial populations.

Now let's reconsider the assumptions about ε_{hij}. If the assumptions about mean and variance can be acceptable, their independence seems not realistic, in particular among the individual errors in a same municipality. Nevertheless these assumptions have only one effect on the modeling: the adoption of the *WLS* method for the models estimations. That means the models have to fit better the provinces with more population, and that seems reasonable.

According with the model (3), the prediction of the municipal *PCDI* should be $\widehat{y}_{ij} = \mathbf{x}'_{ij}\widehat{\boldsymbol{\beta}}$ since the prediction of the municipal error, $\widehat{\varepsilon}_{ij}$, is generally assumed equal to zero. Nevertheless the provincial average of the municipal errors, $\widehat{\varepsilon}_j$, is known before predicting the municipal errors, $\widehat{\varepsilon}_{ij}$; indeed it is known by the estimation of the provincial models (4): $\widehat{\varepsilon}_j = y_j - \mathbf{x}'_j\widehat{\boldsymbol{\beta}}$.

A way to take into account this information is to predict every municipal errors in a province equal to their provincial average: $\widehat{\varepsilon}_{ij} = \widehat{\varepsilon}_j$. Consequently the municipal *PCDI* prediction becames:

$$\widehat{y}_{ij} = \mathbf{x}'_{ij}\widehat{\boldsymbol{\beta}} + (y_j - \mathbf{x}'_j\widehat{\boldsymbol{\beta}}) = y_j + (\mathbf{x}'_{ij} - \mathbf{x}'_j)\widehat{\boldsymbol{\beta}} \tag{6}$$

Therefore the prediction of the municipal *PCDI* can be viewed as an adjustment of the provincial *PCDI* on the basis of the differences between the municipal indicators and the provincial ones. Moreover, the formula (6) assures that the provincial average of all municipal predictions is equal to the known provincial *PCDI*:

$$\sum_i \widehat{y}_{ij}\frac{n_{ij}}{n_j} = y_j \tag{7}$$

3 From a Single Model to *k* Models

The detection of a suitable provincial model (4) (and its estimation) only on the basis of the data of the eight Piedmont provinces would have led to an overfitting model. To get over this problem, the model was initially generalized to the Italian provinces and was therefore estimated using the data of 87 Italian provinces (some

Table 1 Regressors and coefficients

Regressor	Coefficient	Sign.
Intercept	5,710.91	***
Per capita taxable income	0.59	***
Employment rate	69.38	***
Per capita banc deposit	0.18	***
Rate of graduates	−266.40	***
Oldness index	14.94	***

Table 2 Quality indices

Index	Value
R^2	0.962
$\overline{R}^2$	0.959
$\widehat{\sigma}$	486,879.5

provinces were excluded from the analysis because not all the requested data were available). The regression results are reported in Tables 1 and 2.

We can note an unexpected results: the negative contribution of "rate of graduates". It doesn't mean that the relationship between *PCDI* and "rate of graduates" is negative, indeed their correlation is positive, although very low (0.152). It means that the contribution of the "rate of graduates" to the prediction of *PCDI* with the others predictors is negative; it concerns the role of the "rate of graduates" in explaining what it is not explained by the others predictors.

The R^2 and the $\overline{R}^2$ are very high, and that is understandable since the high correlation between *PCDI* and the regressor "per capita taxable income" (0.958). All regressors are significant at 1% level (***) and each one of them improves the Akaike's Information Criterion (AIC) and the Schwarz' Criterion (SC) if added after the other regressors. Nevertheless, even if the R^2 and the $\overline{R}^2$ are very high, we can't state that the model fits the data very well. Indeed the value of the standard regression error, $\widehat{\sigma}$, is not realistic (486,879 euros!). According to the assumption in the Sect. 2, σ is the standard deviation of ε_{hij} and can be viewed as a measure of the average difference between the individual disposal income and the expected *PCDI* in the correspondent municipality. If the model fits well the data, the value of $\widehat{\sigma}$ should be realistic. Therefore, an overall model like (4) is not good for every Italian province. On the other hand, K groups (*clusters*) of provinces may be fitted quite well by K local regression models, like:

$$y_{jk} = \mathbf{x}'_{jk}\boldsymbol{\beta}_{\mathbf{k}} + \varepsilon_{jk} \tag{8}$$

with $\varepsilon_{jk} \sim N(0, \sigma_k^2/n_{jk}), k = 1, .., K$.

The detection of such locals model and the corresponding partition concerns the *clusterwise regression*.

3.1 The Clusterwise Regression

The aim of the *clusterwise regression*, (CR), also named *regression clustering* by other authors (Zhang, 2003), is segmenting a number of units in some clusters in order to detect a good regression model in each cluster. Then clusterwise regression is suitable when the population is not homogeneous, and a single regression model doesn't fit well all the units, but different regression models might fit well partitions of the data. The origins of CR can be founded in the works of Bock (1969) and Spaeth (1979), whose original algorithms can be viewed as a special case of k-means clustering with a criterion based on the minimization of the squared residuals instead of the classical within-class dispersion (Preda and Saporta, 2005).

More specifically, if $G = \{G(1), G(2), \ldots, G(n)\}$ identifies a partition of n units of a population in K clusters, and:

$$V(K, G, \boldsymbol{\beta}_1, \ldots, \boldsymbol{\beta}_K) = \sum_{k} \sum_{G(i)=k} (y_i - \mathbf{x}'_i \boldsymbol{\beta}_k)^2 \tag{9}$$

is the sum of the squared residuals of the K local regressions, the basic algorithm of CR consists on iterating the following two steps:

(a) For given G, $V(K, G, \boldsymbol{\beta}_1, \ldots, \boldsymbol{\beta}_K)$ is minimized by the LS-estimators of the $\boldsymbol{\beta}_1, \ldots, \boldsymbol{\beta}_K$;
(b) For given $\boldsymbol{\beta}_1, \ldots, \boldsymbol{\beta}_K$, $V(K, G, \boldsymbol{\beta}_1, \ldots, \boldsymbol{\beta}_K)$ is minimized by assigning each unit to the cluster where the corresponding regression error is minimum; that identifies a new partition G.

Like in k-means clustering (MacQueen, 1967) the algorithm in converging, because the sequence of $V(K, G, \boldsymbol{\beta}_1, \ldots, \boldsymbol{\beta}_K)$ is, clearly, monotonically non-increasing. But, unlike k-means clustering, the algorithm converges to a local optimal solution, that depends on the initial partitions and not necessarily is the global optimal solution. Therefore, it would be better to simulate several initial partitions in order to choose the best final partition! Since its development, numerous adaptations and extensions of CR have been proposed; DeSarbo et al. (1989) extended clusterwise regression to the case of multiple response variables and repeated measures on subjects and proposed a simulated annealing algorithm for solving the resulting optimization problem. As reported in (Brusco et al., 2008), mixture-model formulations of CR have been proposed by numerous authors (DeSarbo and Cron, 1988; Henning, 2000) that assume the response variable measures are obtained from a mixture of K conditional densities (usually normal) that arise in unknown proportions. Obviously, the bigger the number of clusters, the better the fit of data, but that doesn't mean necessary better partition of data. About this issue, DeSarbo and Cron (1988) suggest to adopt the Akaike's Information Criterion, while Henning (2000) suggest to adopt the Schwarz' Criterion.

A correlated issue is the problem of overfitting, that has been analyzed recently by Brusco et al. (2008).

Table 3 Quality indices for each partition

Num.clusters	1	2	3	4	5
AIC	2,531.3	2,440.8	2,373.2	2,314.2	2,299.2
SC	2,546.1	2,472.9	2,422.5	2,380.8	2,383.1
logL	−1,259.7	−1,207.4	−1,166.6	−1,130.1	−1,115.6
min $\widehat{\sigma}$	486,679	187,114	128,378	104,720	98,436
max $\widehat{\sigma}$	486,679	254,027	181,329	155,416	146,775

Table 4 The four local regressions

	Cluster 1	Cluster 2	Cluster 3	Cluster 4	Overall
Intercept	3,488.45	4,662.35	4,543.81	4,113.12	5,710.91
Per capita taxable income	0.42	0.50	0.40	0.22	0.59
Employment rate	146.23	87.08	106.45	182.05	69.38
Per capita bank deposit	0.05	0.19	0.25	0.21	0.18
Rate of graduates	−144.06	−139.31	−179.61	−183.39	−266.40
Oldness index	16.09	14.51	19.31	21.11	14.94
Provinces	18	31	19	20	88
R^2	1.00	1.00	1.00	1.00	0.99
$\overline{R}^2$	0.99	1.00	1.00	1.00	0.959
$\widehat{\sigma}$	143,368	111,928	104,720	155,416	486,879

3.2 Four Models for PCDI Prediction

To detect the local models (8) for *PCDI* prediction, the basic algorithm of CR, with WLS estimation method, was adopted. According with DeSarbo and Cron (1988) and Henning (2000), partitions in 2, 3, 4, 5 clusters were tried, in order to detect the most suitable solution. For every partition in K clusters, several random initial partition were used.

The Table 3 reports some quality indices of the final (optimal) partitions in different number of clusters.

The partition in four clusters is better according to Schwarz' Criterion, while the partition in five clusters is better according to Akaike's Criterion. In such partition the local regression standard errors are less than in 4-clusters partition, but the improvement is not very significant, so the partition in four clusters was preferred. The Table 4 reports the local regression results of that partition.

Now the standard regression errors of the local regressions are clearly lower, and consequently more realistic than the standard regression error of the overall regression. Both the R^2s and the $\overline{R}^2$s are very high and could be a sign of overfitting, but it is not the case. Indeed the same indices of the overall model are high too and not for the presence of overfitting, as explained in Sect. 2.

3.3 The Municipal PCDI Prediction

Properly, the clusterwise regression described in the last subsection has concerned the provincial models, not the municipal ones. Then the extension of the clustering to the municipal predictions requires the assumption that the *PCDIs* of all municipalities of a province are explained by the model of their province:

$$y_{ijk} = \mathbf{x}'_{ijk}\boldsymbol{\beta}_k + \varepsilon_{ijk} \tag{10}$$

with $\varepsilon_{ijk} \sim N(0, \sigma_k^2/n_{jk})$.

Therefore, the *PCDIs* of the municipality ith of the province jth belonging to the cluster kth will be predict by the following formula:

$$\widehat{y}_{ijk} = y_{jk} + (\mathbf{x}'_{ijk} - \mathbf{x}'_{jk})\widehat{\boldsymbol{\beta}}_k \tag{11}$$

Obviously some municipal *PCDI*s might be explained better by models of other clusters than by their one. Nevertheless there isn't way to known exactly which model is the best for every municipality. Then, in absence of further information, the assumption in (10) can be reasonable at least for middle-big municipalities that are not too different from the profile of their province.

4 Final Considerations

The paper describe a case where the clusterwise regression can be useful to detect a number of suitable regression models in a heterogeneous population. All the methodology can be viewed like a way to predict the municipal *PCDIs* in case of: (1) the *PCDIs* are explainable by some regressors; (2) the *PCDIs* are not known at municipal level, but are known for territorial aggregations; (3) the territorial aggregations are heterogeneous. Obviously the number of territorial aggregations has to be enough numerous for being segmented in clusters where regression models are drawn.

The explained methodology joins in a series of proposals about the Italian municipal disposal incomes, that includes Marbach (1985), Frale (1998) and Bollino and Pollinori (2005), quoting only some authors. Here, as in Marbach, the municipal disposal income is derived from the provincial disposal income, but in Marbach the provincial disposal income is object of prediction; here is exogenous. As in Bollino and Pollinori, the regressive models are heteroscedastic and the estimated provincial errors are used for the prediction of the municipal errors. Those proposals illustrate procedures very articulated, but don't handle the problem of the heterogeneity by means of a model-based approach. The present proposal does it by clusterwise regression.

Finally the provincial *PCDIs* are exogenous data in the models as well as all the regressors. The present paper doesn't consider how they are calculated. Actually the most of them are estimated. For example, the Bank of Italy estimates the *PCDIs* at regional level by a sample survey; private research institutes provide estimations of the *PCDIs* at provincial level, but their methods are not exactly known.

Obviously the quality of the municipal predictions (11) depends on the quality of exogenous data too!

References

Bock, H. H. (1969). The equivalence of two extremal problems and its application to the iterative classification of multivariate data. In *Lecture note*. Matematisches Forschungsinstitut, Oberwolfach, Germany.

Bollino, C. A., & Pollinori, P. (2005). Il valore aggiunto su scala comunale: La Regione Umbria 2001–2003. Quaderni del Dipartimento di economia, Finanza e Statistica (No. 15/2005). Universitá di Perugia.

Brusco, M. J., Cradit, J., Steinley, D., & Fox, G. J. (2008). Cautionary remarks on the use of clusterwise regression. *Multivariate Behavioral Reseach, 43*, 29–49.

DeSarbo, W. S., & Cron, W. L. (1988). A maximum likelihood methodology for clusterwise linear regression. *Journal of Classification, 5*, 249–282.

DeSarbo, W. S., Oliver, R. L., & Rangaswamy, A. (1989). A simulated annealing methodology for clusterwise linear regression. *Psychometrika, 54*, 70–736.

Frale, C. (1998). Stime comunali del reddito disponibile: La provincia di Udine, *Osservatorio permanente dell'economia del Friuli venezia Giulia* (No. 3).

Hennig, C. (2000). Identifiability of models for clusterwise linear regression. *Journal of Classification, 17*, 273–296.

MacQueen, J. B. (1967). Some methods for classification and analysis of multivariate observations. In *Proceedings of 5-th Berkeley Symposium on Mathematical Statistics and Probability* (Vol. 1, pp. 281–297). Berkeley: University of California Press.

Marbach, G. (Ed.). (1985). Il reddito nei comuni italiani 1982. In *Quaderni del Banco di Santo spirito*. Torino: UTET.

Preda, C., & Saporta, G. (2005). Clusterwise PLS regression on a stochastic process. *Computational Statistics and Data Analysis, 49*, 99–108.

Spaeth, H. (1979). Clusterwise linear regression. *Computing, 22*, 367–373.

Unioncamere Piemonte. (2009). *Geografia dei redditi 2009, Osservatorio sul reddito disponibile e prodottoin Piemonte*. Unioncamere Piemonte, Turin, Italy.

Zhang B. (2003). Regression clustering. In *ICDM03, Third IEEE International Conference on Data Mining*, Melbourne, Florida, USA, (p. 451).

A Continuous Time Mover-Stayer Model for Labor Market in a Northern Italian Area

Fabrizio Cipollini, Camilla Ferretti, Piero Ganugi, and Mario Mezzanzanica

Abstract A new and powerful source of information concerning the Italian Labor Market is represented by C.OBB datasets, which record the kind of job contract (with its successive modifications) of all the workers in many Italian Provinces. By means of this information and focusing on the Province of Cremona, we analyze the mobility of employees among different kinds of job contracts (and unemployment also): in particular, from contracts characterized by modest packages of securities toward more structured working relations, ending with Unlimited Time Duration Contracts. The statistical tool used for this analysis is Continuous Time Mover-Stayer Model. Our analysis reveals low mobility from Limited Time Duration to Unlimited Time Duration contracts.

1 Motivation

Thanks to Law 196/1997 ("Pacchetto Treu"), to Law 276/2003 ("Legge Biagi") and to other legislative measures, the Italian Labor Market acquired a degree of flexibility unknown in the previous decades. Currently, in fact, employments can be

F. Cipollini
Department of Statistics, Università di Firenze, Florence, Italy
e-mail: cipollini@ds.unifi.it

C. Ferretti (✉)
Department of Economics and Social Sciences, Università Cattolica del Sacro Cuore, Piacenza, Italy
e-mail: camilla.ferretti@unicatt.it

P. Ganugi
Department of Industrial Engineering, Universitá, degli Studi di Parma, Parma, Italy
e-mail: piero.ganugi@unipr.it

M. Mezzanzanica
Department of Statistics, Milano Bicocca University, Milano, Italy
e-mail: mario.mezzanzanica@unimib.it

A. Giusti et al. (eds.), *Classification and Data Mining*, Studies in Classification, Data Analysis, and Knowledge Organization, DOI 10.1007/978-3-642-28894-4_22,

Table 1 Classification of states. (Thanks to Pietro Antonio Varesi and Michele Bricchi of UCSC, Piacenza)

State	Description	Group
1	Unlimited duration and full time	Unlimited time duration
2	Unlimited duration and partial time	
3	Expiry job and full time	Limited time duration
4	Expiry job and partial time	
5	Apprenticeship	
6	Co.co.pro. and Co.co.co	
7	Self-employment	
8	Unemployment	

ruled by 35 different kinds of agreements, among which only two have Unlimited Time Duration (UTD). One of the most important (and till now unanswered) questions risen by this new Legislation concerns the degree of mobility of individuals from Limited Time Duration (LTD), and other precarious job agreements, toward UTD contracts. Another important issue concerns possible differences between genders.

Traditionally, the evolution of individuals among states is modelled using a Markov Chain, which has the drawback to overestimate the mobility (Blumen et al., 1955; Spilerman, 1972). In this sense, the Mover-Stayer model has the relevant advantages to overcome this pitfall. Its continuous time version (Fougere and Kamionka, 2003; Frydman and Kadam, 2004) seems to be a suitable statistical model because individuals can change state at every instant of time.

2 The C.OBB Data and the Job States

C.OBB data are a new and powerful source of information, fed by the compulsory communications which Italian firms have to transmit to the Labour Office of Province for every new or expired engagement.

The dataset employed in the analysis includes all workers having a contract in the Province of Cremona in any day of January 2007 and of December 2009 (thus, in the middle of this interval a person can be also unemployed for some time). Once records lacking of fundamental variables are cleaned from, the dataset includes 45,898 workers. Considering that Labour Forces in the Province of Cremona amount to 164,000 units, and that C.OBB data includes only a very small fraction of self-employees, this dataset represents a relevant portion of the whole job market of the Province.

To simplify the analysis, we grouped the original 35 job contracts into 8 possible states, sorted according to their package of securities (see Tursi and Varesi, 2010) as shown in Table 1.

3 The Continuous Time Mover Stayer Model

The (discrete time) Mover-Stayer Model (MS) has been introduced in Blumen et al. (1955) as an extension of the classical Markov Chain Model. Further developments are in Goodman (1961), Frydman (1984) and Spilerman (1972); interesting empirical applications can be found, among others, in Quah (1993), Fougere and Kamionka (2003) and Cipollini et al. (2012).

The model can be viewed both as a *mixture* model and as a *latent class* model. In fact, MS is a mixture of two Markov chains because, given a population of units and a set of mutually exclusive states $S = \{1 \ldots, k\}$, the population itself is partitioned in two groups: the *Movers* and the *Stayers*. Every individual starting from a state i can be a Stayer, with probability s_i, or a Mover, with probability $1-s_i$. Movers move across S according to a classical Markov chain with transition matrix M; Stayers do not move from their starting state and so follow a degenerate chain with transition matrix I_k. MS is also a latent class model because, among the observed $n_i^{(s)}$ units never moving from state i, an individual cannot be identified as a genuine Stayer or "not-yet-moved" Mover.

The parameters of the model are the vector $s = (s_1, \ldots, s_k)$ (where each $s_i \in [0,1]$) and the matrix M (where each $M_{i,j} \geq 0$ and $\sum_{j \in S} M_{i,j} = 1$). The 1-step global transition matrix P, thus, has elements $P_{i,j} = s_i \mathrm{I}(i = j) + (1 - s_i) M_{i,j}$ where $\mathrm{I}(\cdot)$ denotes the indicator function.

The MS has been extended also to a *continuous time* framework (CTMS) by Singer and Spilerman (1976) and Frydman and Kadam (2004). Continuous time means that Movers can move across S at each instant of time and that there exists a *generating matrix* Q such that the transition matrix on a time interval $[0,t]$ can be expressed as $M(0,t) = e^{tQ}$ (the exponential matrix function, see Golub and Van Loan 1996). In such a case, the parameters of the model are s and Q. The main advantage of the CTMS, with respect to the MS, is the time flexibility, since it makes possible to estimate a transition matrix referred to any time interval $[0,t]$. Its possible pitfalls are *embeddability* and *aliasing* (see Singer and Spilerman 1976 for details).

4 Bayesian Inference on the CTMS

Inferential methods for the parameters of the CTMS are proposed in Frydman and Kadam (2004), Fougere and Kamionka (2003) and Inamura (2006). In this work we follow Fougere and Kamionka (2003) by readapting their Bayesian approach to our data and estimating the parameters via *Gibbs Sampler*. We summarize here its essential points considering a 2-dimensional r.v. The method can however be extended to more general cases: we refer to Casella and Robert (2009) for a deeper handling.

Let (X, Y) a r.v. whose conditional p.d.f.'s, $f_{X|Y}$ and $f_{Y|X}$, are available. Given a starting value y_0, random draws x_j and y_j are iteratively sampled from

$X_j|Y = y_{j-1}$ and $Y_j|X = x_j$, respectively, according to their p.d.f.'s $f_{Y|X}$ and $f_{X|Y}$. The sequence (X_j, Y_j) is proved to be a Markov Chain converging in distribution to f_{XY}, so that a suitable sequence $\{(x_j, y_j) : j = m_0 + 1, \ldots, m_0 + m\}$ can be taken as an m-dimensional sample from (X, Y). m_0 is the burn-in period needed to reach convergence, that in practical applications must be properly checked (Casella and Robert, 2009).

Referring now to the CTMS (Sect. 3), we aim to obtain the posterior distribution of the parameters s and Q, using the Gibbs Sampler and the whole set of observed data. In order to do this, we introduce the following notation:

- T is the last time period;
- Z_i is the (unknown) actual number of Stayers in state i and $Z = \{Z_i : i = 1, \ldots, k\}$;
- $n_i^{(0)}$ is the observed number of individuals in state i at time 0 and $n^{(0)} = \{n_i^{(0)} : i = 1, \ldots, k\}$;
- $n_i^{(s)}$ is the observed number of individuals never moving from state i and $n^{(s)} = \{n_i^{(s)} : i = 1, \ldots, k\}$;
- N_{ij} is the observed number of transitions from state i to state j and $N = \{N_{ij} : i, j = 1, \ldots, k\}$;
- $O = \{n^{(s)}, N\}$ is the whole set of observed data describing the dynamics.

About the a priori distribution of s, we assume that each s_i follows a *Beta* distribution, namely $s_i|n_i^{(0)} \sim \text{Beta}(a_i)$, where a_i is a 2-dimensional vector of positive elements.

As per the a priori distribution of $Q = \log(M)$, it is convenient to express it by working on M. Hence, we assume that each row of M is conditionally distributed as a *Dirichlet*, namely $M_{i..}|n_i^{(0)} \sim \text{Dir}(b_i)$, where b_i is a k-dimensional vector of positive elements.

Finally the a priori distribution of $Z|n_i^{(0)}$ is $\text{Bin}(n_i^{(0)}, s_i)$. Note that all distributions are defined given the starting value $n^{(0)}$.

By means of standard arguments of Bayesian inference (Berger, 1985), we obtain the following conditional distributions

$$s_i|Z_i = z, n_i^{(0)}, O \sim \text{Beta}(a_{i,2} + Z_i, a_{i,2} + n_i^{(0)} - Z_i), \tag{1}$$

$$M_{i..}|Z_i = z, n_i^{(0)}, O \sim \text{Dir}(b_{i,1} + N_{i,1}, \ldots, b_{i,i} + N_{i,i} - Tz, \ldots, b_{i,k} + N_{i,k}) \tag{2}$$

$$Z_i|s, M, n_i^{(0)} O \sim \text{Bin}\left(n_i^{(s)}, p_i = \frac{s_i}{s_i + (1 - s_i) M_{i,i}^T}\right) \tag{3}$$

from which we can implement the Gibbs Sampler and eventually draw random samples from the distribution of (s, Q) given O (note that the algorithm first generates M and then calculates $Q = \log(M)$).

Table 2 Relative error $e(0, t)$ for the whole dataset, varying t

t	Quarter	Half year	Year	Two years
$e(0, t)$	0.027	0.073	0.202	0.221

Table 3 Estimated $\hat{P}(0, 8)$ (2-years matrix on the whole data, expressed with percentages)

State	1	2	3	4	5	6	7	8
1	86.66	2.77	5.25	2.02	0.59	0.53	0.05	2.13
2	6.62	88.93	2.30	0.87	0.28	0.37	0.04	0.59
3	23.32	4.16	50.05	11.16	2.05	1.68	0.23	7.34
4	8.41	5.48	10.50	71.51	0.96	0.97	0.19	1.97
5	14.30	2.39	7.12	3.61	68.45	0.70	0.13	3.30
6	19.58	7.45	20.62	12.60	3.00	27.65	0.39	8.71
7	20.90	7.21	15.79	9.47	3.52	4.11	33.14	5.86
8	31.83	9.47	29.31	15.81	4.20	3.00	0.31	6.08

5 Main Results

The estimation procedure illustrated in Sect. 4 has been employed for estimating s and Q on the whole sample and by gender, considering a time unit corresponding to a quarter. Once checked the convergence of the algorithm, the point estimate of each parameter is obtained by taking the average of the corresponding sample from the posterior distribution. As in Fougere and Kamionka (2003) we choose hyperparameters $a_{i,j}$ and $b_{i,j}$ equal to two, having tested that estimates are stable with respect to different choices.

Goodness-of-fit tests based on χ^2 distribution, as suggested by Cipollini et al. (2012), are of limited applicability in this case, because of the presence of very small and null values. Then let $P_{obs}(0, t)$ and $\hat{P}(0, t)$ be the observed and the estimated transition matrix, respectively, after time t. As suggested in Frydman et al. (1985) we evaluate the model fit by means of the quantity

$$e(0, t) = ||P_{obs}(0, t) - \hat{P}(0, t)||_2/||P_{obs}(0, t)||_2$$

where $||.||_2$ is the 2-norm of matrices. The quantity $e(0, t)$ represents the relative error committed considering $\hat{P}(0, t)$ instead of $P_{obs}(0, t)$ (see Golub and Van Loan 1996). Table 2 summarizes our results for different spans of time.

Estimated $\hat{Q}$ and $\hat{s}$ are useful tools for evaluating persistence and transitions of individuals among states. In this work, however, we simplify the exposition of results, focusing on the transition probabilities $\hat{P}(0, t)$ estimated from the model. Table 3 shows $\hat{P}(0, 8)$, the estimated 2-years transition matrix, for the whole population.

Estimates evidence a low mobility of workers from LTD Contracts (states 3, 4, 5, 6) to the UTD Contracts (states 1 and 2). For example, the probability of Expiry Full Time Contracts to be changed in UTD is about 28 % within 2 years. Transition probabilities referred to other LTD Contracts, in the same direction, are even lower.

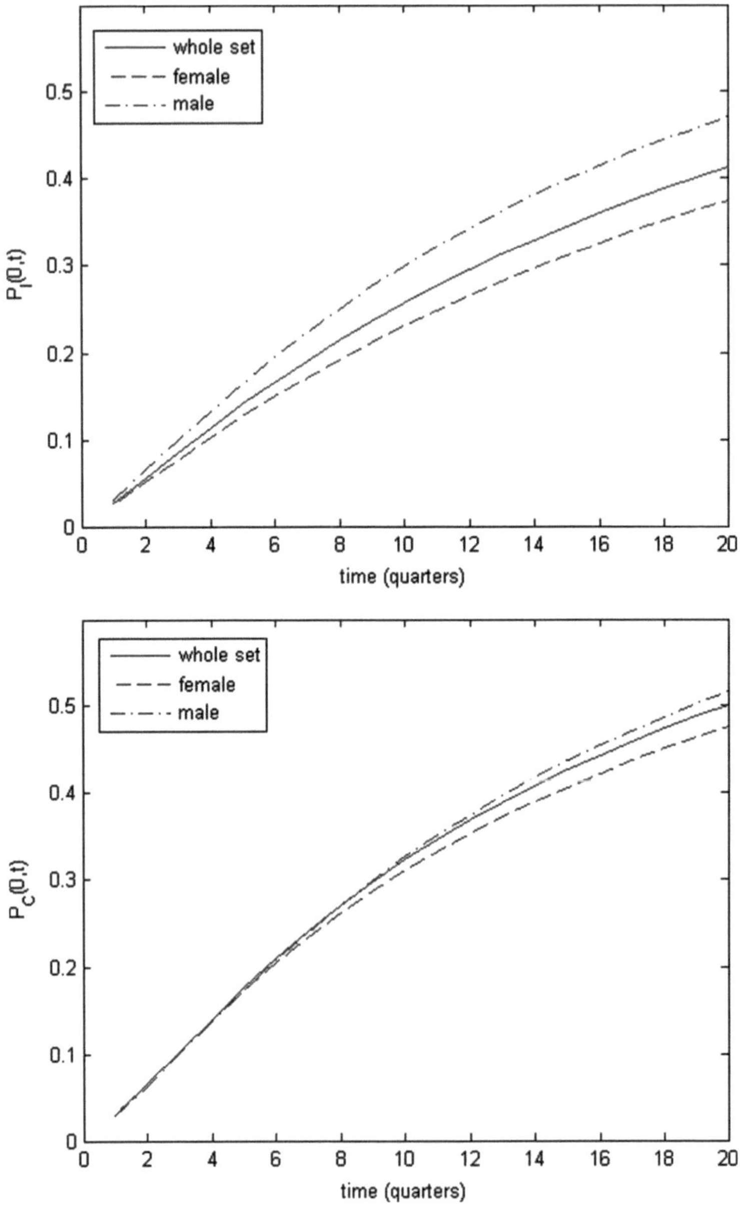

Fig. 1 Estimated transition probabilities towards UTD w.r.o. time: from expiry job (*upper*) and from co.co. (*lower*)

In analyzing results, for sake of space, we focus on the transition probabilities of individuals starting from Co.co.pro and Expiry Job, towards UTD. Let $p_C(0,t)$ and $p_I(0,t)$, respectively, be the probabilities of having a co.co.pro and a expiring working contract at time 0, and an UTD contract (at full or partial time) at time t.

Table 4 Estimated unconditional distributions among contracts, expressed as percentages

		State							
Group	t	1	2	3	4	5	6	7	8
Whole set	p_0	54.94	18.90	9.60	7.69	6.57	1.15	0.07	1.08
	p_8	53.27	19.50	9.97	8.40	5.23	0.99	0.11	2.52
	p_{20}	51.89	20.47	10.64	9.22	4.07	1.00	0.13	2.57
	p_∞	48.34	24.88	10.76	9.77	2.61	1.01	0.13	2.50
Female	p_0	40.37	29.53	10.05	11.59	5.89	1.41	0.10	1.05
	p_8	39.70	30.38	9.05	12.54	4.75	1.10	0.14	2.34
	p_{20}	38.99	31.61	8.90	13.28	3.75	1.05	0.15	2.28
	p_∞	36.21	36.41	8.51	13.15	2.45	1.00	0.15	2.14
Male	p_0	70.69	7.30	9.10	3.43	7.30	0.85	0.05	1.27
	p_8	67.91	7.66	10.96	3.87	5.75	0.89	0.09	2.87
	p_{20}	65.98	8.29	12.74	4.42	4.43	0.98	0.11	3.06
	p_∞	63.59	10.48	13.83	4.92	2.88	1.06	0.11	3.13

Figure 1 shows the estimated $\hat{p}_C(0,t)$ and $\hat{p}_I(0,t)$ as functions of t, comparing results for genders.

Strong differences among the different groups are evident. According to gender the difference looks neat: mobility towards UTD contracts is higher for men than for women.

In Table 4 we report the unconditional distributions p_t in the different states, computed at several times t (p_0 is the starting distribution, p_∞ is the equilibrium distribution) and estimated from the model. We observe that the weight of full-time UTD is the highest for every group, but the difference with respect to the other states is more relevant for males, and weaker for females. Furthemore we note that on one side the full time UTD tends to decrease, and on the other side the partial time UTD tends to increase.

6 Conclusion

The strong motivation of this paper is to investigate the mobility of Italian workers among job agreements based on different packages of securities. With this aim, we analyze C.OBB data for the Province of Cremona by means of a CTMS. The results evidence a modest mobility of workers from LTD to UTD contracts. In particular, females have about half probability of achieving UTD position than males.

Further research will be directed to define an index of mobility for CTMS, using the axiomatic approach proposed by Shorrocks (1978) and Geweke et al. (1986). Secondly we aim to extend the same analysis to other Italian Provinces.

References

Berger, J. O. (1985). *Statistical decision theory and Bayesian analisys* (2nd ed.). New York: Springer.

Blumen, I., Kogan, M., & McCarthy, P. J. (1955). *The industrial mobility of labor as a probability process*. Ithaca: Cornell University Press.

Casella, G., & Robert, C. P. (2009). *Introducing MonteCarlo methods with R*. New York: Springer.

Cipollini, F., Ferretti, C., & Ganugi, P. (2012). Firm size dynamics in an industrial district: the Mover Stayer Model in action, A. Di Ciaccio et al. (eds), *Advanced statistical methods for the analysis of large datasets*, Studies in Theoretical and Applied Statistics, pp. 443–452, Springer Verlag Berlin Heidelberg.

Fougere, D., & Kamionka, T. (2003). Bayesian inference for the Mover-Stayer model of continuous time. *Journal of Applied Economics, 18*, 697–723.

Frydman, H. (1984). Maximum-Likelihood estimation in the Mover-Stayer model. *Journal of the American Statistical Association, 79*, 632–638.

Frydman, H., & Kadam, A. (2002). Estimation in the continuous time Mover Stayer model with an application to bond rating migration. *Applied Stochastic Models in Business and Industry, 20*, 155–170.

Frydman, H., Kallberg, J. K., & Kao, D. L. (1985). Testing the adequacy of Markov Chain and Mover-Stayer models as representation of credit behavior. *Operations Research, 33*(6), 1203–1214.

Geweke, J., Robert, C. M., & Gary, A. Z. (1986). Mobility indices in continuous time Markov chains. *Econometrica, 54*(6), 1407–1423.

Golub, G. H., & Van Loan, C. F. (1996). *Matrix computation* (3rd ed.). Baltimore: Johns Hopkins University Press.

Goodman, L. A. (1961). Statistical methods for the Mover-Stayer model. *Journal of the American Statistical Association, 56*, 841–868.

Inamura, Y. (2006). Estimating continuous time transition matrices from discretely observed data. Bank of Japan Working Paper Series No. 06-E 07, Tokyo.

Quah, D. (1993). Empirical cross section dynamics in economic growth. *Empirical Economic Review, 37*, 426–434.

Shorrocks, A. F. (1978). The measurement of mobility. *Econometrica, 46*(5), 1013–1024.

Singer, B., & Spilerman, S. (1976). The representation of social processes by Markov models. *American Journal of Sociology, 82*(1), 1–54.

Spilerman, S. (1972). Extensions of the Mover-Stayer model. *American Journal of Sociology, 78*, 599–626.

Tursi, A., & Varesi, P. A. (2010). *Lineamenti di Diritto del Lavoro*. Padua: CEDAM.

Model-Based Clustering of Multistate Data with Latent Change: An Application with DHS Data

José G. Dias

Abstract Finite mixture modeling has been used extensively as a model-based clustering technique. This research addresses the application of mixture models to multistate data (sequences of states) under the Markov assumption. By assuming a latent or hidden Markov process, the model incorporates the estimation of the misclassification error. The data are from the life history calendar from the Brazilian Demographic and Health Survey (DHS) 1996 in which contraceptive use dynamics are surveyed retrospectively. The results show that the dynamics are heterogeneous with two subpopulations.

1 Introduction

Finite mixture modeling has been a powerful tool for capturing unobserved heterogeneity in a wide range of social and behavioral science data (Clogg, 1995). Nowadays increasing rich data sets with longitudinal structure have created a need for the development of more advanced statistical models that take into account the dynamics of social phenomena (e.g. random effect models, latent growth models). We introduce a specific finite mixture model for modeling multistate data that takes into account unobserved heterogeneity in line with Dias and Vermunt (2007), but extends it by allowing for misclassification error.

The paper is organized as follows: Sect. 2 presents the mixture model with latent change; Sect. 3 reports an application in demography in which the model is applied in the modeling of contraceptive use dynamics. The paper concludes with a summary of the main findings.

J.G. Dias (✉)
Instituto Universitário de Lisboa (ISCTE-IUL), Business Research Unit (BRU-IUL)
e-mail: jose.dias@iscte.pt

A. Giusti et al. (eds.), *Classification and Data Mining*, Studies in Classification, Data Analysis, and Knowledge Organization, DOI 10.1007/978-3-642-28894-4_23,

2 The Finite Mixture Model with Latent Change

Let us assume that we have n sequences of T observations. Let y_{it} represent the observed state response for individual i at time point t, where $i \in \{1, \ldots, n\}$, $t \in \{1, \ldots, T\}$, and y_{it} can take M different values (states). In addition to the observed "response" variable y_{it}, we assume two different latent variables: a time-constant discrete latent variable and a time-varying discrete latent variable. The former, which is denoted by $w \in \{1, \ldots, S\}$, captures the unobserved heterogeneity across the sample of individuals; that is, individuals are clustered based on differences in their dynamics. The time-varying latent variable is denoted by $z_t \in \{1, \ldots, K\}$.

Let $f(\mathbf{y}_i; \boldsymbol{\varphi})$ be the probability function associated with observation i, where $\boldsymbol{\varphi}$ is the vector of parameters in the model. The model is defined as:

$$f(\mathbf{y}_i; \boldsymbol{\varphi}) = \sum_{w=1}^{S} \sum_{z_1=1}^{K} \cdots \sum_{z_T=1}^{K} f(w) f(z_1|w) \prod_{t=2}^{T} f(z_t|z_{t-1}, w) \prod_{t=1}^{T} f(y_{it}|z_t). \quad (1)$$

As in any mixture model, the observed data density $f(\mathbf{y}_i; \boldsymbol{\varphi})$ is obtained by marginalizing over the latent variables. Because the latent variables are discrete, this simply involves the computation of a weighted average of class-specific probability functions, where the (prior) class membership probabilities or mixture proportions serve as weights (McLachlan and Peel, 2000). We assume that within cluster w the sequence $\{z_1, \ldots, z_T\}$ is in agreement with a first-order Markov chain. Moreover, we assume that the observed state at a particular time point depends only on the latent state at this time point; i.e., conditionally on the latent state z_t, the response y_{it} is independent of states at other time points, which is often referred to as the local independence assumption. As far as the first-order Markov assumption for the latent state conditional on cluster membership w is concerned, it is important to note that this assumption is not as restrictive as one may initially think. It does clearly not imply a first-order Markov structure for the responses y_{it}. The standard hidden Markov model (HMM) (Baum et al., 1970; Rabiner and Juang, 1986) turns out to be a special case of this model by eliminating the time-constant latent variable w from the model, that is, by assuming that there is no unobserved heterogeneity across respondents.

The characterization of the model is given by:

- $f(w)$ is the prior probability of belonging to a particular cluster w with multinomial parameter $\pi_w = P(W = w)$;
- $f(z_1|w)$ is the initial latent state probability; that is, the probability of having a particular initial state conditional on belonging to cluster w with multinomial parameter $\lambda_{kw} = P(Z_1 = k|W = w)$;
- $f(z_t|z_{t-1}, w)$ is a latent transition probability; that is, the probability of being in a particular state at time point t conditional on the state at time point $t - 1$ and cluster membership; assuming a time-homogeneous transition process, we have

$p_{jkw} = P(Z_t = k|Z_{t-1} = j, W = w)$ as the relevant multinomial parameter. Thus, the model allows that each cluster has its specific transition dynamics;
- $f(y_{it}|z_t)$, the probability density of having a particular observed state in sequence i at time point t, conditional on the regime occupied at time point t, is assumed to have a multinomial distribution characterized by parameters $p_{km} = P(y_{it} = m|Z_t = k)$. Note that these parameters are assumed invariant across clusters, an assumption that may, however, be relaxed.

Since $f(\mathbf{y}_i; \boldsymbol{\varphi})$, defined by Eq. (1), is a mixture of hidden Markov models, it defines a flexible model that can accommodate unobserved heterogeneity (mixture component) and misclassification error (discrete hidden Markov model). The model has $(S + K + M)(K - 1) + S - 1$ parameters to be estimated, including $S - 1$ class sizes, $S(K - 1)$ initial state probabilities, $K(K - 1)$ transition probabilities, and $M(K - 1)$ p_{km} probabilities. As the number of latent states K equals the number of observed states M, p_{km} identifies the degree of misclassification.

Maximum likelihood (ML) estimation of the model parameters involves maximizing the log-likelihood function: $\ell(\boldsymbol{\varphi}; \mathbf{y}) = \sum_{i=1}^{n} \log f(\mathbf{y}_i; \boldsymbol{\varphi})$, a problem that can be solved by means of the Expectation-Maximization (EM) algorithm (Dempster et al., 1977). The E step computes the joint conditional distribution of the latent variables given the data and the current provisional estimates of the model parameters. In the M step, standard complete data ML methods are used to update the unknown model parameters using an expanded data matrix with the estimated densities of the latent variables as weights. Since the EM algorithm requires us to compute and store the $S \cdot K^T$ entries in the E step, this makes this algorithm impractical or even impossible to apply with more than a few time points. However, for hidden Markov models, a special variant of the EM algorithm has been proposed that is usually referred to as the forward-backward or Baum-Welch algorithm (Baum et al., 1970). The Baum-Welch algorithm circumvents the computation of this joint posterior distribution making use of the conditional independencies implied by the model. The Baum-Welch algorithm for HMMs can easily be adapted to estime this model.

An important modeling issue is the setting of S, the number of clusters needed to capture the unobserved heterogeneity across individuals. The selection of S is typically based on information statistics such as the Bayesian Information Criterion (BIC) (Schwarz, 1978). In our application we select S that minimizes the BIC value defined as:

$$BIC_S = -2\ell_S(\hat{\boldsymbol{\varphi}}; \mathbf{y}) + N_S \log n, \tag{2}$$

where N_S is the number of free parameters of the model and n is the sample size.

3 Application

Life history calendar (LHC) is a major and relatively new instrument for the collection of retrospective data (Belli et al., 2001). We illustrate the approach using data from the Brazilian Demographic and Health Survey (BDHS) conducted

Table 1 Model selection

S	$\ell(\varphi; \mathbf{y})$	#par	BIC	$\Delta BIC\%$
1	−13784.51	44	27886.33	–
2	−13211.06	69	26919.73	−3.47%
3	−12845.88	94	26369.66	−2.04%
4	−12699.63	119	26257.44	−0.43%
5	−12599.53	144	26237.51	−0.08%

between March 1996 and June 1996 (BENFAM and Macro International, 1997). The BDHS includes a calendar of monthly data on contraceptive use and pregnancy status. The BDHS is a nationally representative, stratified two-stage sample. The calendar covers the period from January 1991 to the month of interview. We selected for this analysis the São Paulo region, corresponding to 1,355 women. We aggregate the original state space into five states, namely: (1) Non-use of contraception, (2) Sterilization (Female sterilization, Male sterilization), (3) Pregnancy (Pregnancy, Birth, Terminated pregnancy/non-live birth), (4) Pill, and (5) Other temporary methods (IUD - Intrauterine device, Injections, Diaphragm/foam/jelly, Condom, Periodic abstinence/rhythm, Withdrawal, and Other traditional methods).

Women are grouped into clusters on the basis of similarity of their behavior. We estimate the model with a different number of clusters from 1 to 5, using 200 different starting values to avoid local maxima. We notice that mixture model log-likelihood surfaces tend to be complex with several local optima (Dias and Wedel, 2004). This is particularly true for latent class models with nominal manifest data in which the number of parameters to be estimated tends to increase faster with the number of latent states. Therefore, despite model estimation being time consuming; a large number of starting values has to be set to provide confidence in the parameter estimates being reported.

Table 1 provides the values of the log-likelihood, the number of parameters in the model, the value of BIC, and the variation in BIC. The results suggest that three clusters should be taken into account, given that the BIC value tends to stabilize from three to four clusters. However, in the solution with three clusters, the third cluster size is 0.61%, which suggests the two-cluster solution as the best one. I notice that from a public policy point of view, clusters or segments must be big enough to be worth to be targeted. Thus, we selected the two-latent class solution. Cluster sizes are 52.7% and 47.3% for cluster 1 and 2, respectively.

Table 2 reports the misclassification estimated probabilities for both aggregate and two-cluster based solutions. Despite potential problems of memory recall in retrospective studies, we observe that the estimated misclassification error is small in both cases. The clustering solution improves the classification for all states, in particular for pregnancy state. There is no measurement error for the sterilization state.

Table 2 Misclassification probability estimates

Solution	Observed state	Latent state 1	2	3	4	5
Aggregate						
	Non-use of contraception	0.9998	0.0000	0.0386	0.0034	0.0005
	Sterilization	0.0000	1.0000	0.0000	0.0000	0.0000
	Pregnancy	0.0000	0.0000	0.9613	0.0000	0.0000
	Pill	0.0001	0.0000	0.0000	0.9914	0.0002
	Other temporary methods	0.0001	0.0000	0.0001	0.0051	0.9994
Cluster-based						
	Non-use of contraception	0.9999	0.0000	0.0002	0.0019	0.0002
	Sterilization	0.0000	1.0000	0.0000	0.0000	0.0000
	Pregnancy	0.0000	0.0000	0.9998	0.0001	0.0000
	Pill	0.0001	0.0000	0.0000	0.9931	0.0000
	Other temporary methods	0.0001	0.0000	0.0000	0.0049	0.9998

Table 3 Initial probability estimates

States	Clusters 1	2	Aggregate
1	0.072	0.878	0.452
2	0.346	0.002	0.183
3	0.085	0.028	0.060
4	0.315	0.072	0.200
5	0.182	0.021	0.106

Tables 3 and 4 depict the two-cluster dynamics of the latent process. Results for the homogeneous population (aggregate results) are reported as well. From Table 3 we observe that 45.2% of the women did not use contraception at the beginning of the LHC. However, the proportion is quite heterogeneous at cluster level. Indeed, whilst only 7.2% were not using contraception in cluster 1, that proportion increased to 87.8% in cluster 2. Overall, cluster 2 contains women that were not using contraception at the beginning of the LHC, in opposition to cluster 1 that contains women in states sterilized, pregnancy, using pill, or other temporary methods.

Table 4 provides the transition probability estimates at aggregate and cluster levels. Results assuming homogeneity show a strong persistence of staying in the same state. Indeed, excluding pregnancy (0.88), the probability that the process remains in the same state is always above 0.95. Note that sterilization is an absorbing state (1.00). This description of the dynamics of contraceptive use is not very informative, because all women are assumed to follow exactly the same pattern over time.

For the two-cluster analysis we observe that in cluster 1 the probability of staying in state non-use of contraception is no longer so persistent (0.78) with higher probabilities of moving to state pregnancy and pill than in cluster 2.

Table 4 Transition probability estimates

	Origin	Destination 1	2	3	4	5
	1	0.9807	0.0002	0.0096	0.0060	0.0035
	2	0.0000	1.0000	0.0000	0.0000	0.0000
Aggregate	3	0.0535	0.0171	0.8836	0.0239	0.0220
	4	0.0125	0.0014	0.0054	0.9746	0.0062
	5	0.0071	0.0013	0.0107	0.0063	0.9745
	1	0.7808	0.0028	0.1045	0.0767	0.0351
	2	0.0000	1.0000	0.0000	0.0000	0.0000
Cluster 1	3	0.0776	0.0196	0.8766	0.0117	0.0145
	4	0.0152	0.0019	0.0034	0.9744	0.0050
	5	0.0070	0.0016	0.0078	0.0028	0.9809
	1	0.9892	0.0001	0.0047	0.0033	0.0027
	2	0.0000	1.0000	0.0000	0.0000	0.0000
Cluster 2	3	0.0800	0.0118	0.8805	0.0132	0.0145
	4	0.0165	0.0000	0.0036	0.9699	0.0099
	5	0.0200	0.0000	0.0132	0.0241	0.9426

On the other hand, for instance, cluster 2 shows more dynamism between states pill and other temporary methods than cluster 1. This clearly shows that clusters identify women at different stages of the life course and consequently with different needs of information in terms of family planning.

Finally, Table 5 provides a description of the clusters based on background or profiling variables. We perform a binary logistic regression to control simultaneously the ability of the profiling variables in discriminating the two cluster. We confirm that São Paulo region is rather homogenous, as place of residence, education of respondent, and occupation of respondent are not able to distinguish the two clusters. Cluster allocation was based on the maximum posterior probability estimate. For these variables, differences are rather small. Thus, the results show that place of residence is not an important driver of the unobserved heterogeneity as it happens in Northeast region of Brazil (Dias and Willekens, 2005).

Regarding cluster 1, only 10.1% of the women have less than 24 years in opposition to cluster 2 with 59.5%. The same happens with the category never married with 5.1% and 58.0% in clusters 1 and 2, respectively. We conclude that the age and current marital status are the best variables to identify the clusters with different needs and both play a pivotal role in identifying family planning needs for this population, as they differ at different stages of the life course. These results have implications for family planning policies and programmes. In particular, communication and other media supports should be developed taking into account age groups as they define segments with distinct unmet needs of information (Wedel and Kamakura, 2000).

Table 5 Clusters profiling

Background characteristics		Aggregate	Clusters 1	2
Age of respondent (in years)*				
	15–19	0.180	0.015	0.379
	20–24	0.145	0.086	0.216
	25–29	0.162	0.186	0.132
	30–34	0.164	0.241	0.072
	35–39	0.136	0.195	0.065
	40–44	0.119	0.157	0.073
	45–49	0.094	0.120	0.063
Place of residence				
	Capital, large city	0.444	0.450	0.437
	Small city	0.335	0.339	0.330
	Town	0.139	0.134	0.145
	Countryside	0.082	0.077	0.088
Education of respondent				
	No education	0.027	0.027	0.026
	Primary	0.277	0.335	0.208
	Secondary	0.607	0.549	0.678
	Higher	0.089	0.089	0.088
Current marital status*				
	Never married	0.292	0.051	0.580
	Married	0.511	0.715	0.267
	Living together	0.108	0.136	0.075
	Widowed	0.015	0.014	0.018
	Divorced	0.018	0.026	0.010
	Not living together	0.055	0.058	0.050
Occupation of respondent				
	Employee	0.699	0.658	0.745
	Self-employee	0.279	0.314	0.240
	Employer	0.022	0.027	0.015

* $p < 0.001$

4 Conclusion

We provide a new approach to modeling demographic multistate data incorporating unobserved heterogeneity and misclassification error. This research extends results reported in Dias and Willekens (2005), in which contraceptive use dynamics were modeled as a manifest process rather than a latent one. We applied this model to Life History Calendar data from the Brazilian Demographic and Health Survey 1996 to identify groups of women with similar contraceptive use and pregnancy careers. We found two clusters with differential contraceptive use and dynamics. These results have important reproductive health implications as they help identify subpopulations with distinct unmet needs to be addressed using different programmes.

Acknowledgements The author would like to thank the Fundação para a Ciência e a Tecnologia (Portugal) for its financial support (PTDC/CS-DEM/108033/2008). I also wish to express my gratitude to the two anonymous referees for their valuable comments on this paper.

References

Baum, L. E., Petrie, T., Soules, G., & Weiss, N. (1970). A maximization technique occurring in the statistical analysis of probabilistic functions of Markov chains. *Annals of Mathematical Statistics, 41*, 164–171.

Belli, R. F., Shay, W. L., & Stafford, F. P. (2001). Event history calendars and question list surveys. *Public Opinion Quarterly, 65*, 45–74.

BENFAM & Macro International (1997). *Pesquisa Nacional Sobre Demografia e Sade (PNDS), Brasil, 1996*, Rio de Janeiro.

Clogg, C. C. (1995). Latent class models. In G. Arminger, C. C. Clogg, & M. E. Sobel (Eds.), *Handbook of statistical modeling for the social and behavioral sciences* (pp. 311–359). New York: Plenum Press.

Dempster, A. P., Laird, N. M., & Rubin, D. B. (1977). Maximum likelihood from incomplete data via the EM algorithm (with discussion). *Journal of the Royal Statistical Society, Series B, 39*, 1–38.

Dias, J. G., & Vermunt J. K. (2007). Latent class modeling of website users' search patterns: Implications for online market segmentation. *Journal of Retailing and Consumer Services, 14*, 359–368.

Dias, J. G., & Wedel, M. (2004). On EM, SEM and MCMC performance for problematic Gaussian mixture likelihoods. *Statistics and Computing, 14*(4), 323–332.

Dias, J. G., & Willekens, F. (2005). Model-based clustering of sequential data with an application to contraceptive use dynamics. *Mathematical Population Studies, 12*(3), 135–157.

McLachlan, G. J., & Peel, D. (2000). *Finite Mixture Models*. New York: Wiley.

Rabiner, L. R., & Juang, B. H. (1986). An introduction to hidden Markov models. *IEEE Acoustics, Speech and Signal Processing Magazine, 3*, 4–16.

Schwarz, G. (1978). Estimating the dimension of a model. *Annals of Statistics, 6*, 461–464.

Wedel, M., & Kamakura, W. (2000). *Market segmentation: Conceptual and methodological foundations*. Boston: Kluwer Academic Publishers.

An Approach to Forecasting Beanplot Time Series

Carlo Drago and Germana Scepi

Abstract Visualization and Forecasting of time series data is difficult when the data are very numerous, with complex structures as, for example, in the presence of high volatility and structural changes. This is the case of high frequency data or, in general, of financial data, where we cannot clearly visualize the single data and where the necessity of an aggregation arises. In this paper we deal with the specific problem of forecasting beanplot time series. We propose an approach based firstly on a parameterization of the beanplot time series and successively on the chosen best forecasting method with respect to our data. In particular we experiment with a strategy to use combination forecast methods in order to improve the forecasting performance.

1 Introduction

In the analysis of complex time series, such as high frequency data or financial data with a peculiar data structure, it is not always possible to have effective visualization and reliable forecasting. This problem was first studied in the Symbolic Data Analysis literature (Diday and Noirhomme-Fraiture, 2008) where important results in Symbolic Forecasting were obtained by Maia et al. (2008) on Interval-data, then by Arroyo et al. (2010) on Histogram data. The problem of the visualization of complex time series has been explored in a previous paper proposed by Drago and Scepi (2011). This paper verified the necessity of searching and analysing an aggregate behaviour for complex data and defined a new peculiar aggregated time series called beanplot time series. Here we propose an approach for forecasting beanplot time series. In particular, following a short introduction

C. Drago (✉) · G. Scepi
University of Naples Federico II, Via Cinthia, Monte Sant'Angelo (NA), Italy
e-mail: carlo.drago@unina.it; scepi@unina.it

A. Giusti et al. (eds.), *Classification and Data Mining*, Studies in Classification, Data Analysis, and Knowledge Organization, DOI 10.1007/978-3-642-28894-4_24,

on the definition and main characteristics of beanplot time series (in Sect. 2), we deal with the problem of searching an appropriate parameterization for defining an external model with the aim of forecasting beanplot time series. We propose our forecasting approach in Sect. 4, while we illustrate it using real data in the application (Sect. 4.1). In the specific case of financial and high frequency data, we are trying to predict the associated risk or the volatility that can occur over time. Size and shape can represent either the internal variation over the interval temporal, so, in this sense there is a specific link between the data collection and quantitative methods used: we try to forecast the market "instability" or the internal variation in the data.

2 Beanplot Time Series, Internal and External Modeling

The Beanplot time series $\{b_{Y_t}\} t = 1 \ldots T$ is an ordered sequence of beanplots or densities over time. The time series values can be viewed as realizations of an X beanplot variable in the temporal space T, where t represents the single time interval. The choice of the length of the single time interval t (day, month, year) depends on the specific data features and objectives the analyst wants to study. A beanplot realization at time t is a combination between a 1-d scatterplot and a density trace. It is defined (Kampstra, 2008) as:

$$\hat{f}_{h,t} = \frac{1}{nh} \sum_{n}^{i=1} K\left(\frac{x - x_i}{h}\right) \tag{1}$$

where x_i $i = 1 \ldots n$ is the single observation in each t, K is a Kernel and a h is a smoothing parameter defined as a bandwidth. K can be a Gaussian function with mean zero and variance 1. The Kernel as we know is a non-negative and real-valued function $K(z)$ satisfying: $\int K(z)dz = 1, \int zK(z)dz = 0, \int z^2 K(z) = k_2 < \infty$ with the lower and upper limits of integration being $-\infty$ and $+\infty$. It is possible to use various Kernel functions: uniform, triangle, Epanechnikov, quartic (biweight), tricube (triweight), Gaussian and cosine. The choice of the kernel in the beanplot time series seems not particularly relevant (Racine, 2008) and our simulations show the different kernels tend to fit similarly the underlying phenomena. The choice of the h value is more important than the choice of K (Silverman, 1986). To select the h is commonly used the MISE, the mean integrated squared error:

$$MISE(h) = E \int (\hat{f}_h(x) - f(x))^2 \, dx. \tag{2}$$

With small values of h, the estimate looks "wiggly" and spurious features are shown. On the contrary, high values of h give a too smooth estimate and it may not reveal structural features, as for example bimodality, of the underlying density. In literature, several methods for choosing the bandwidth were proposed (Jones et al., 1996). The choice of the bandwidth is relevant (Drago adn Scepi, 2010) also because some observations or internal models can be considered outliers and they need to be handled or weighted differently. In our approach we choose the value of h from whose computed with the Sheather Jones method (Sheather and Jones, 1991) that defines the optimal h in a data-driven approach. In the beanplot the variability or size is represented by the interval related to the minima a_{L_t} and the maxima a_{U_t}. The beanplot interval $[a_t]$, is the ordered pair $[a_t] = [a_{L_t}, a_{U_t}]$ where a_{L_t}, a_{U_t} are the interval bounds such as $a_{L_t} \leq a_{U_t}$. Inside the interval $[a_{L_t}, a_{U_t}]$ are represented the single primary observations (represented as a 1-dimensional scatterplot or strip chart) so we are able to understand the location of the single observations. The measure of size in the beanplot $\{b_{Y_t}\}$ is:

$$a_{S_t} = a_{U_t} - a_{L_t} \qquad t = 1 \ldots T. \tag{3}$$

At the same time it is possible to consider the interval composed by the two consecutive sub-intervals (or half-point) through the beanline (the radii of the beanplot $\{b_{Y_t}\}$): $[a] = \langle a_{C_t}, a_{R_t} \rangle$ with a_{C_t} the centre and a_{R_t} the radii. At this point, we need to define an external model before forecasting our data. Our idea is to parameterize each beanplot by using specific attributes of each beanplot and by considering their realization over the time. So we define a Beanplot Attribute Time Series a realization of a single beanplot $\{b_{Y_t}\}$ $t = 1 \ldots T$ descriptor over the time. We decide to consider the coordinates $\bar{x}$ and $\bar{y}$ as descriptors of the beanplot. We refer to them as descriptor points because we show they measure the beanplot structure. In particular, the $\bar{x}$ time series show the location and the size of the beanplot over the time while the $\bar{y}$ represent the shape over the time (see Figs. 1–3). To specifically parameterize the beanplot we first choose the number n of descriptor points and then we obtain the coordinates $\bar{x}$ and $\bar{y}$. If we consider an high number of points n in the procedure, we obtain a more precise approximation of the beanplot. In our approach, we consider the values of $\bar{x}$ and $\bar{y}$ corresponding to the 25th, 50th and 75th percentile. We define this procedure internal modeling. For each descriptor, we obtain an attribute time series considering all parameters over time.

3 An Approach to Forecasting Beanplot Time Series

The second step is analysing the structure of the external models. Therefore we have to test the structure of each attribute time series. As we know, they represent the beanplot dynamics over the time, so we can use a specific method to forecast the attribute time series to obtain the prediction at time $t + 1$, $t + 2$ and so on. Successively we must decide which model has to be used for the forecasting approach. We can use univariate methods or multivariate methods. In the first case we are assuming there is no specific relationship between the attribute time series, in the second case we are assuming a relationship exists. So it is important first to test

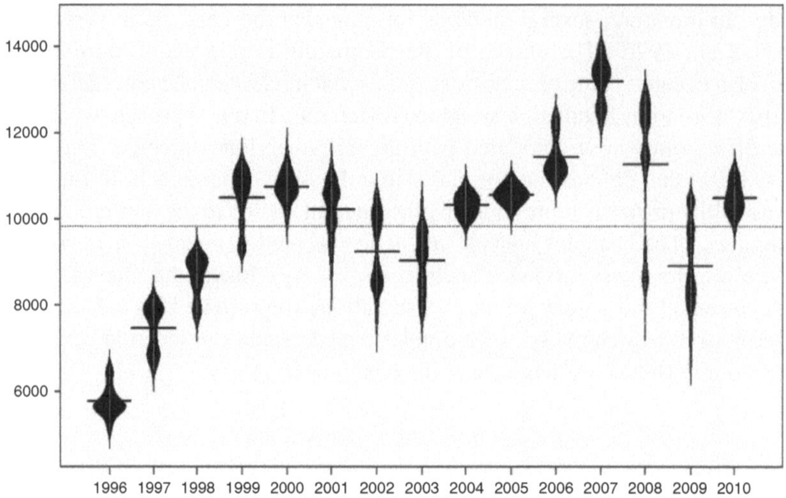

Fig. 1 Beanplot time series

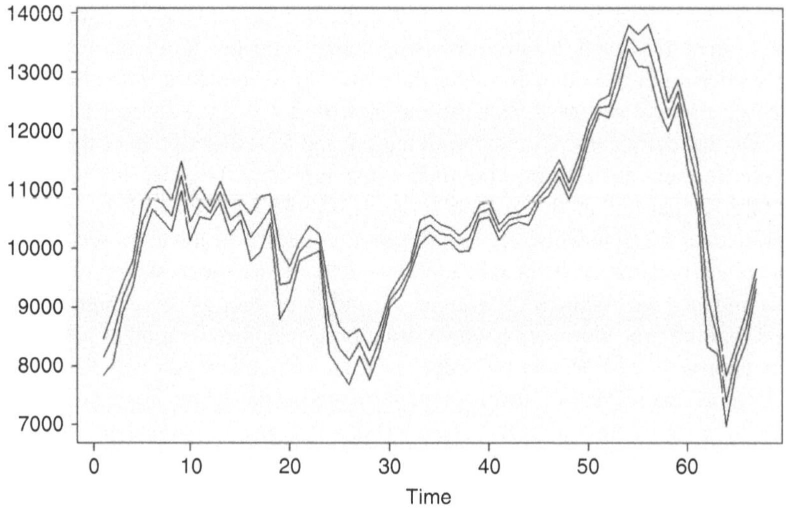

Fig. 2 *X* attribute time series

the stationarity of the attribute time series and successively to define the possible cointegration between the series. Then, it is important to verify the autocorrelation of the attribute time series and the possible structural changes. Only at the end of this analysis do we decide which model can be used for forecasting our attribute time series. After the identification of the forecasting models, it is necessary to estimate the different models for obtaining the forecasts and, finally, to evaluate the reliability of our forecasts. The diagnostic procedure is important because we

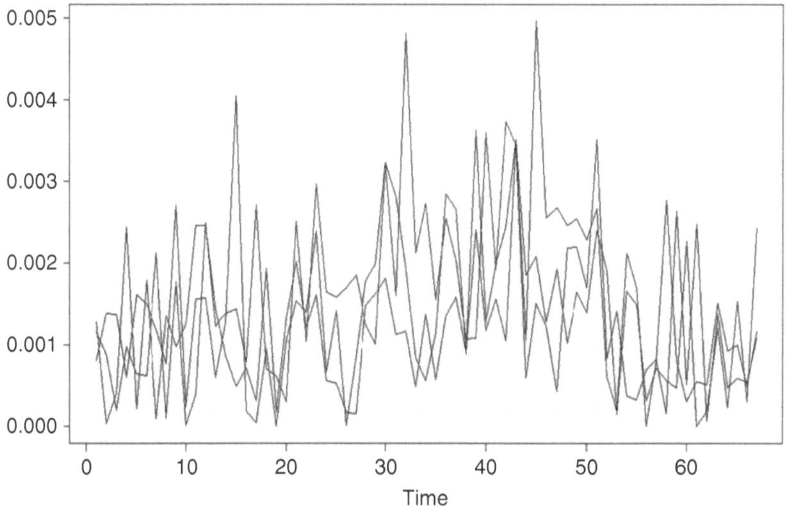

Fig. 3 *Y* attribute time series

can evaluate the models and eventually to re-specify them. At the same time, it is very useful to consider the performance of the different forecasting models by considering some evaluation indexes. Anyway, using more than one forecast may be necessary to obtain better predictions (Timmermann, 2006). So we have:

$$F_t = \gamma_1(t)F_t^1 + \gamma_2(t)F_t^2 + \cdots + \gamma_m(t)F_t^m + \zeta_t. \tag{4}$$

In the combination of the forecasts we use $F_1 \ldots F_m$ forecasts coming from differing forecasting techniques (i.e. Exponential Smoothing, Splinef, Theta methods). Various strategies of the weights estimation of the combination model $\gamma_1 \ldots \gamma_m$ are used. Firstly, there can be some structural changes and it could be necessary to take into account more than one forecasting model. Secondly, the use of different models can reduce the risk associated with choosing one single model. The complete procedure is depicted in Algorithm 1.

4 Forecasting the Beanplot Time Series Related to the Dow Jones Market

The application is related to the Dow Jones market data set from the years 1928–2010, and its purpose is to forecast the beanplot time series for the specific period 2009–2010. In particular we consider in the models the data from the 1998–2008. The specific aim in the forecasting process is to predict the instability in the market over time. Methods used in the forecasting models are Smoothing Splines and

Algorithm 1: Forecasting beanplots algorithm

Data: A set of attribute time series for the beanplot series $\{b_{Y_t}\}$ $t = 1 \ldots T$ each one representing a different model parameter $k \in K$

Result: A set of F forecasts of the attribute time series representing the predictions of the models

begin
- Choice of the forecasting horizon considered I
- Choice of the number of observations n to consider in the models
- Is there a relationship between the attribute time series?
- **if** *so, then* **then**
  - are the series stationary? Model the series using a VAR
  - Are the series cointegrated? Model the series using a VECM
  - Obtain F^1 to F^m forecasts for each k
- **end**
- **for** $k \in K$ **do**
  - Model the series using different univariate methods
  - Obtain F^1 to F^m forecasts for each k
- **end**
- Model Diagnostics
- **if** *Residuals aren't white noise* **then**
  - Re-specify the models
- **end**
- Compute accuracy forecast measurements
- Forecasts combination
- **if** *A combination can improve the forecasts* **then**
  - Choice the alternative forecast strategy
- **end**
- Search Algorithm
- Is the set of information (the n observation used) optimal?
- **if** *The set of information is not optimal* **then**
  - Seek for the best interval maximizing the forecasts accuracy
- **end**

end

Automatic Arima (Hyndman and Khandakar, 2008; Hyndman, 2011) primarily, and also VAR (Vector Autoregressive Models) and VECM (Vector Error Correction Models) in the case of the Y attribute time series. In a second forecasting model we use a combination approach, by combining different forecasts obtained by different methods. Lastly, we compare all the results, using as a benchmark the Naive model (that is representing the prediction obtained by considering the last observation). Here we compute the set of X attribute time series (years 1996–2010), by starting from the parameterization of coordinates. Here we can consider a different temporal interval than the interval used in the data visualization (for considering the statistical features of the beanplots). At the end of the parameterization, we obtain 6 attribute time series (3 for the X and 3 for the Y) around 60 observations (years 1996–2010 approximately). The three X attribute time series are related to the 25th, 50th and the 75th percentile where each Y is associated to the X. We call these intervals as extreme risk intervals, with minima or lower risk, median risk and maxima or upper risk. These intervals are directly related to the beanplot structure. The attribute time series for the X show the long run dynamics of the beanplots and also the impact of the financial crisis. By considering the Y attribute time series (year 1996–2010), we need to remind that we are considering a different temporal interval (not the yearly temporal interval) but only 2 months (around 40 observations in a beanplot data). In this representation, we can observe the complexity of the initial series, observed in particular in the changes from time to time in the beanplot shapes (represented by the Y). This behaviour is related to the short run dynamics of the series.

4.1 Diagnostic Models an Accuracy of the Forecasts

We start to use as an univariate forecasting model the Smoothing Splines for the three X attribute time series. Results are compared with the real value, where the previous observations represent the naive forecasting model. We outperform the naive model. In particular, we perform a MAPE of 5–6 %. The result seem to be good, and anyway the lower part of the beanplots (the lower risk interval) it is more difficult to predict in conditions of high volatility, so we expect in these cases lower performance in the prediction models. By considering an alternative and competitive forecasting univariate model (by Automatic Arima) we outperform the naive forecasting model but also outperform the smoothing splines approach. In particular the automatic Arima algorithm selects the best Arima model by minimizing the AIC (Akaike Information Criteria index). We consider the Y attribute time series associated to the X attribute time series in the stationary framework as example of the VAR forecasting model. Here as well, we outperform the naive model, but anyway the results are not as good as in the X case. Here, for predicting the Y attribute time series we use the Smoothing Splines approach two times out of three, the models outperform the naive method. In any event, the performances of the methods for the Y are not as good as the X case (and that could be expected by understanding the nature of the attribute time series). The Smoothing Splines approach seems to be the best approach for the Y case in forecasting of the Y attribute time series. We consider another relevant element: an original algorithm that optimizes the forecasting model by selecting the relevant temporal information (by minimizing the MAPE in the validation period of the model). Therefore, the procedure is divided into distinct parts: running the algorithm to minimize the MAPE in the validation period and using the temporal interval for the forecasting. In this way we are explicitly selecting the relevant set of information (without structural breaks) in our data. The results seems to be good (at least with respect to the other models used as benchmarks). The decision to use a selection algorithm of the optimal interval for the Y attribute time series can be explained because we are dealing explicitly with very volatile attribute time series (the Y) with

Table 1 Accuracy of the forecasting models on the attribute time series

Attribute time series	Method	1	2	3
X	Smoothing Splines	6.79	7	3.28
X	Auto Arima	7.23	0.87	4.22
X	Combination forecasts	2.18	2.72	2.12
Y	Smoothing splines with search	24.11	34.92	24.54

[a] The considered forecast accuracy measure in the table is the Mean Absolute Percentage Error (MAPE)

dynamics with occurrence of frequent structural changes. See Table 1 to compare the MAPE for the Y as different attribute time series. We use combinations for two reasons: we can take into account precisely the uncertainty in choosing a specific model (in fact sometimes the choice of a specification in a single forecasting model can be very hard) and we can specifically take in account the structural changes that could be captured (or we try to capture) by considering combinations of forecasts. Therefore, we consider predictions obtained by using various methods: Smoothing Splines, Auto-Arima, the mean of the period, the Theta method and the Exponential Smoothing. See Table 1 to compare the different MAPE for the X. We do not use any special weighting structure but we consider only the average between results obtained by the different methods. The forecasting performances outperform the single other models based on the single forecasting model chosen. Here we are considering an interval of prediction of five periods ahead (for this reason we do not apply this method to the Y).

5 Conclusions

In this paper, we propose a new procedure to forecast beanplot time series. This type of data seems relevant to taking into account the intra-period variability where the time series are overwhelming and they need to be aggregated. In particular, the results are related to obtaining the data from the original time series, to parameterizing these data (internal model) and applying these methods in forecasting financial data (the external model) using combination forecasts. Future research will be devoted to improving the forecasting processes considering different approaches both in method (for example using methods as the K-Nearest Neighbor), and in combinations of forecasting by considering other different approaches.

References

Arroyo, J., González-Rivera, G., & Maté, C. (2010). Forecasting with interval and histogram data: Some financial applications. In Ullah, A., & Giles, D. E. (Eds.), *Handbook of empirical economics and finance* (pp. 247–280). London: Chapman & Hall/CRC.

Diday, E., & Noirhomme-Fraiture, M. (2008). *Symbolic data analysis and the SODAS software*. London: Wiley.
Drago, C., & Scepi, G. (2010). Forecasting by Beanplot time series. In *Electronic Proceedings of Compstat* (pp. 959–967). Berlin/Heidelberg: Springer.
Drago, C., & Scepi, G. (2011). Visualizing and exploring high frequency financial data. In S. Ingrassia, R. Rocci, & M. Vichi (Eds.), *New perspectives in statistical modeling and data analysis* (pp. 283–290). Dordrecht: Springer.
Hyndman R.J. (2011) forecast: Forecasting functions for time series. R package version 3.14. http://CRAN.R-project.org/package=forecast
Hyndman, R. J., & Khandakar, Y. (2008). Automatic time series forecasting: The forecast package for R. *Journal of Statistical Software, 27*(3), 1–22.
Jones, M. C., Marron, J. S., & Sheather, S. J. (1996). A brief survey of bandwidth selection for density estimation. *Journal of the American Statistical Association, 91*(433), 401–407.
Kampstra, P. (2008). Beanplot: A boxplot alternative for visual comparison of distributions. *Journal of Statistical Software, 28*, 1–9.
Maia, A. L., De Carvalho, F., & Lurdermir, T. B. (2008). Hybrid approach for interval-valued time series forecasting. *Neurocomputing, 71*(16–18), 3344–3352.
Racine, J. S. (2008). Nonparametric econometrics: A primer. *Foundation and Trends in Econometrics, 3*(1), 1–88.
Sheather, S. J., & Jones, M. C. (1991). A reliable data-based bandwidth selection method for Kernel density estimation. *Journal of the Royal Statistical Society, Series B, 53*, 683–690.
Silverman, B. W. (1986). *Density estimation for statistics and data analysis*. London: Chapman and Hall.
Timmermann, A. (2006). Forecast combinations. In G. Elliott, C. Granger, A. Timmermann (Eds.), *Handbook of economic forecasting* (pp. 135–196). Amsterdam/Boston: Elsevier/North-Holland.

Shared Components Models in Joint Disease Mapping: A Comparison

Emanuela Dreassi

Abstract Two models for jointly analysing the spatial variation of incidences of three (or more) diseases, with common and uncommon risk factors, are compared via a simulation experiment. In both models, the linear predictor can be decomposed into shared and disease-specific spatial variability components. The two models are the shared model on the original formulation that use exchangeable Poisson distribution as response multivariate variable and shared components model that use a Multinomial one. The simulation study, performed using three different degree of spatial unstructured poisson over-dispersion, shows that models behave similarly. However, they perform differently for the shared clustering terms when a different level of spatial unstructured over-dispersion is present.

1 Introduction

A great amount of the literature deals with disease mapping, as the statistical analysis of geographical patterns of disease. Any spatial variation may be explained by different risk factors, therefore disease mapping allows to state hypotheses concerning their aetiology. Interest in joint disease mapping increased over recent years: joint statistical modelling of several diseases on the same spatial location, with different and common aetiologies. Joint analysis highlights common and uncommon geographical patterns of risk and obtains more precise and convincing results.

Various attempts to consider simultaneously more than one disease have been made: by a multilevel model as Langford et al. (1999) and Leyland et al. (2000), or by an ecological regression approach where a disease represent a covariate of

E. Dreassi (✉)
Department of Statistic "G. Parenti" University of Florence, Florence, Italy
e-mail: dreassi@ds.unifi.it

A. Giusti et al. (eds.), *Classification and Data Mining*, Studies in Classification, Data Analysis, and Knowledge Organization, DOI 10.1007/978-3-642-28894-4_25,

the model as Bernardinelli et al. (1997). However, the joint modelling approach seems to be more naive, as all diseases enter as response variables with reference to unobserved latent risk factors. More recently, joint modelling following a Multivariate Gaussian Markov random field has been proposed; see Gelfand and Vounatsou (2003) and Jin et al. (2005). Dabney and Wakefield (2005) remade a proportional mortality model to the joint mapping of two diseases.

In this paper, we focus on a particular class of models: shared component models. Originally introduced by Knorr-Held and Best (2001), these models have been extended to more than two diseases by Held et al. (2005) and from exchangeable Poisson response to a Multinomial one by Dreassi (2007). In Dreassi (2007) a Multinomial model (PL) is presented and compared with exchangeable Poisson model (SC) by a real example. In this paper, a simulation study is conducted to evaluate and compare more deeply the performances of both models.

The paper is organized as follows. Section 2 introduces the joint analysis with shared components model following exchangeable Poisson model (SC) and Multinomial models (PL). Section 3 describe the simulation experiment. Results are showed in Sect. 4 and conclusion in Sect. 5.

2 Shared Components Models

Shared components models highlight common and specific spatial components, allowing the linear predictor to be decomposed into shared and disease-specific spatial variability terms.

2.1 Shared Components Exchangeable Poisson Model

Let y_{ik} denote the number of death cases for k-th disease ($k = 1, \ldots, K$) and i-th area ($i = 1, \ldots, I$). Each y_{ik} is assumed to follow a Poisson distribution with parameters $E_{ik}\theta_{ik}$, where E_{ik} represent the expected cases in i-th area and k-th disease and θ_{ik} the relative risk. Following the standard model of Besag et al. (1991) on consider a log link for θ_{ik}

$$\log(\theta_{ik}) = \alpha_k + u_{ik} + v_{ik} \tag{1}$$

where α_k represents a cause-specific intercept, such as an overall risk level, u_{ik} is a spatially structured term, and v_{ik} a spatially unstructured term.

The prior distribution for the model parameters is as follows. The intercept α_k has a flat non-informative distribution. The heterogeneity terms v_{ik} are independent, each v_{ik} being Normal $(0, \lambda_{vk}^{-1})$ (λ_{vk} represents the precision parameter). Using Gaussian Markov random fields (GMRFs) models in order to cope the spatial structure, the clustering terms u_{ik} are modeled conditionally on $u_{l\sim ik}$ terms

($\sim i$ indicates adjacent areas to i-th ones, $l = 1, \ldots, I$ and n_i their number; where adjacent means that two areas share an edge or, for islands, that exists a boat connection), as Normal$(\bar{u_{ik}}, (\lambda_{uk} n_i)^{-1})$ where $\bar{u_{ik}} = \sum_{l \sim i} \frac{u_{lk}}{n_i}$.

The hyperprior distributions of the precision parameters λ_{vk} and λ_{uk} are assumed to be Gamma (0.5, 0.0005).

Following Knorr-Held and Best (2001) and Held et al. (2005), a model on the shared components formulation is considered: the structured spatial terms (clustering) u_{ik} in (1) are decomposed into a shared and a disease-specific effect. So, for example, when $K = 3$ each disease's clustering term could be

$$\begin{aligned} u_{i1} &= us1_i \times \omega_1 + us2_i \times \delta_1 + up_{i1} \\ u_{i2} &= us1_i \times \omega_2 + us2_i \times \delta_2 + up_{i2} \\ u_{i3} &= us1_i \times \omega_3 \end{aligned} \tag{2}$$

where $us1_i$ and $us2_i$ represent the shared clustering components (the know risk factors pattern) and up_{i1} and up_{i2} the specific ones. The scale parameters $\omega_1, \ldots, \omega_3$ and δ_1, δ_2 allow the shared components to vary per cause by a constant factor.

Terms $\log \omega_1, \ldots, \log \omega_3$ and $\log \delta_1, \log \delta_2$, constrained to $\sum_{k=1}^{3} \log \omega_k = 0$ and $\sum_{k=1}^{2} \log \delta_k = 0$, are assumed to be multivariate normal distributed with zero mean and variance covariance matrix respectively

$$\Sigma_\omega = \sigma_\omega^2 \begin{pmatrix} 1 & -1/2 & -1/2 \\ -1/2 & 1 & -1/2 \\ -1/2 & -1/2 & 1 \end{pmatrix} \quad \text{and} \quad \Sigma_\delta = \sigma_\delta^2 \begin{pmatrix} 1 & -1 \\ -1 & 1 \end{pmatrix} \tag{3}$$

Knorr-Held and Best (2001) consider $\sigma_\omega^2 = \sigma_\delta^2 = 0.17$. The $us1_i$, $us2_i$, up_{i1} and up_{i2} terms are modelled following a GMRF as described before.

2.2 *Shared Components Multinomial Model*

In Dreassi (2007), following suggestions by proportional mortality model (Dabney and Wakefield, 2005) and shared component model (Knorr-Held and Best 2001 and Held et al. 2005) to highlight similarity and dissimilarity on spatial patterns, an other shared component model is introduced: a Multinomial (or polytomous logit) (PL) model. In this model, a disease is regarded as reference category, and for each predictor the shared components model formulae can be adopted. In the model proportionality is assumed and then a model for death for each disease, area and age-stratum j is considered without knowledge of the population at risk.

Let $y_{ij} = (y_{ij1}, \ldots, y_{ijk}, \ldots, y_{ijK})\prime$ be distributed according to a multinomial with parameters m_{ij} and probability vector $\pi_{ij} = (\pi_{ij1}, \ldots, \pi_{ijk}, \ldots, \pi_{ijK})\prime$, where $m_{ij} = \sum_{k=1}^{K} y_{ijk}$ and $\sum_{k=1}^{K} \pi_{ijk} = 1$. A polytomous logit model is considered: each category probability is modeled as

$$\pi_{ijk} = \phi_{ijk} / \sum_{r=1}^{K} \phi_{ijr} \tag{4}$$

where each log odd

$$\log(\phi_{ijk}) = \alpha_k^\diamond + a_{jk}^\diamond + u_{ik}^\diamond + v_{ik}^\diamond \tag{5}$$

is decomposed additively into a disease-specific intercept $\alpha_k^\diamond$ (representing overall difference between k-th disease and K-th reference disease), $a_{jk}^\diamond$ a time-structured term by age and disease representing difference between k-th disease and reference category, and structured $u_{ik}^\diamond$ and unstructured $v_{ik}^\diamond$ spatial effects (again representing difference on the spatial structured and unstructured spatial terms between the disease k considered and the reference disease).

For the $a_{jk}^\diamond$ term a first order random walk with independent Gaussian increments is assumed. For the other terms prior are equal to SC model.

Representing, for example, the third disease the reference category (when $K = 3$), $\alpha_3^\diamond = 0$, $a_{j3}^\diamond = 0$ (for each age-class $j = 1, \ldots, 13$), $u_{i3}^\diamond = 0$ and $v_{i3}^\diamond$ (for each area $i = 1, \ldots, I$) are defined, as constraint for identifiability.

Note that terms $u_{i1}^\diamond$ and $u_{i2}^\diamond$ represent differences between first disease and reference category disease clustering, and between third disease and reference category disease clustering, respectively,

$$u_{i1}^\diamond = u_{i1} - u_{i3} \qquad \text{and} \qquad u_{i2}^\diamond = u_{i2} - u_{i3} \tag{6}$$

We consider a model where the difference structured spatial terms (clustering) in Eq. (5) are decomposed into a shared and a disease-specific effect (Held et al. 2005). We can represent each clustering term for the first and second disease, respectively, as

$$u_{i1}^\diamond = us2_i \times \delta_1 + up_{i1} \qquad \text{and} \qquad u_{i2}^\diamond = us2_i \times \delta_2 + up_{i2} \tag{7}$$

where $us2_i$ is the shared clustering component and up_{i1} and up_{i2} is the disease specific one; both are distributed according GMRF models. Prior distributions are the same described before for SC model. Note that Eqs. (6) and (7) imply

$$u_{i1} = u_{i3} + ua_i \times \delta_1 + up_{i1} \qquad \text{and} \qquad u_{i2} = u_{i3} + ua_i \times \delta_2 + up_{i2} \tag{8}$$

which is different from Eq. (2) because we are forcing to be $\omega_1 = \omega_2 = \omega_3$.

3 Simulation Study

To evaluate the proposed shared components models, considering exchangeable Poisson (SC) or Multinomial (PL) models, we conducted a simulation study.

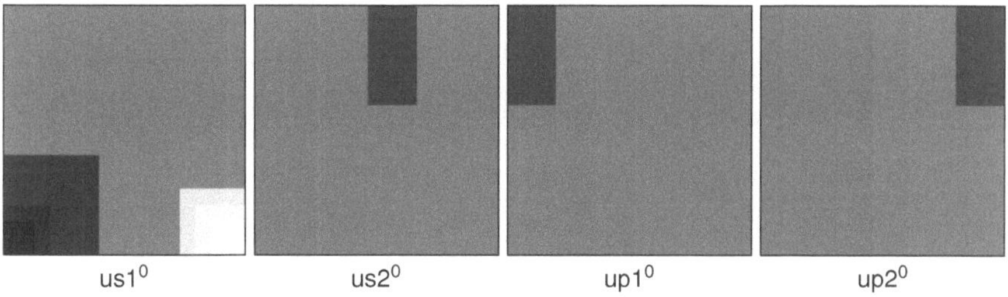

Fig. 1 True clustering terms

The shared components models used for the simulation experiment has been conceived with reference to a specific application: three diseases, a common risk factor and another risk factor shared by two diseases only. In the present application, the incidence of the disease that shared only one risk factor represents the reference category for the Multinomial model (PL). Then a shared component is considered, representing the second risk factor common only for the two diseases adjusted for the first risk factor. Finally, including disease specific terms in the predictors, the possibility of other different risk factors is investgated.

We used three different disease maps (each map with $n = 225$ areas) taken square areas over a 15×15 grid. For each i-th area $i = 1, \ldots, 225$, and for each k disease $k = 1, 2, 3$, we generated 100 deaths counts from Poisson (100 θ^0_{ik}). We assumed that each $\log \theta^0_{ik}$ for $k = 1, 2, 3$ is equal to

$$\begin{aligned} \log \theta^0_{i1} &= us1^0_i + us2^0_i + up1^0_i + v_{i1} \\ \log \theta^0_{i2} &= us1^0_i + us2^0_i + up2^0_i + v_{i2} \\ \log \theta^0_{i3} &= us1^0_i + v_{i3} \end{aligned} \tag{9}$$

$us1^0$ range from -0.20 to 0.22, instead $us2^0$, $up1^0$ and $up2^0$ range from 0 to 0.15. We fix ω_1, ω_2, ω_3, δ_1 and δ_2 equal to 1 and α and heterogeneity v_{ik} equal to zero.

Figure 1 shows the map of each clustering terms: shared $us1^0$ and $us2^0$, and specific $up1^0$ and $up2^0$ respectively. Figure 2 describes the disease true map θ^0_k for the latter parameter setting.

Then, we consider different specification for heterogeneity terms v_{ik}: a normal random effect with zero mean and σ^2 equal to 0.10 or 0.20. The performances of different models considering no heterogeneity, or heterogeneity at different level are evaluated with respect to two basic criteria: the Bias (BIAS) and the root mean squared error (RMSE).

We estimate $us1$ $us2$ $up1$ $up2$ using (SC) shared poisson model and $us2$ $up1$ and $up2$ using (PL) multinomial model.

The marginal posterior distributions of the parameters of interest for both models are approximated by Monte Carlo Markov Chain methods.

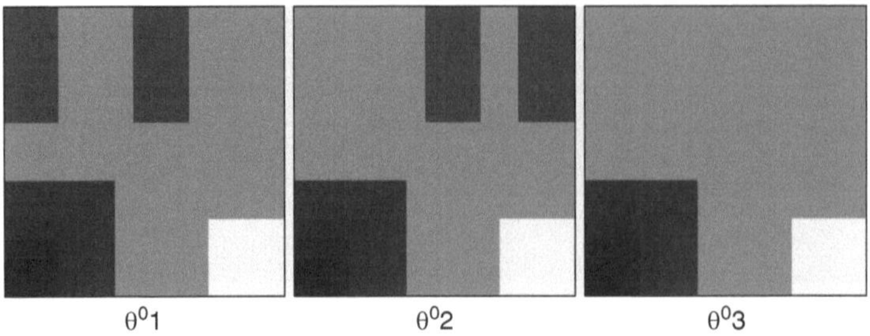

Fig. 2 True map of disease

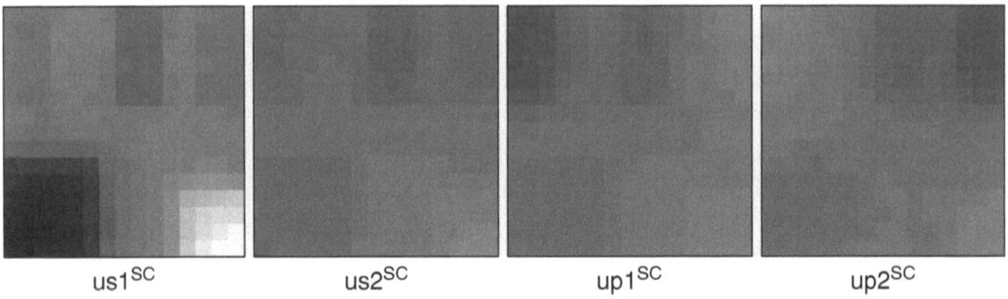

Fig. 3 Estimated clustering terms with SC

The estimates for SC model are obtained using specific MCMC software. It uses joint updates of the latent spatial fields and it is able to incorporate sum to zero constraints in spatial fields explicitly in the prior and in the MCMC algorithm.

For PL model we used *Brugs* library of R software in order to perform the MCMC analysis. The convergence of the algorithm has been evaluated for a subset of identifiable parameters (precision hyperparameters) for some simulation iteration. The algorithm seems to converge after a few thousand iterations. However, given also the very high number of (non monitored) parameters in the model, we decided to discard the first 200,000 iterations (burn-in) and to store for estimation 2,000 samples (one each 100) of the following 200,000 iterations.

4 Results

Figures 3 and 4 show the average over the 100 simulated data of clustering terms estimates by the two different models. Each average map is on a 15×15 grid, with the same levels of gray (from -0.177 to 0.168).

Results about mean BIAS and mean RMSE over areas are set out in Tables 1–3. They suggest a similar behavior of the models for specific clustering terms $up1$ and $up2$, while PL performs lower than SC for shared clustering term $us2$. However, when σ^2 increases, so spatial unstructured poisson over-dispersion is higher, PL

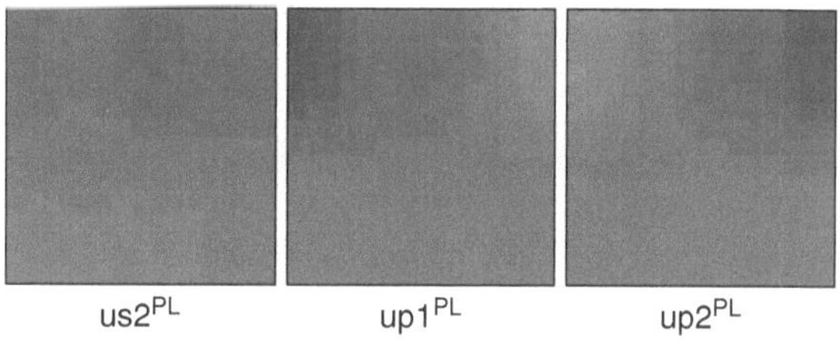

Fig. 4 Estimated clustering terms with PL

Table 1 Mean BIAS and mean RMSE from shared component Poisson model (SC) and shared components Polytomous model (PL) for shared and specific clustering components. No heterogeneity terms

	Mean BIAS		Mean RMSE	
Clustering	SC	PL	SC	PL
$us1$	−0.017		0.069	
$us2$	−0.012	−0.046	0.038	0.099
$up1$	−0.012	−0.012	0.016	0.018
$up2$	−0.012	−0.013	0.017	0.019

Table 2 Mean BIAS and mean RMSE from shared component Poisson model (SC) and shared components Polytomous model (PL) for shared and specific clustering components. Heterogeneity terms using $\sigma^2 = 0.10$

	Mean BIAS		Mean RMSE	
Clustering	SC	PL	SC	PL
$us1$	−0.017		0.306	
$us2$	−0.012	−0.050	0.124	0.094
$up1$	−0.012	−0.013	0.016	0.020
$up2$	−0.012	−0.013	0.018	0.020

Table 3 Mean BIAS and mean RMSE from shared component Poisson model (SC) and shared components Polytomous model (PL) for shared and specific clustering components. Heterogeneity terms using $\sigma^2 = 0.20$

	Mean BIAS		Mean RMSE	
Clustering	SC	PL	SC	PL
$us1$	−0.017		0.439	
$us2$	−0.102	−0.060	0.125	0.090
$up1$	−0.012	−0.013	0.017	0.020
$up2$	−0.012	−0.012	0.018	0.020

performs better than SC for shared clustering term $us2$ in terms of mean RMSE. For higher level of this over dispersion SC model seem to be more unstable to identify the two shared clustering terms.

5 Conclusion

As stated in Dreassi (2007), the SC model for joint disease mapping is perhaps more 'natural' and 'elastic' than PL model: both risk factors are considered as shared components of the model, and both common clustering terms are allowed to vary per cause for a multiplicative constant factor. In turn, the PL model gives some advantages: it allows to analyse mortality data without knowing the population at risk and to consider variability on age effect estimates in the model. Using a particular disease as reference category, we can omit a GMRF for the shared terms common to all the diseases with some advantages on computational time; moreover using a Multinomial model instead than exchangeable Poisson for a multivariate problem seem to be more convenient. Advantages and disadvantages for each model have been disregarded using an unrealistic, but particular simulation experiment; accordingly results from simulation give us information about the performances on estimating clustering terms. The simulation study suggests that both models provide similar estimates; however, PL model behaves lower for shared clustering term when spatial unstructured poisson over-dispersion is not present.

References

Bernardinelli, L., Pascutto, C., Best, N. G., & Gilks, W. R. (1997). Disease mapping with errors in covariates. *Statistics in Medicine, 16*, 741–752.

Besag, J., York, J., & Mollié, A. (1991). Bayesian image restoration, with two applications in spatial statistics (with discussion). *Annals of the Institute of Statistical Mathematics, 43*, 1–59.

Dabney, A. R., & Wakefield, J. C. (2005). Issues in the mapping of two diseases. *Statistical Methods in Medical Research, 14*, 83–112.

Dreassi, E. (2007). Polytomous disease mapping to detect uncommon risk factors for related diseases. *Biometrical Journal, 49*(4), 520–529.

Gelfand, A., & Vounatsou, P. (2003). Proper multivariate conditional autoregressive models for spatial data analysis. *Biostatistics, 4*, 11–25.

Held, L., Natário, I., Fenton, S. E., Rue, H., & Becker, N. (2005). Towards joint disease mapping. *Statistical Methods in Medical Research, 14*, 61–82.

Jin, X., Carlin B. P., & Banerjee, S. (2005). Generalized hierarchical multivariate CAR models for areal data. *Biometrics, 61*(4), 950–961.

Knorr-Held, L., & Best, N. (2001). A shared component model for detecting joint and selective clustering of two diseases. *Journal of the Royal Statistical Society Series A (Statistics in Society), 164*, 73–86.

Langford, I. H., Leyland, A. H., Rasbash, J., & Goldstein, H. (1999). Multilevel modelling of the geographical distributions of diseases. *Journal of the Royal Statistical Society Series C (Applied Statistics), 48*, 253–268.

Leyland, A. H., Langford, I. H., Rasbash, J., & Goldstein, H. (2000). Multivariate spatial models for event data. *Statistics in Medicine, 19*(17–18), 2469–2478.

Piano and Guitar Tone Distinction Based on Extended Feature Analysis

Markus Eichhoff, Igor Vatolkin, and Claus Weihs

Abstract In this work single piano and guitar tones are distinguished by means of various features of the music time series. In a first study, three different kinds of high-level features and MFCC are taken into account to classify the piano and guitar tones. The features are called high-level because they try to reflect the physical structure of a musical instrument on temporal and spectral levels. In our study, three spectral features and one temporal feature are used for the classification task. The spectral features characterize the distribution of overtones, the temporal feature the energy of a tone. In a second study as many low level and the high level features as possible proposed in the literature are combined for the classification task.

1 Introduction

What characterizes the sound of a musical instrument? Because single tones of different instruments may have the same pitch and loudness, it is important to consider timbre represented by the distribution of overtones in periodograms to distinguish instruments. This distribution depends on the physical structure of the musical instrument, see Fletcher (2008). Also the non-harmonic timbre characteristics play an important role as stated in Livshin and Rodet (2006). On the other side the energy of every single tone has got a temporal envelope that differs from one musical instrument to another. Both ideas are used in this paper to build so-called high-level features for instrument characterization.

M. Eichhoff (✉) · C. Weihs
Chair of Computational Statistics, TU Dortmund, Dortmund, Germany
e-mail: eichhoff@statistik.tu-dortmund.de; weihs@statistik.tu-dortmund.de

Igor Vatolkin
Chair of Algorithm Engineering, TU Dortmund, Dortmund, Germany
e-mail: igor.vatolkin@tu-dortmund.de

A. Giusti et al. (eds.), *Classification and Data Mining*, Studies in Classification, Data Analysis, and Knowledge Organization, DOI 10.1007/978-3-642-28894-4_26,

Moreover, MFCC have already shown to be useful for classification tasks in speech processing (Rabiner and Juang, 1993) as well as in musical instrument recognition (Krey and Ligges, 2010). One of the first studies with the deeper analysis of audio feature impact for automatic instrument identification was provided in Brown et al. (2001). We designed two different studies for identification of piano and guitar tones. The first one takes four concrete groups of high-level features into account and provides a classifier hyperparameter tuning for the building of models within the R software package (Bischl, 2011). Another one considers a very large feature set from the actual research and implies the feature selection by evolutionary strategies implemented in AMUSE (Vatolkin et al., 2010). Though the results of the both studies can not be directly compared because of the different focus points (less and well-designed features with more classifier tuning against many features with sophisticated selection but no hyperparameter analysis), they give some important insights for the handling of the partial steps of the complex instrument identification task.

2 First Study: Characterization by High Level Features

Each tone consists of an audio signal $x[n]$, $n \in \{1, \ldots 52920\}$, the first 1.2 s are used for the calculation of the feature vectors. The signal has a sampling rate $sr = 44,100$ Hz. Each piano or guitar tone is windowed by half-overlapping segments w_s, $s \in \{1, \ldots, 25\}$ of a size of 4,096 samples. It can be divided into four phases: Attack, decay, sustain and release (see Fig. 1). In order to distinguish such phases, blocks of five consecutive time windows each are constructed (see Fig. 2) and so-called block-features are calculated for each block by taking medians over the window-wise calculated characteristics.

2.1 Groups of Features

Taking the upper and lower shape of the energy envelope of a tone into account the absolute values $|x[n]|$ define the so-called **Absolute Amplitude Envelope** $\mathrm{e} \in IR^{1\times 132}$ as follows by using non-overlapping frames of size 400:

$$e = \left(\max_{1 \le i \le 400} \{|x[i]|\}, \max_{401 \le i \le 800} \{|x[i]|\}, \ldots, \max_{l-399 \le i \le 52800} \{|x[i]|\} \right), l = \left\lfloor \frac{N}{400} \right\rfloor \cdot 400.$$

In the study, overall 132 Absolute Amplitude Envelope block features are used. A visualization of the Absolute Amplitude Envelope is given in Fig. 3 for a piano tone.

The periodogram P_X of each window is calculated at fixed Fourier-coefficients $\{k_1, \ldots, k_{2048}\}$ of a signal X. Additionally for each window the fundamental frequency (called $\hat{f}_0$) is estimated by using tuneR (Ligges 2010) so that overtones can be calculated as $\hat{f}_i = (i+1) \cdot \hat{f}_0,\ i \in \{0, \ldots, 13\}$.

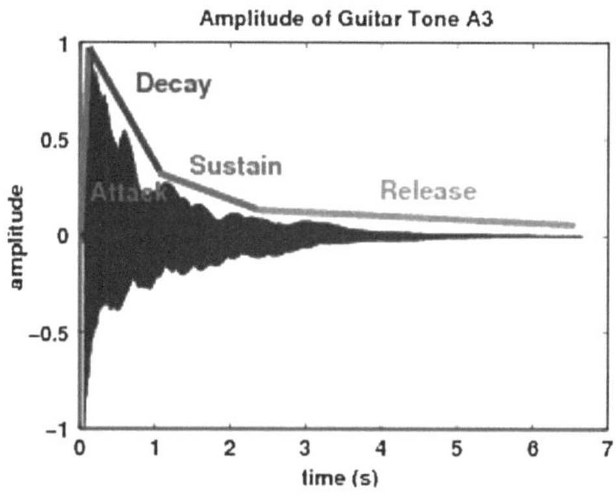

Fig. 1 ADSR-curve of a musical signal

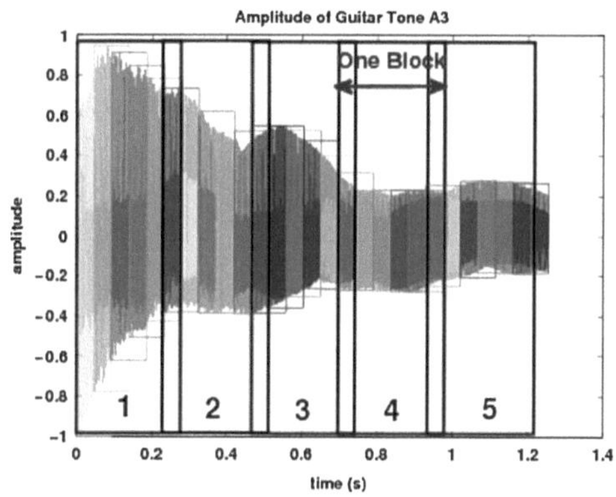

Fig. 2 Blocks of a tone

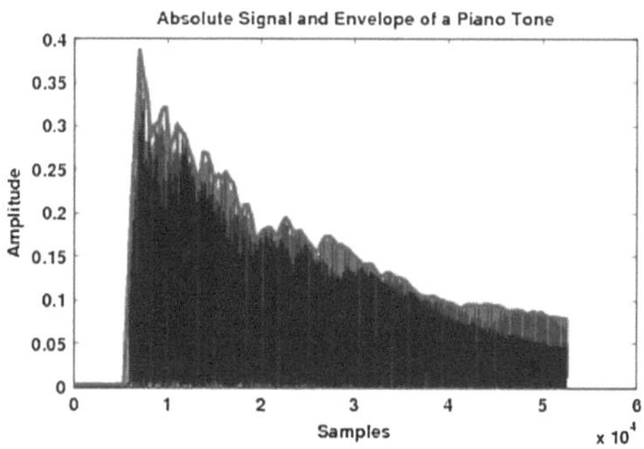

Fig. 3 Musical signal and its envelope

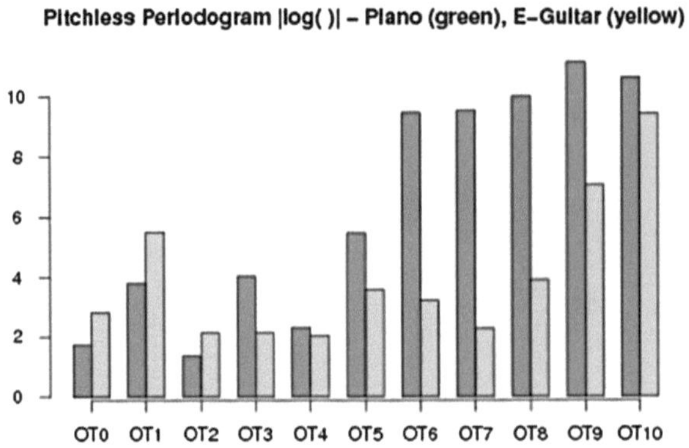

Fig. 4 Pitchless periodogram

At first the fundamental frequencies $\hat{f}_0^{w_s}$ are estimated per window w_s, $s \in \{1,\dots,25\}$ and as described in the beginning of the second section the estimated block-fundamental-frequencies $\hat{f}_0^{b_{\lceil \frac{\xi}{5} \rceil}}$ and block-overtones $\hat{f}_i^{b_{\lceil \frac{\xi}{5} \rceil}}$ are calculated as $\hat{f}_0^{b_{\lceil \frac{\xi}{5} \rceil}} = \text{median}\left(\hat{f}_0^{w_r}, \hat{f}_0^{w_{r+1}}, \dots, \hat{f}_0^{w_{r+4}}\right)$ and $\hat{f}_i^{b_{\lceil \frac{\xi}{5} \rceil}} = (i+1)\,\hat{f}_0^{b_{\lceil \frac{\xi}{5} \rceil}}$ for $r \in \{1, 6, 11, 16, 21\}, i \in \{0, 1, \dots, 13\}$.

After calculating the thirteen block-overtones and block-fundamental-frequency the **Pitchless Periodogram (PiP)** $p \in I\!R^{70}$ can be calculated. It is defined as

$$p = \left(p_1^{k_0}, p_1^{k_1}, \dots, p_1^{k_{13}}, p_2^{k_0}, \dots, p_2^{k_{13}}, \dots, p_5^{k_0}, \dots, p_5^{k_{13}}\right),$$

$$p_{\lceil \frac{\xi}{5} \rceil}^{k_i} := \text{median}\left(P_{x_{w_r}}(k_i), P_{x_{w_{r+1}}}(k_i), \dots, P_{x_{w_{r+4}}}(k_i)\right),$$

with k_i defined by $\left|\hat{f}_i^{b_{\lceil \frac{\xi}{5} \rceil}} - k_i/4096 \cdot sr\right| = \min_{1 \le j \le 2048} \left|\hat{f}_i^{b_{\lceil \frac{\xi}{5} \rceil}} - j/4096 \cdot sr\right|$, $i \in \{0, 1 \dots, 13\}, r \in \{1, 6, 11, 16, 21\}$.

The periodogram is called *pitchless* because the pitch is ignored, only the periodogram heights are considered on an equidistant scale $i \in \{0, \dots, 13\}$. In the study, overall 70 Pitchless Periodogram block features are used. In Fig. 4 the Pitchless Periodogram of one piano and one guitar tone (first block) can be seen for 10 overtones. A log-transformation of the original pitchless periodogram feature vector is carried out to improve visualization.

The power spectrum is calculated by a Discrete Fourier Transformation (DFT) using Hamming windows and a subsequent log-transformation. After mapping the powers of the spectrum onto the mel scale by using triangular filters the discrete cosine transformation is applied yielding the **MFCC coefficients** (see Rabiner and Juang 1993). In the study, overall 80 MFCC block features are used. Figure 5 shows an example of the MFCC.

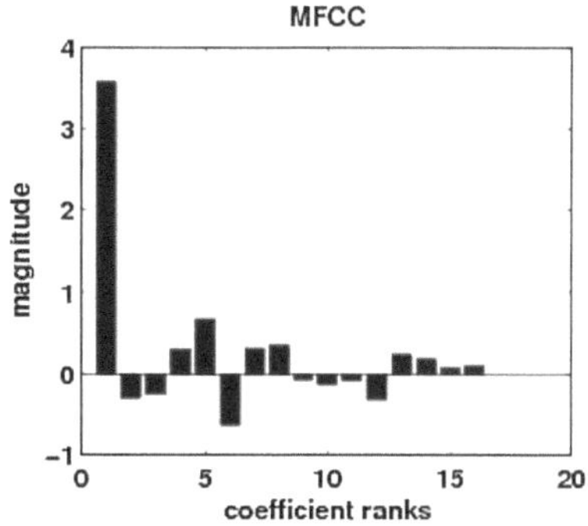

Fig. 5 MFCC

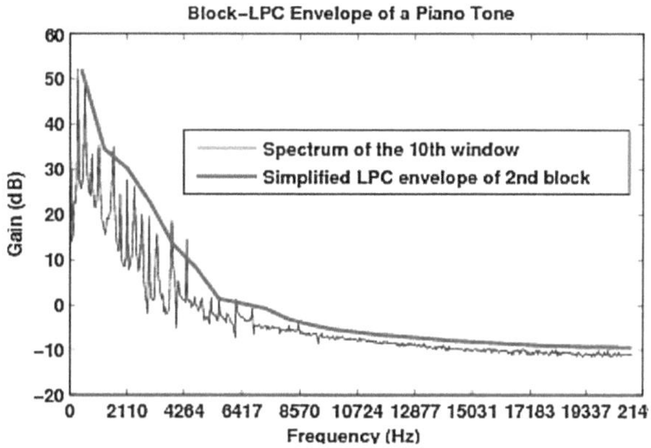

Fig. 6 LPC simplified spectral envelope

The **Linear prediction Coefficients (LPC)** Simplified Envelope is a smoother of the spectral envelope (see Makhoul 1975). In the study, overall 125 LPC block features are used. Figure 6 shows an example of the LPC Simplified Spectral Envelope of a piano tone.

2.2 *Classification and Evaluation*

Classification was carried out using 270 guitar and 275 piano tones to train the models by means of three randomly chosen tenfold cross-validations and 4,309 guitar and 1,345 piano tones for evaluation. For variable selection in each iteration of each cross-validation a logistic model is calculated and by stepwise forward selection those variables are selected that minimize the AIC-criterion (Akaike, 1974). Three vectors v, w, z each containing the frequencies of selection per variable and per cross-validation are calculated. Only those variables may be selected which have been chosen at least, e.g., nine of ten times in the median by the criterion ($V \geq 9$). These are those entries of the vector $v_{med} = (median(v_i, w_i, z_i))_{i=1,\ldots,d}$, d = number of variables, that are equal to nine.

Table 1 Evaluation results (mmce) in % with or without variable filtering

Dimensions: methods	70 P	276 PLM	212 PLM (Mnw)	6 PLM ($V \geq 9$)	6 PLM (Mnw, $V \geq 8$)
SVM	79.06	59.25	50.05	6.44	**3.78**
ADABOOST	**16.66**	**3.50**	**1.72**	**6.3**	4.60
KNN	26.70	16.03	18.63	8.57	6.41
RANDOM FOREST	17.85	4.82	4.15	6.05	5.87

P, PiP; L, LPC (downsampled to 22,050 Hz); M, MFCC; Mnw, MFCC not windowed

Table 2 Classification results by using sequential forward selection

Methods	PLM	Dim.	PLM (Mnw)	Dim.
LDA	**6.75**	7	6.40	6
LOGREG	7.56	7	5.05	7
ADABOOST	7.12	6	**3.35**	10

The example tones are taken from the McGill University Master DVD set (McGill University, 2010), from the RWC database (Goto et al., 2003) and the music instrument samples of Electronic Music Studios, Iowa (University of Iowa, 2010). In Tables 1 and 2 the evaluation results of the classification with the eight different statistical classification methods linear discriminant analysis (LDA), logistic regression (LOGREG), decision trees (RPART), support vector machines (SVM, Gauss-kernel), boosting, k-nearest-neighbour (*k*NN) and random forests are shown. The last five methods have hyperparameters. For these methods, in the tables the mean misclassification error corresponds to that hyperparameter combination that leads to the lowest mean misclassification error over the three different tenfold cross-validations. The computational calculation was done with MATLAB (Lartillot and Toiviainen, 2007) and the R-packages tuneR and mlr, see Ligges (2010) and Bischl (2011).

From Table 1 one can see that the best result is reached for the non-windowed version of PLM with the Adaboost method. It can also be seen that variable selection by a logistic regression pre-step may drastically reduce the error rate, e.g. in case of SVM and kNN. Table 2 shows that the results by using sequential forward selection embedded in the classification step itself are similar to the ones carried out by a variable selection in a pre-step (see Table 1).

2.3 *Interpretation of Selected Features*

Table 3 shows examples of selected features coded as (name_blocknumber_feature-number). Obviously, the first and the fifth block are the most important. The first block contains the first 0.279 s and corresponds to the attack phase of the tone that is rather specific for musical instruments. The fifth block starts after 1.117 s containing, thus, the declining part of the energy (sustain or decay-phase) which appears to be also useful to discriminate between piano and guitar tones.

Table 3 Classification results by using different forward selection methods

Methods	Detected features
PLM (Mnw, $V \geq 8$)	lpc_block5_6, lpc_block5_8, mfcc_1, mfcc_2,chroma_block1_1stOT
PLM ($V \geq 9$)	lpc_block5_8, mfcc_block1_1, mfcc_block1_4,mfcc_block5_1 mfcc_block5_2
PLM (Mnw),	mfcc_2, mfcc_1, chroma_block1_6thOT,lpc_block4_8
ADABOOST (sfs)	lpc_block5_6, lpc_block1_10, lpc_block5_11, chroma_block2_7thOT lpc_block1_9, lpc_block4_3
PLM, LDA (sfs)	mfcc_block5_2, mfcc_block1_1, mfcc_block5_1, chroma_block4_1stOT lpc_block5_8, lpc_block1_6,lpc_block4_24

Table 4 Mean values of hold out errors and selected features after 1,500 evaluations

ES parameters		Error				No. of selected features			
p_{01}	γ	C4.5	Random forest	NB	SVM	C4.5	Random forest	NB	SVM
0.1	32	16.98	14.97	11.38	**07.00**	338.2	343.9	258.6	357.6
0.01	32	19.57	13.68	10.89	07.01	335.1	339.1	198.9	344.1
0.001	32	16.76	12.93	12.25	07.27	332.9	332.6	245.8	348.1
0.001	128	16.47	13.42	13.42	08.31	133.5	199.6	37.1	222.0

3 Second Study: Classification with a Large Feature Set

In a second study we started with a rather large feature set and applied feature selection by means of an evolutionary strategy searching for the best characteristics for building of classification models. We used a set of audio descriptors introduced in (Theimer et al, 2008) extended with many MIR Toolbox functions (Lartillot and Toiviainen, 2007). The extraction was done within the framework AMUSE (Vatolkin et al., 2010). The initial number of the feature dimensions was equal to 265. Since we distinguished between the features extracted from the middle of the attack interval, onset frame and the middle of the release interval, the number of the features to select was finally set to $265 \cdot 3 = 795$. Classification was carried out by decision tree C4.5, random forest, naive Bayes (NB) and SVM (with linear kernel).

Evolutionary Strategies (ES) were applied for feature selection as developed in Bischl et al. (2010). The solution representation was a binary vector with length M equal to the number of features: $\mathbf{f} = \{f_0, \ldots, f_M\}$ ($f_i = 1$ if the i-th feature is selected; otherwise $f_i = 0$). We applied asymmetric mutation in order to reduce the number of selected features. The asymmetric switch probabilty p_{01} was set to $\{0.1, 0.01, 0.001\}$. The general mutation probabilty γ was set to $32/M$, but $128/M$ for $p_{01} = 0.001$. We carried out 10 statistical runs with 1,500 evaluations for each classification algorithm and each ES parameter configuration. As the fit criterion we used the tenfold cross validation training error.

Table 4 illustrates the mean error and the mean number of the remained features over 10 ES runs on the test set. Figure 7 summarizes the results for the different classifiers for both training and test errors. It can be clearly seen, that the classifier choice has a larger impact on the performance than the ES parametrization.

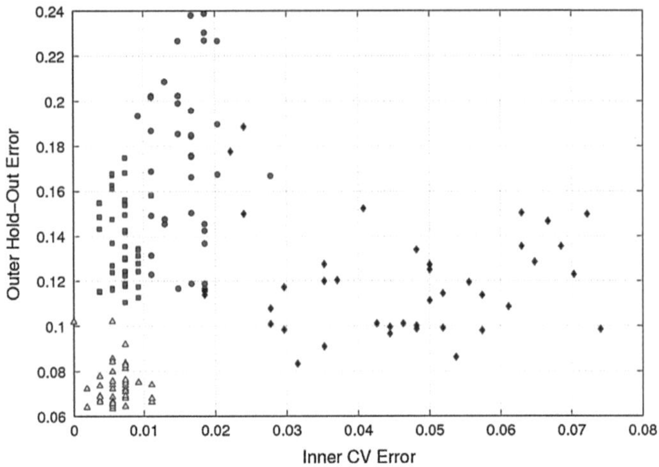

Fig. 7 Both errors after optimization runs (*circles*: C4.5; *squares*: random forest; *diamonds*: NB; *triangles*: SVM)

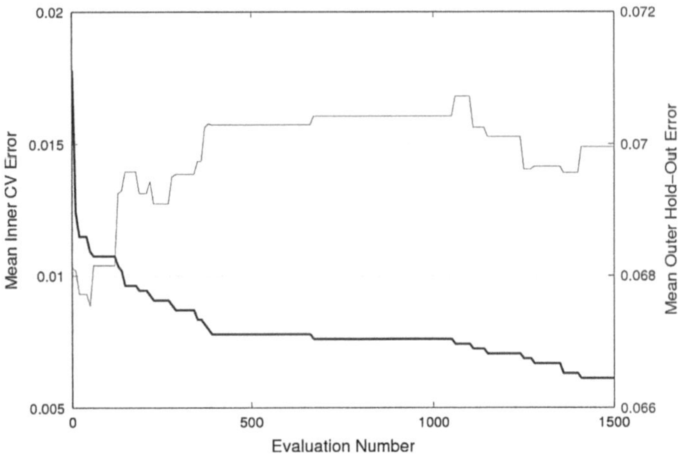

Fig. 8 Progress of an exemplary optimization run for SVM

SVM achieves the smallest errors on the test set. Random forest outperforms C4.5 in most cases. It is interesting to see, that naive Bayes is slightly better than random forest on the test set despite of the rather poor performance on the training set. The drawback of the optimization can be seen in the Fig. 8: though the inner CV error continuously falls during the optimization, the outer hold out error can not be predicted and increases during the evaluation progress due to overfitting.

Another interesting observation is the number of selected features after the optimization (cf. Table 4): after 1,500 evaluations for SVM runs still more than 200 features are used, whereas naive Bayes optimization decreases the number of the selected features to less than 40 for ES with $\gamma = 128$ and $p_{01} = 0.001$. It can also be seen that decreasing p_{01} is not strongly reducing the feature number. However, an extended ES parameter analysis is beyond the scope of this work.

4 Conclusions

As the classification of piano and guitar tones show good results – especially in case of high level features and variable selection – we next would like to consider the classification of other musical instruments like strings and reed instruments together with piano and guitar.

References

Akaike, H. (1974). A new look at the statistical model identification. *IEEE Transactions on Automatic Control*, *19*(6), 716–723. doi:10.1109/TAC.1974.1100705.

Bischl, B. (2011). Machine learning in R, R-package, TU Dortmund. http://r-forge.r-project.org/projects/mlr/.

Bischl, B., Vatolkin, I., & Preuss, M. (2010). Selecting small audio feature sets in music classification by means of asymmetric mutation. In: *Proceedings of the 11th International Conference on Parallel Problem Solving From Nature (PPSN)*, Krakow, pp. 314–323.

Brown, J. C., Houix, O., & McAdams, S. (2001). Feature dependence in the automatic identification of musical woodwind instruments. *Journal of the Acoustical Society of America*, *109*, 1064–1072.

Fletcher, N. H. (2008). *The physics of musical instruments*. New York: Springer.

Goto, M., Hashiguchi, H., Nishimura, T., & Oka, R. (2003). RWC music database: Music genre database and musical instrument sound database. In: *ISMIR 2003 Proceedings*, Baltimore, pp. 229–230.

Krey, S., & Ligges, U. (2010). SVM based instrument and timbre classification. In Locarek-Junge, H., & Weihs, C. (Eds.), *Classification as a tool for research*. Berlin/Heidelberg/New York: Springer.

Lartillot, O., & Toiviainen, P. (2007). A matlab toolbox for musical feature extraction from audio. In: *International Conference on Digital Audio Effects*, Bordeaux.

Ligges, U. (2010). tuneR–analysis of music. http://r-forge.r-project.org/projects/tuner.

Livshin, A., & Rodet, X. (2006). The significance of the non-harmonic "Noise" versus the harmonic series for musical instrument recognition. In: *ISMIR 2006 Proceedings*, Victoria, pp. 95–100.

Makhoul, J. (1975). Linear prediction: A tutorial review. *IEEE*, *63*, 56.

McGill University. (2010). McGill master samples collection on DVD. http://www.music.mcgill.ca/resources/mums/html.

Rabiner, L., & Juang B. H. (1993). *Fundamentals of speech recognition*. Englewood Cliffs: Prentice Hall PTR.

Theimer, W., Vatolkin, I., & Eronen, A. (2008). *Definitions of audio features for music content description* (Tech. Rep. TR08-2-001) University of Dortmund, Chair of Algorithm Engineering.
University of Iowa. (2010). Electronic music studios. Musical instrument samples. http://theremin.music.uiowa.edu.
Vatolkin, I., Theimer, W., & Botteck, M. (2010). AMUSE (Advanced MUSic Explorer)–A multitool framework for music data analysis. In *Proceedings of the 11th International Society for Music Information Retrieval Conference (ISMIR)*, Utrecht, pp. 33–38.

Auralization of Auditory Models

Klaus Friedrichs and Claus Weihs

Abstract Computational auditory models describe the transformation from acoustic signals into spike firing rates of the auditory nerves by emulating the signal transductions of the human auditory periphery.
The inverse approach is called auralization, which can be useful for many tasks, such as quality measuring of signal transformations or reconstructing the hearing of impaired listeners. There have been few successful attempts to auditory inversion each of which deal with relatively simple auditory models.
In recent years more comprehensive auditory models have been developed which simulate nonlinear effects in the human auditory periphery. Since for this kind of models an analytical inversion is not possible, we propose an auralization approach using statistical methods.

1 Introduction

An auditory model is a computer model of the human auditory system. It requires an acoustic signal as input and outputs the spike firing rates of the auditory nerve fibers. The human auditory system consists of roughly 30,000 auditory nerve fibers but in auditory models this is usually simplified by a much smaller quantity. In most models the auditory system is coded by a multichannel bandpass filter where each channel represents one specific nerve fiber. As in the human system each channel has its specific center frequency by which the perceptible frequency range is defined.

K. Friedrichs (✉) · C. Weihs
Chair of Computational Statistics, TU Dortmund, Dortmund, Germany
e-mail: friedrichs@statistik.tu-dortmund.de; weihs@statistik.tu-dortmund.de

A. Giusti et al. (eds.), *Classification and Data Mining*, Studies in Classification, Data Analysis, and Knowledge Organization, DOI 10.1007/978-3-642-28894-4_27,

The inversion procedure to resynthesize the original signal from the auditory model output can be used for quality measurements of signal algorithms. Therefore, it is not essential to get the original signal but it is sufficient to get a signal which sounds like the original one. This procedure is called auralization. One exemplary potential of auralization is reconstructing the hearing of impaired listeners which is a significant task for improving hearing aids. Jepsen has introduced how cochlear hearing loss can be modeled in an auditory model (Jepsen et al., 2006). To interpret the output of such a modified model the output simply has to be auralized through the original auditory model.

There have already been successful attempts of auralization by analytical inversion. However, they all deal with relatively simple auditory models. Slaney et al. used an auditory model which contains of a filter bank, a half-wave rectifier (HWR) and an automatic gain control. In their model only the HWR stage results in information loss and can not be inverted directly. Hence, they introduced techniques to reconstruct this information by using knowledge about each channel's signal (Slaney et al., 1994). Hohmann presented an approach to invert the gammatone filter bank, a filter which is used in many auditory models (Hohmann, 2002). Feldbauer et al. analyzed the problem from another direction. They developed an auditory model with the intention that it can be inverted with a relatively low computational effort. Their model consists of the gammatone filter bank, a HWR, a power-law compressor and an adaptive subsampling mechanism. While the effect of the power-law compressor can be inverted directly, the HWR and the adaptive subsampling mechanism are undone by bandpass filtering. Additionally, they used two correction steps to compensate energy loss of the pulses (Feldbauer et al., 2005).

In recent years more comprehensive auditory models have been developed. The most common model is the one of Meddis and Sumner which is based on animal observations and psychoacoustic phenomena (Sumner et al., 2002). This model contains of several stages:

1. Response of the middle ear, modeled by a second-order linear bandpass Butterworth filter
2. Filtering of the basilar membrane, modeled by a dual-resonance-nonlinear (DRNL) filter
3. Inner-hair-cell model
   (a) Transduction of basilar membrane motion into receptor potential
   (b) Calcium controlled transmitter release function
   (c) Quantal and probabilistic model of synaptic adaptation
4. Auditory-nerve response

Thereby, cochlear-nonlinearities are modeled which are important with respect to many perceptual experiments. Unfortunately, the resulting problem is impossible to invert analytically. Therefore, in this paper an auralization approach using statistical methods is introduced.

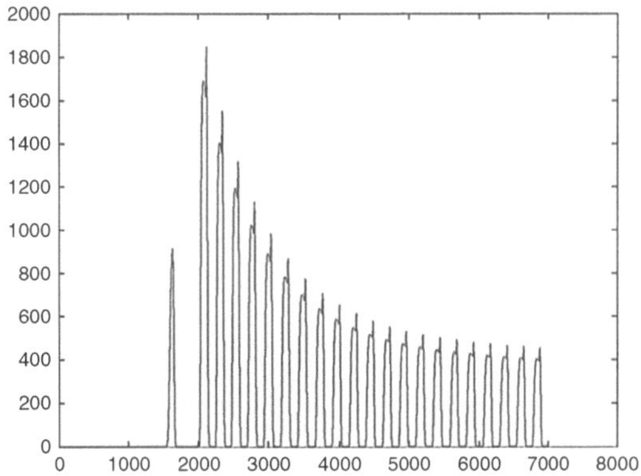

Fig. 1 Exemplary spike firing rate of one channel

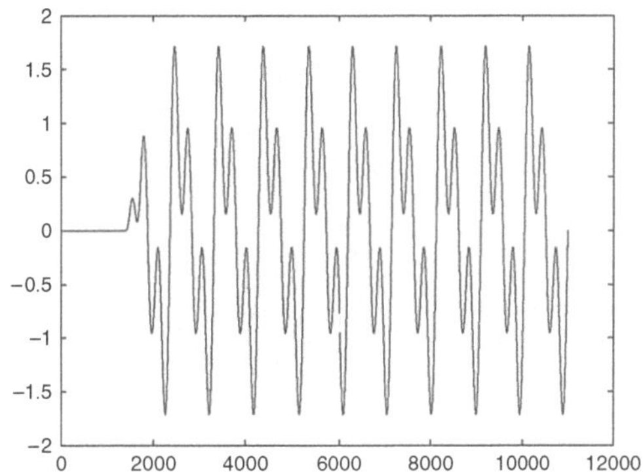

Fig. 2 Harmonic tone

2 Auralization by Using Statistical Methods

In this study auralization is done for an auditory model of Meddis, which is modeled by a 30 channels filter bank with center frequencies between 100 and 3,000 Hz. Figure 1 shows an exemplary output of one channel. To simplify the problem in this study it is assumed that the input stimulus is a harmonic tone. Harmonic tones are typical for musical sounds. They consist of a fundamental frequency, the key tone and integer multiples of this frequency, which are called overtones. The sound of a harmonic tone is defined by its involved frequencies and the power of each frequency. Figure 2 shows an exemplary harmonic tone which contains a key tone of 100 Hz (91 dB) and the overtones of 200 Hz (83 dB) and 300 Hz (89 dB).

To resynthesize a harmonic tone in this paper a two-stage concept is proposed. In a first step the overtones have to be detected by classification and in a second step the power of each overtone has to be estimated by regression. A crucial method for the whole task is to use the phase locking effect which was introduced by Moissl and Meyer-Base (Moissl et al., 2000). This effect phase-locks the impulse rate of the channels to the stimulus. In our problem this implies that each frequency, which is part of the input signal, will also occur in some channels.

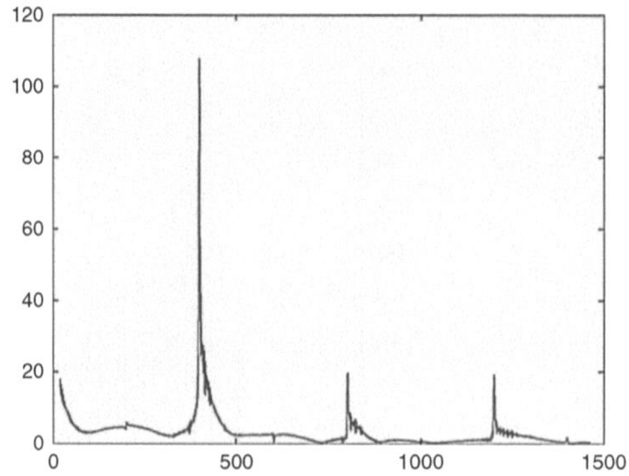

Fig. 3 DFT of channel 13: 400 Hz is part of the tone

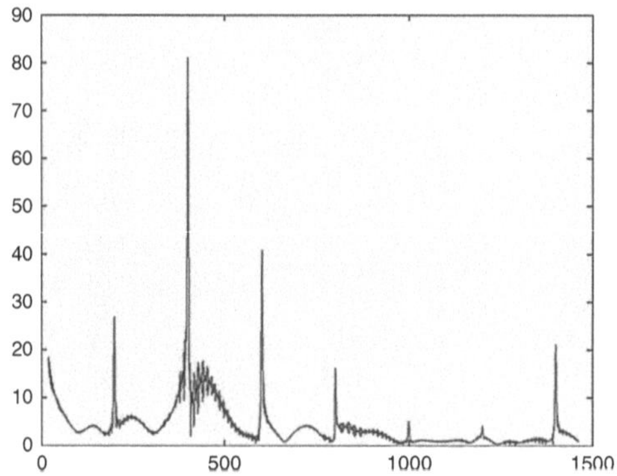

Fig. 4 DFT of channel 13: 400 Hz is not part of the tone

2.1 Frequency Detection

In frequency detection the periodicities of each channel chart have to be analyzed. Therefore, the discrete fourier transform (DFT) of each channel output is generated. Because of the phase locking effect in each of these charts peaks should occur at frequencies which are part of the acoustic stimulus. Such a peak gets stronger the smaller the difference is between this frequency and the center frequency of the channel. From this it follows that it is sufficient to detect in each channel only the frequencies which are between the center frequencies of both surrounding channels. Furthermore, the restriction on harmonic tones ensures that in each of the 30 channels at most one frequency has to be detected.

Figure 3 shows the DFT of channel 13 for a harmonic tone which includes an overtone with the frequency of 400 Hz. This frequency is also the maximum peak of this chart. A first approach is detecting the main peaks of all channels and, in this way, getting all frequency components. Unfortunately, Fig. 4 shows that the

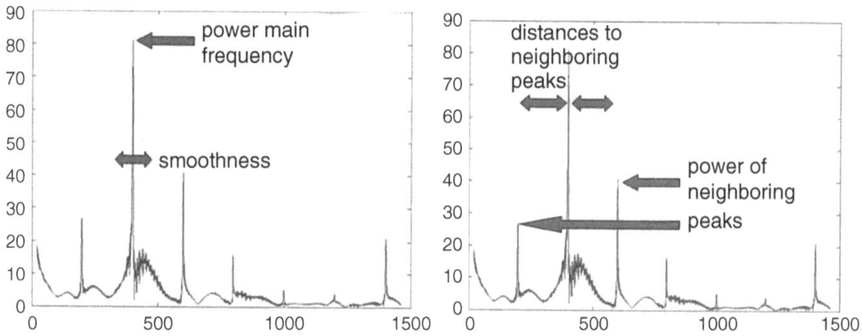

Fig. 5 Features

maximum peak does not always define a frequency which is part of the acoustic stimulus. Here the maximum peak is also at 400, but in this example 400 Hz is not part of the input tone. After having detected the maximum peak of a channel it has to be classified if this frequency is in fact part of the original signal in order to construct a classification rule. In this study this is done by classification trees. Therefore, a training set of harmonic tones is required. Because there are differences between low and high frequencies each channel needs its own decision tree. For each tree seven features are used: The suggested frequency itself, the power of the main peak, the smoothness of this peak, the distance to the neighboring peaks and the power of these peaks. These features are visualized in Fig. 5.

The main frequency is considered because the distance to the center frequency of the channel could have an impact and a higher power of this peak should increase its probability for being part of the input. Additionally, its smoothness as well as the information about the neighboring peaks could yield essential information about the relative power compared to the surrounding frequencies. Figure 6 shows the features of a harmonic tone which consists of the frequencies 200, 400, 600 and 800 Hz. In this example only the feature vectors of the bold typed channels are used. For all other channels the main frequency can be neglected directly because it is outside the particular channel range.

To test the approach a training set of 20,000 tones is used, generated by the following rules. Each tones consists of up to 8 overtones and has a fundamental frequency uniformly distributed between 80 and 1,080 Hz. Each overtone (up to 3 kHz) is chosen as a component with probability 0.7. Each component is generated with a power uniformly distributed between 82 and 92 dB and the tone is built by summarizing all components. Figure 7 shows an exemplary classification tree. The error rates of the 30 decision trees are calculated by tenfold cross validation. As it can be seen in Table 1 these error rates are between 0 and 0.4 %. This result is even further improved when bringing the results of all channels together since most frequencies can be detected in two channels.

channel	center frequency	main frequency	power main frequency	smoothness	lower frequency peak	power lower frequency	higher frequency peak	power higher frequency	frequency is tone component
1	100	200,7	63,3	23,4	53,1	92,3	32,5	200,7	-
2	112	200,7	74,2	19,0	46,8	79,1	38,8	200,7	-
3	126	200,7	74,4	24,9	37,0	58,6	20,5	200,7	-
4	142	200,7	79,5	85,0	20,0	183,1	23,3	399,9	-
5	160	200,7	91,7	36,6	28,5	30,8	24,5	401,4	-
6	**180**	**200,7**	**106,3**	**54,2**	**21,8**	**183,1**	**26,0**	**399,9**	yes
7	**202**	**202,1**	**113,4**	**41,0**	**22,4**	**184,6**	**27,5**	**399,9**	yes
8	227	202,1	84,1	70,3	22,9	184,6	26,7	399,9	-
9	256	200,7	81,5	20,5	22,8	183,1	29,8	46,9	-
10	287	202,1	86,9	13,2	22,4	184,6	33,8	79,1	-
11	323	202,1	86,5	17,6	21,4	184,6	30,3	112,8	-
12	363	202,1	84,2	23,4	19,4	184,6	26,1	143,6	-
13	**409**	**401,4**	**81,1**	**43,9**	**26,9**	**199,2**	**40,8**	**200,7**	no
14	459	602,1	88,9	16,1	28,1	200,7	26,3	599,1	-
15	517	600,6	95,6	60,1	19,9	584,5	29,6	600,6	-
16	**581**	**602,1**	**104,4**	**41,0**	**20,4**	**585,9**	**30,7**	**599,1**	yes
17	**653**	**602,1**	**99,8**	**33,7**	**21,0**	**585,9**	**23,2**	**599,1**	yes
18	**734**	**801,3**	**104,8**	**36,6**	**20,7**	**785,2**	**30,4**	**799,8**	yes
19	**826**	**801,3**	**101,8**	**36,6**	**20,8**	**785,2**	**28,5**	**799,8**	yes
20	928	602,1	77,8	26,4	29,9	200,7	35,0	199,2	-
21	1044	801,3	101,9	26,4	22,7	199,2	28,1	799,8	-
22	1174	801,3	103,8	26,4	21,0	785,2	32,2	799,8	-
23	1320	801,3	103,8	29,3	24,4	199,2	35,0	799,8	-
24	1484	801,3	103,6	26,4	32,4	199,2	36,6	799,8	-
25	1669	801,3	101,7	26,4	41,9	199,2	38,9	799,8	-
26	1877	801,3	96,8	26,4	51,4	199,2	38,1	799,8	-
27	2110	801,3	89,3	23,4	55,4	199,2	41,0	600,6	-
28	2373	801,3	79,3	23,4	54,4	199,2	41,3	600,6	-
29	2668	801,3	67,9	13,2	50,1	199,2	38,7	600,6	-
30	3000	801,3	57,4	13,2	44,2	199,2	34,3	600,6	-

Fig. 6 Feature generation of an exemplary harmonic tone. In the first two columns the channel number and its corresponding center frequency are listed. The third column shows the detected main frequency of each channel chart. If it is similar to the center frequency the whole row is marked bold and only in this case it has to be classified if the detected frequency is part of the tone. The features, which are used for this classification task, are listed in columns 4–9. Finally, the last column shows the target variable which enables supervised learning

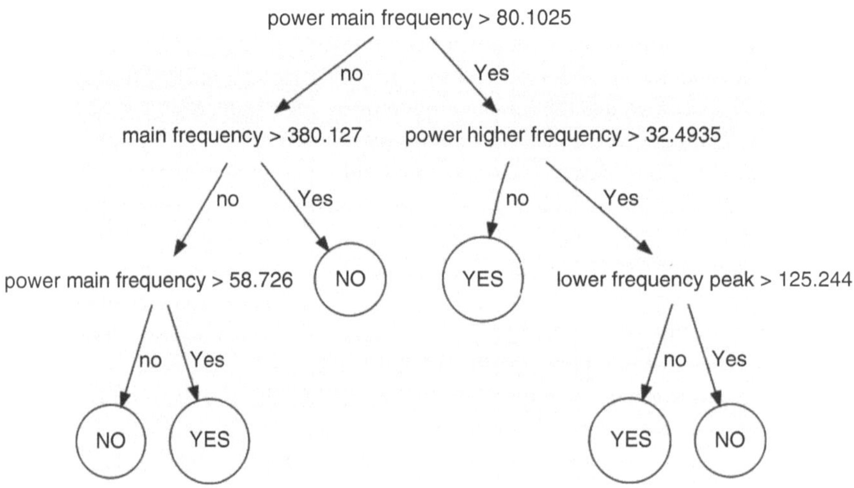

Fig. 7 Classification tree of channel 13

Table 1 Error rates of frequency detection

Channel number	1	2	3	4	5	6	7	8	9	10	11	12	13	14	15
Error rate in %	0.4	0.0	0.0	0.0	0.0	0.0	0.0	0.0	0.0	0.0	0.0	0.0	0.0	0.2	0.3
Channel number	16	17	18	19	20	21	22	23	24	25	26	27	28	29	30
Error rate in %	0.3	0.3	0.2	0.3	0.3	0.3	0.4	0.2	0.2	0.0	0.0	0.0	0.0	0.0	0.0

Table 2 Error rates of the power estimation of the 1st training set

Channel number	1	2	3	4	5	6	7	8	9	10	11	12	13	14	15
Average squared error	1.4	1.1	0.8	0.9	1.8	2.1	2.2	2.3	2.1	2.2	1.9	2.3	2.4	2.2	2.0
Channel number	16	17	18	19	20	21	22	23	24	25	26	27	28	29	30
Average squared error	2.3	2.0	2.2	1.8	2.1	2.3	2.3	2.1	1.7	1.3	1.2	1.0	0.9	0.8	0.7

2.2 *Power Estimation of Each Frequency*

After having detected all frequencies which are components of the signal, the power of each frequency has to be estimated. Therefore, a similar approach as for the frequency detection is applied. The same features are used but here the target variable is the power in dB of each frequency component. Again, we use decision trees for this regression problem. This approach is further improved by additional features: The average firing activity in the analyzed channel as well as the power of the analyzed frequency in the other 29 channels are considered since it is supposed that these two features are correlated with the sound volume of the frequency in the input signal. Furthermore, there are strong interactions between the frequency components of the input signal. Hence, also the overtone number of the analyzed frequency and the frequencies of the other over tones are applied as features by using the results of the frequency detection.

Again, a training set is used which is generated similar to the one used in Sect. 2.1. In a second experiment a simplified training set is used in which each harmonic tone consists of up to 4 overtones instead of 8. The error rates are calculated again by tenfold cross validation. The average squared error of the 30 regression trees are listed in Table 2 for the first training set and in Table 3 for the second training set. These error rates should be compared to the range of the power of each frequency which is between 82 and 92 dB. Since the range is uniformly distributed the naive approach, which estimates always the mean (87 dB), has an estimated average squared error of 8.3 dB.

For the simplified tones of the second training set these error rates are between 0.2 and 0.7 dB. This is virtually not hearable for humans and ensures almost an accurate reconstruction of the original signal. Unfortunately, the error rates for the more complex tones are inferior. Here the error rates are up to 2.4 dB. The reason is the following. With more possible overtones there are much more combinatorial possibilities to generate harmonic tones which contain a specific frequency. But all frequencies influence the firing activities of all channels.

Table 3 Error rates of the power estimation of the 2nd training set

Channel number	1	2	3	4	5	6	7	8	9	10	11	12	13	14	15
Average squared error	0.4	0.2	0.2	0.4	0.3	0.5	0.5	0.4	0.5	0.5	0.6	0.4	0.7	0.5	0.6
Channel number	16	17	18	19	20	21	22	23	24	25	26	27	28	29	30
Average squared error	0.6	0.7	0.5	0.6	0.6	0.4	0.5	0.3	0.3	0.4	0.3	0.2	0.3	0.2	0.2

3 Conclusion

Using classification and regression methods for auralization of auditory models seems to be a promising approach. While the frequency detection is mostly solved the power estimation of each frequency component still needs improvement. One simple improvement could be a larger training set. Some pretests have shown that smaller training sets than the ones used in the previous chapter lead to inferior results. Therefore, it can be expected a larger training set will lead to smaller error rates. Furthermore, other regression methods than decision trees might perform better in the power estimation task. Another improvement could result from using additional features and a criteria for variable selection.

Future studies have to show the ability to generalize the proposed auralization approach to non-harmonic sounds. Finally, in order to reconstruct a series of acoustic signals a mechanism for detecting temporal changes has to be developed.

Acknowledgements This work was supported by the Collaborative Research Center "Statistical modeling of nonlinear dynamic processes" (SFB 823) of the German Research Foundation (DFG).

References

Feldbauer, C., Kubin, G., & Kleijn, W. B. (2005). Anthropomorphic coding of speech and audio: A model inversion approach. *EURASIP Journal on Applied Signal Processing, 2005*, 571618.

Hohmann, V. (2002). Frequency analysis and synthesis using a Gammatone filterbank. *Acta Acustica United with Acustica, 88*(3), 433–442.

Jepsen, M. L., Dau, T., & Ewert, S. (2006). *A model of the normal and impaired auditory system*. Academic dissertation, Technical University of Denmark.

Moissl U., & Meyer-Base U. (2000). Decoding of neural firing to improve cochlear implants. *Proceedings of SPIE, 4055*, 337–348.

Slaney, M., Naar, D., & Lyon, R. F. (1994). Auditory model inversion for sound separation. In *Proceedings of IEEE Intnational Conference Acoustics, Speech, Signal Processing*, Adelaide.

Sumner, C. J., O'Mard, L. P., Lopez-Poveda, E. A., & Meddis, R. (2002). A revised model of the inner-hair cell and auditory nerve complex. *Journal of the Acoustical Society of America, 111*(5), 2178–2188.

Visualisation and Analysis of Affiliation Networks as Tools to Describe Professional Profiles

Cristiana Martini

Abstract The analysis of professional profiles is both a crucial and a challenging issue: professional profiles evolve over time, especially in the forefront fields. The interrelations between working activities and professional profiles can be seen as an affiliation network, where a set of actors (activities) co-participates in a set of events (professions); the same is true with the relation between competencies and professions. The techniques developed to analyse and represent affiliation networks can then be applied to the analysis of professional profiles. This paper discusses the application of some techniques developed to analyse and represent affiliation networks to the analysis of professional profiles, with an example on the professional profiles operating in the Research and Development (R&D) field.

1 The Analysis of Professional Profiles

The labour market is a mutable reality, where professional profiles change, and new professional roles emerge and/or replace the old ones; the International Labour Office (ILO) periodically updates the International Standard Classification of Occupation (ISCO), as far as most of the national statistical agencies. The change is especially fast when we consider innovative services or developing sectors, and even more for the highly qualified jobs; in this case, the new profiles are often unclear also to the agencies that are called to educate people for these professional roles.

An effective way to depict a job is by means of the work activities that are performed, analysing an activity-by-job ($A \times J$) matrix $\mathbf{A}$, whose (a, j)-th element represents the relevance of the a-th activity in the j-th job. As an alternative,

C. Martini (✉)
University of Modena and Reggio Emilia, Reggio Emilia, Italy
e-mail: cmartini@unimore.it

A. Giusti et al. (eds.), *Classification and Data Mining*, Studies in Classification, Data Analysis, and Knowledge Organization, DOI 10.1007/978-3-642-28894-4_28,

a job can be described by means of the competencies (knowledge, skills and attitudes) which are required to cover that position; different mixtures of competencies give rise to different professional profiles, which can be described by analysing an analogous competency-by-job ($C \times J$) matrix **C**, whose (c, j)-th element represents the importance of the c-th competency for the j-th job.

The importance of each competency and the pertinence of each activity can be gathered from different informants, e.g. job incumbents, employers, line managers, experts in charge of personnel hiring within companies or other experts. According to the kind of information obtained from the respondents, the competency-by-job matrix **C** and the activity-by-job matrix **A** contain the frequency of employees in the j-th job who possess the c-th competency or perform the a-th activity, or the average score of the c-th competency or the a-th activity for the j-th job.

Job/task analysis and competency modeling are widely used in human resource management to get job descriptions, job specifications and further analyses which could lead to improvement in all aspects of management practice (Schneider and Konz, 1989; Mirabile, 1997), but affiliation networks have never been applied in this context. Aim of this contribution is to show that the structure of relationships among activities, competencies and jobs can effectively be described by affiliation networks, and the techniques developed to analyse these networks can be applied to the analysis and visual representation of professional profiles (Sect. 2). Section 3 contains an application to data collected in the Research and Development field, while in Sect. 4 some final considerations are drawn.

2 Jobs, Competencies and Activities as Affiliation Networks

Affiliation networks are 2-mode networks, consisting of a set of actors and a set of events (Wasserman and Faust, 1994); a distinctive feature of 2-mode affiliation networks is duality, i.e. events can be described as collections of individuals affiliated with them, while actors can be described as collections of events with which they are affiliated.

The activity-by-job and competency-by-job matrices described above can be interpreted as affiliation networks, where a set of actors (competencies or activities) participate in defining a set of events (professional profiles). Affiliation networks are usually applied to social circles in the traditional sense (individuals' affiliation with collectivities or social events), while no applications to the job analysis are found in the international literature.

Affiliation networks can be straightforwardly represented with the affiliation network matrix, i.e. a matrix of zeros and ones only, where each row describes the actor's affiliation with the events, and each column describes the memberships of the event; this matrix can be easily derived from the **A** or the **C** matrix. If the cells of the **A** or the **C** matrices are pertinence/importance scores, scores indicating an adequate or high importance will be substituted by 1, while scores indicating scarce or no importance will be replaced by 0. If the cells of the original matrix are frequencies, and the non-zero cells are relatively few, they can all be replaced by 1, otherwise the lowest frequencies will be set to zero.

Substantive applications of affiliation networks often focus on just one of the modes; 1-mode analyses use adjacency matrices derived from the affiliation matrix, where ties between pairs of actors are based on the connections implied by events and originate co-membership (or co-attendance, or co-affiliation) relations, while the ties between pairs of events derive from the linkages generated by actors, and give rise to overlapping (or interlocking) relations; the derived 1-mode networks are then non-directional and valued.

Different 1-mode networks can be obtained from jobs, activities and competencies, which analyse complementary aspects of the relations among these entities:

- A 1-mode activity network; the corresponding adjacency matrix **AP** is a square $(A \times A)$ matrix, where the (i, j)-th cell reports the number of professional profiles where the i-th and j-th activities are jointly performed.
- A 1-mode competency network; the corresponding adjacency matrix **CP** is a square $(C \times C)$ matrix, where the (i, j)-th cell indicates the number of professional profiles where the i-th and j-th competencies are jointly required.
- A 1-mode job network based on activities; the corresponding adjacency matrix **PA** is a square $(P \times P)$ matrix, where the (i, j)-th cell indicates the number of activities shared by the i-th and the j-th jobs.
- A 1-mode job network based on competencies; the corresponding adjacency matrix **PC** is a square $(P \times P)$ matrix, where the (i, j)-th cell indicates the number of competencies shared by the i-th and the j-th jobs.

All these adjacency matrices can be visually represented and described through network analysis. The density measures computed for the job networks **PA** and **PC** indicate the general overlapping degree among professional profiles in terms of performed activities and required competencies, while for the activity network **AP** and the competency network **CP** they indicate the general degree of tranverseness of the considered sets of activities and competencies; density measures are meaningful mainly to compare different sets of jobs, competencies and activities. However, scholars do not agree on the procedure to calculate density measures for valued networks (Scott, 1991). A possible solution is to transform the valued network in a binary network, by determining a threshold and substituting 1 to the values of the adjacency matrix above this threshold and 0 to the ones below; obviously, this transformation implies a loss of information, and the choice of the threshold should be driven by substantive reasons.

Centrality measures provide overlapping indices for each professional profile in the job networks **PA** and **PC**, or transverseness indices for each activity or competency in the corresponding networks **AP** and **CP**; this is particularly true for the degree centrality of a node, i.e. the number of edges incident upon that node, which highlights those professional profiles that share many activities or competencies with other profiles, while betweennes and closeness centrality measures have a less illuminating meaning in the present context.

A further tool of network analysis that gives interesting information for job analysis is the search for cliques, i.e. the largest subsets of network nodes connected to all other nodes in the subset. For valued networks, a clique at level c is the complete sub-graph of maximum size whose ties have weight at least equal to c; from a substantive point of view, a clique at level c is a set of particularly similar jobs (i.e. sharing at least c activities or competencies), or, in the analysis of matrices **AP** and **CP**, the set of activities or competencies more transversal to the analysed jobs.

Simultaneous representation of actors and events is also possible, but these methods are less developed; the most straightforward approach to give a 2-mode representation of an affiliation network is a bipartite graph, i.e. a graph in which the nodes are partitioned into two subsets and the lines connect pairs of nodes belonging to different subsets. However, this representation can be unwieldy when used to depict large affiliation networks; Borgatti and Everett (1997) point out the complexity of a bipartite graph with 18 and 14 points.

Correspondence analysis is another method for representing both rows and columns of a 2-mode affiliation matrix in a plot where points representing actors are placed close to each other if they mostly attended the same events, while points representing events are plotted close together if they were attended by mostly the same actors (Wasserman et al., 1990); closeness between actor-points and event-points indicate that those actors attended those events. Correspondence analysis includes an adjustment for marginal effects, therefore, actors are plotted close to events which they attended and which were attended by few other people, and events are plotted close to actors who attended those events and few other events (Borgatti and Everett, 1997; Wasserman and Faust, 1994).

3 An Application to the R&D Field

This application is focused on the professional profiles operating in the Research and Development (R&D) field. Jobs, activities and competencies were surveyed by interviewing the directors of a random sample of 31 (out of 66) R&D companies with at least three operators in the Veneto Region.[1] The face-to-face interviews focused on the professional profiles employed in the firm (at least at a technical level); for each profile, directors were asked how many people in such position were employed in the firm, what activities they had to perform in their working tasks, and what competencies were needed to cover each position. Answers were given in open form, and then categorised and coded; this operation gives rise to

[1]Companies with less than three operators were excluded because clear professional profiles and roles separation are hard to achieve with only one or two operators; the random sample was stratified by companies' size and sub-sector of activity (R&D in natural sciences and engineering and R&D in social sciences and humanities).

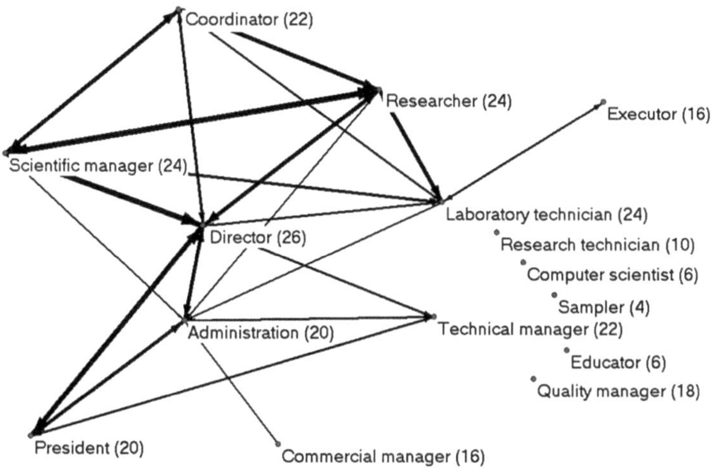

Fig. 1 Graph representation of the 1-mode network of jobs based on activities. Density = 0.5733 (in brackets the degree centrality of each node)

15 professional profiles (covering the whole range from technical to managerial positions), 41 activities and 48 technical-specific competencies. The final dataset contains data on 450 persons who work in the R&D field; the (a, j)-th element of the matrix **A** is the frequency of operators in the j-th job who are required to perform the a-th activity, and given the large number of zero-cells,[2] all non-zero frequencies are replaced by 1 in the affiliation matrix.

The interconnection between jobs and activities is first analysed by means of a network approach: the 2-mode network is transformed according to a 1-mode approach, giving rise to separate 1-mode networks for activities, competencies, and jobs. Since the main interest is on the description of professional profiles, only the two 1-mode job networks are reported and analysed.

The 1-mode job network based on activities, where connections indicate common activities across jobs, is a weakly connected network, indicating that different roles may require the same competencies, but the tasks distribution is quite clear; Fig. 1 shows all the ties valued at least 3, i.e. connections indicate pairs of jobs sharing at least 3 out of the 41 activities[3]; ties' thickness is proportional to weights. The main overlapping regards managerial and coordinative roles, firstly the scientific manager, who shares the managerial activities with the director and the operational ones with the researcher; the similarity between the director and the president is also remarkable. Conversely, the most separate profiles are computer scientist, educator and sampler, whose tasks are particularly well defined and unique.

---

[2]In the **A** matrix all the frequencies (about 1,000) are concentrated in only 100 of the 615 cells.

[3]To improve the readability of the graph, the adjacency matrix has been dichotomised by setting at zero ties with values 1 and 2.

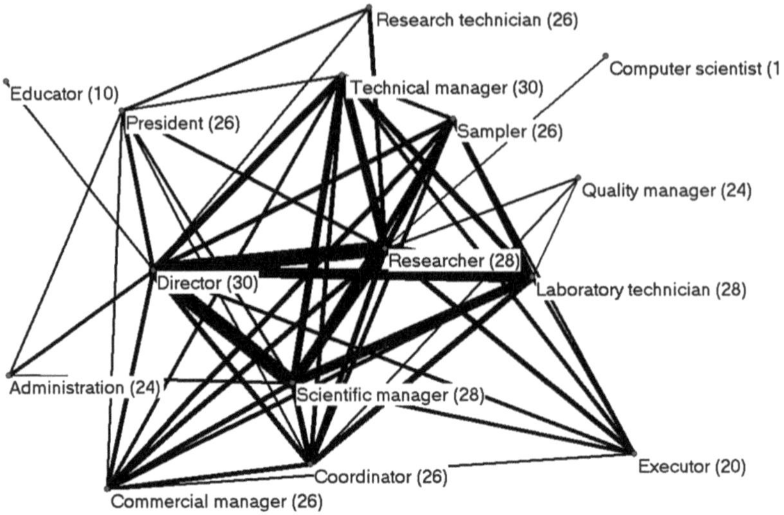

Fig. 2 Graph representation of the 1-mode network of jobs based on competencies. Density = 0.8133 (in brackets the degree centrality of each node)

The 1-mode job network based on competencies (where the connections indicate common required competencies) is much more dense. Figure 2 shows again ties valued at least 3, highlighting a strong interconnection between the director, the scientific manager and the researcher, which constitute a clique at level 14; these profiles are the core of the R&D firm, and share the scientific responsibility of the research, but from different points of view and with complementary work activities: while the scientific manager is in charge of the theoretical and programmatic aspects of the research activity, the researcher has technical tasks, and the director deals with the managerial aspects. Very strong connections are observed also between this first group and the technical manager or the laboratory technician; connections are relevant with samplers, commercial managers, coordinators and executors too, while the educator and computer scientist are completely set apart.

The higher density of the job-competency network indicates that professional roles share competencies more than activities, probably because an efficient tasks distribution implies a limited overlapping of tasks; an example is given by the sampler, who shares a large number of competencies but performs quite different activities.

A simultaneous analysis of actors and events is possible through correspondence analysis, which generates a plot where professional profiles and activities are located according to two main dimensions: the first separates technical and managerial profiles and explains 23.2 % of the total inertia, while the second goes from operative roles to scientific planning, explaining 20.6 % of the total inertia. In order to guarantee readability, Fig. 3 only reports work activities and Fig. 4 professional profiles (obviously, analogous representations can be obtained for competencies).

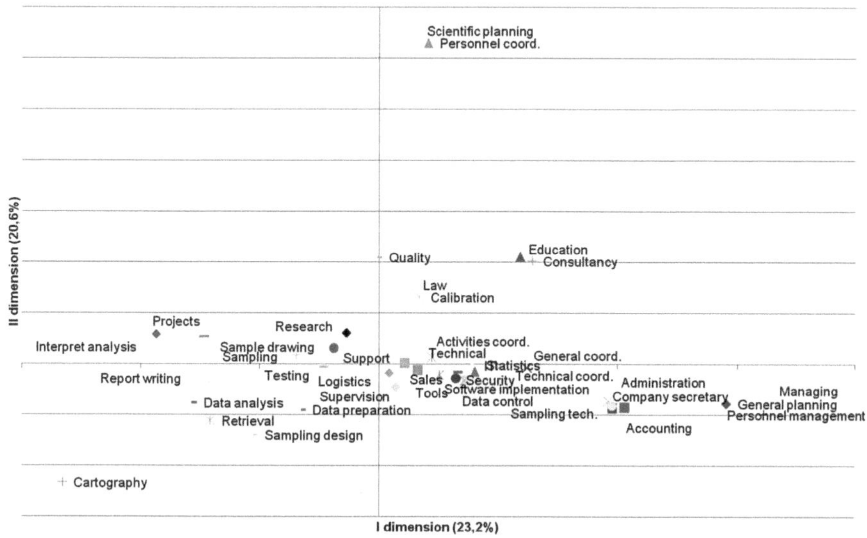

Fig. 3 Plot of correspondence analysis scores for jobs and activities – Representation of activities

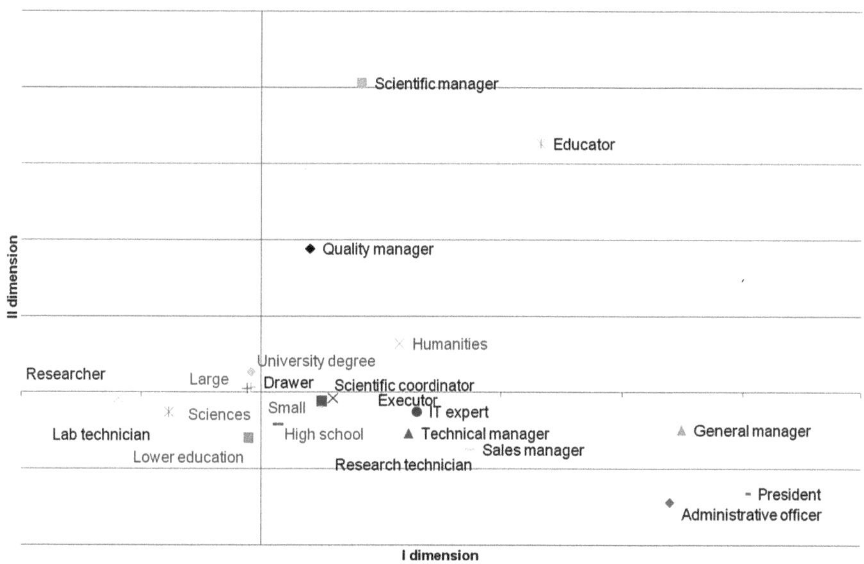

Fig. 4 Plot of correspondence analysis scores for jobs and activities – Representation of jobs

The correspondence analysis underlines the existence of three main professional areas, centred on the three core profiles highlighted by the network analysis:

- A managerial area, with executive and administrative profiles, characterised by governing and accounting competencies, and activities related to lead the group;

- A scientific area, with scientific managers involved in scientific planning and personnel coordination; the required competencies are not only scientific, but also include social endowment and communicativeness;
- A technical-operational area, with different versions of the technical profiles, whose required competencies reflect the wide range of research applications performed by the R&D firms.

4 Final Remarks

The affiliation network approach to the study of professional profiles allows to highlight the mutual relationships within each set of entities (professions, competencies, activities), or between them. The described approaches of analysis are not an exhaustive range of all the indices and techniques developed to study affiliation networks, but a mere example of meaningful applications, with a preference for the most simple and intuitive approaches (*in primis*, visual representations). Further interesting results for the analysis of the relationship between jobs, activities and competencies can be obtained through the analysis of centrality and structural similarity of 2-mode networks (Borgatti and Everett, 1997).

In 1-mode networks derived from affiliation matrices cells are frequencies, e.g. the number of times two jobs require to perform the same activity. Therefore, these are unnormalised measures of similarity between couples of jobs, and jobs requiring a higher number of competencies or activities are more likely to be connected. Co-affiliation matrices can be normalised by dividing each frequency by a maximum possible score, e.g. the total number of activities, the minimum number of activities required by the two jobs, the number of activities required by at least one of the two jobs, etc.... Normalised measures are not frequencies of co-occurrences, but a measure of preference (Borgatti and Halgin, 2011). In the present application data have not been normalised to avoid the risk of emphasizing the similarity between couples of jobs summarily described, but the issue deserves further analyses.

The range of activities and, even more, of competencies which should be used in this kind of analyses is still an open question. The inclusion of soft skills or cross-occupational competencies can enrich the analysis and emphasise non-technical analogies between professional profiles, also operating in different fields, but in a joint analysis of technical and soft skills the risk is to hide the role of technical competencies, since soft-skills are by definition cross-occupational, and tend to connect the network much more strongly than technical skills do. This problem can (at least partly) be solved by normalising the network, but probably a dual comparative analysis of 1-mode networks of jobs generated by technical competencies and by soft skills would be more effective in identifying similarities and differences among jobs.

References

Borgatti, S. P., & Everett, M. G. (1997). Network analysis of 2-mode data. *Social Networks, 19*, 243–269.

Borgatti, S. P., & Halgin, D. S. (2011). Analyzing affiliation networks. In P. Carrington, & J. Scott (Eds.), *The Sage handbook of social network analysis* (pp. 417–433). London: Sage Publications.

Mirabile, R. (1997). Everything you wanted to know about competency modeling. *Training and Development, 51*, 73–77.

Schneider, B., & Konz, A. M. (1989). Strategic job analysis. *Human Resource Management, 28*, 51–63.

Scott, J. (1991). *Social network analysis: A handbook.* London: Sage Publications.

Wasserman, S., & Faust, K. (1994). *Social network analysis: Methods and applications.* Cambridge: Cambridge University Press.

Wasserman, S., Faust, K., & Galaskiewicz, J. (1990). Correspondence and canonical analysis of relational data. *Journal of Mathematical Sociology, 15*, 11–64.

Graduation by Adaptive Discrete Beta Kernels

Angelo Mazza and Antonio Punzo

Abstract Various approaches have been proposed in literature for the kernel graduation of mortality rates. This paper focuses on the discrete beta kernel estimator, proposed in Mazza and Punzo (New perspectives in statistical modeling and data analysis, studies in classification, data analysis and knowledge organization, Springer, Berlin/Heidelberg, 2011), which is conceived to naturally reduce boundary bias and in which age is pragmatically considered as a discrete variable. Here, an attempt to improve its performance is provided by allowing the bandwidth to vary with age according to the reliability of the data expressed by the amount of exposure. A formulation suggested in Gavin et al. (Trans Soc Actuaries 47:173–209, 1995) is used for the local bandwidth. A simulation study is accomplished to evaluate the gain in performance of the local bandwidth estimator with respect to the fixed bandwidth one, and an application to mortality data from the Sicily Region (Italy) for the year 2008 is finally presented.

1 Introduction

Mortality rates are age-specific indicators, commonly used in demography. They are also widely adopted by actuaries, in the form of mortality tables, to calculate life insurance premiums, annuities, reserves, and so on. Producing these tables from a suitable set of crude (or raw) mortality rates is called graduation, and this subject has been extensively discussed in the actuarial literature (see, e.g., Copas and Haberman, 1983). To be specific, the crude rates $\mathring{q}_x$, for each age x, can be seen as arising from a sample of deaths, of size d_x, from a population, initially exposed to the

A. Mazza (✉) · A. Punzo
Dipartimento di Economia e Impresa, Università di Catania, Catania, Italy
e-mail: a.mazza@unict.it; antonio.punzo@unict.it

A. Giusti et al. (eds.), *Classification and Data Mining*, Studies in Classification, Data Analysis, and Knowledge Organization, DOI 10.1007/978-3-642-28894-4_29,

risk of death, of size e_x, and thus they contain random fluctuations. The situation is commonly summarized via the model $d_x \sim \text{Bin}(e_x, q_x)$, where q_x represents the true, but unknown, mortality rate at age x. Because of the dependence structure characterizing the true rates, a common, prior opinion about their form is that each true rate of mortality is closely related to its neighbors. This relationship is expressed by the belief that the true rates progress smoothly from one age to the next. So, the next step is to graduate the crude rates in order to produce smooth estimates, $\widehat{q}_x$, of the true rates. This is done by systematically revising the crude rates, in order to remove any random fluctuations. Nonparametric models are the natural choice if the aim is to reflect this belief. Furthermore, a nonparametric approach can be used to choose the simplest suitable parametric model, to provide a diagnostic check of a parametric model, or to simply explore the data (see (Härdle, 1992), for a detailed discussion on the chief motivations that imply their use, and (Debòn et al., 2006) for an exhaustive comparison of nonparametric methods in the graduation of mortality rates).

Kernel smoothing is one of the most popular statistical methods for nonparametric graduation. Among the various alternatives existing in literature, the attention is here focused on the discrete beta kernel estimator proposed by Mazza and Punzo (2011). Roughly speaking, the genesis of this model starts with the consideration that, although age X is in principle a continuous variable, it is typically truncated in some way, such as age at last birthday, so that it takes values on the discrete set $\mathcal{X} = \{0, 1, \ldots, \omega\}$, ω being the highest age of interest. Note that the discretization of age, from a pragmatical and practical point of view, could also come handy to actuaries that have to produce "discrete" graduated mortality tables starting from the observed counterparts. In the estimator proposed in Mazza and Punzo (2011), discrete beta distributions, as defined in Punzo and Zini (2012) and parameterized according to Punzo (2010), are considered as kernel functions, in order to overcome the problem of boundary bias commonly arising from the use of symmetric kernels. The support $\mathcal{X}$ of the discrete beta, which can be asymmetric, in fact matches the age range and this, when smoothing is made near the boundaries, allows avoiding allocation of weight outside the support (e.g. negative or unrealistically high ages).

In this paper, we attempt to improve the performance of the discrete beta kernel estimator by allowing the bandwidth to vary with x according to the reliability of the data expressed by the e_x. With this aim, we adopt a formulation suggested in Gavin et al. (1995) for the local bandwidth.

The paper can be summarized as follows. In Sect. 2 the (fixed) discrete beta kernel estimator is illustrated and in Sect. 3 an its adaptive version is provided. In Sect. 4 cross-validation estimation of the local bandwidth is described. In Sect. 5 a simulation study is performed, with the aim to ascertain the gain in performance of the local-bandwidth estimator with respect to the fixed-bandwidth one. Finally, in Sect. 6, we present an application to mortality data from the Sicily Region (Italy) for the year 2008.

2 Discrete Beta Kernel Graduation

Given the crude rates $\mathring{q}_y$, $y \in \mathcal{X}$, the Nadaraya-Watson kernel estimator of the true but unknown mortality rate q_x at the evaluation age x is

$$\widehat{q}_x = \sum_{y=0}^{\omega} \frac{k_h(y; m = x)}{\sum_{j=0}^{\omega} k_h(j; m = x)} \mathring{q}_y = \sum_{y=0}^{\omega} K_h(y; m = x) \mathring{q}_y, \quad x \in \mathcal{X}, \tag{1}$$

where $k_h(\cdot; m)$ is the *discrete kernel function* (hereafter simply named *kernel*), $m \in \mathcal{X}$ is the single mode of the kernel, $h > 0$ is the so-called (fixed) *bandwidth* governing the bias-variance trade-off, and $K_h(\cdot; m)$ is the normalized kernel. Since we are treating age as being discrete, with equally spaced values, kernel graduation by means of (1) is equivalent to moving (or local) weighted average graduation (Gavin et al., 1995).

As kernels in (1) we adopt

$$k_h(x; m) = \left(x + \frac{1}{2}\right)^{\frac{m+\frac{1}{2}}{h(\omega+1)}} \left(\omega + \frac{1}{2} - x\right)^{\frac{\omega+\frac{1}{2}-m}{h(\omega+1)}}. \tag{2}$$

The normalized version, $K_h(x; m)$, corresponds to the discrete beta distribution defined in Punzo and Zini (2012) and parameterized, as in Punzo (2010), according to the mode m and another parameter h that is closely related to the distribution variability. Substituting (2) in (1) we obtain the discrete beta kernel estimator that was introduced in Mazza and Punzo (2011).

Roughly speaking, discrete beta kernels possess two peculiar characteristics. Firstly, their shape, fixed h, automatically changes according to the value of m. Secondly, the support of the kernels matches the age range $\mathcal{X}$ so that no weight is assigned outside the data support; this means that the order of magnitude of the bias does not increase near the boundaries. Further details are reported in Mazza and Punzo (2011).

3 An Adaptive Variant

Rather than restricting h to a fixed value, a more flexible approach is to allow the bandwidth to vary according to the reliability of the data. Thus, for ages in which the amount of exposure (sample size) e_x is relatively larger, a low value for h results in an estimate that more closely reflects the crude rates. For ages in which the exposure is smaller, such as at old ages, a higher value allows the estimate of the true rates of mortality to progress more smoothly; this means that at older ages we are calculating local averages over a greater number of observations. This technique is often referred to as a variable or *adaptive kernel estimator* because it is characterized by an adaptive bandwidth $h_x(s)$ which depends on the exposure and is function of a further sensitive parameter s.

Although our knowledge of the amount of exposure can be built into the basic model (1) in a number of ways (see Gavin et al., 1995), here we adopt a natural formulation according to which $h_x(s)$ is simply the global bandwidth h multiplied by a local factor $l_x(s)$, that is

$$h_x(s) = hl_x(s) = h\left(f_x^{-1} \Big/ \max_{x\in\mathcal{X}}\left\{f_x^{-1}\right\}\right)^s, \quad x \in \mathcal{X}, \tag{3}$$

where

$$f_x = e_x \Big/ \sum_{y=0}^{\omega} e_y, \quad x \in \mathcal{X},$$

is the empirical frequency of exposed to the risk of death at age x, and with $s \in [0,1]$. In (3) f_x^{-1} is normalized so that $l_x(s) \in (0,1]$. The observed exposures decide the shape of the local factors $l_x(s)$, while s is necessary to dampen the possible extreme variations in exposure that can arise between young and old ages. Naturally, $l_x(0) = 1$; in this case we are ignoring the variation in exposure, which gives a fixed-width estimator. Finally note that lower values of e_x produce a higher $l_x(s)$; this allows more smoothing to be applied at those ages.

Summarizing, using (3) we are calculating a different bandwidth for each age $x \in \mathcal{X}$ at which the curve is to be estimated, leading model (1) to become

$$\widehat{q}_x = \sum_{y=0}^{\omega} \frac{k_{h_x}(y; m = x)}{\sum_{j=0}^{\omega} k_{h_x}(j; m = x)} \mathring{q}_y = \sum_{y=0}^{\omega} K_{h_x}(y; m = x)\, \mathring{q}_y, \quad x \in \mathcal{X}, \tag{4}$$

where the notation h_x is used to abbreviate $h_x(s)$. Thus, for each evaluation age x, the $\omega + 1$ discrete beta distributions $K_{h_x}(\cdot; m = x)$ vary for the placement of the mode as well as for their variability as measured by h_x.

4 The Choice of *h* and *s*

In (4), two parameters need to be estimated: sensitivity, s, and global bandwidth, h. Although s could be selected by cross-validation, we prefer to choose this parameter subjectively, as in Gavin et al. (1995). Once s has been chosen, cross-validation can be still used to select h.

The cross-validation statistic or score, $CV(h|s)$, for model (4) is

$$CV(h|s) = \sum_{x\in\mathcal{X}} \left(\mathring{q}_x - \widehat{q}_x^{\,t,(-x)}\right)^2, \tag{5}$$

where

$$\widehat{q}_x^{t,(-x)} = \sum_{\substack{y \in \mathcal{X} \\ y \neq x}} \frac{K_{h_x}(y; m = x)}{\sum_{\substack{j \in \mathcal{X} \\ j \neq x}} K_{h_x}(j; m = x)} \mathring{q}_y$$

is the estimated value at age x computed by removing the crude rate $\mathring{q}_x$ at that age. The values of h that minimizes $CV(h|s)$ could be referred to as cross-validation bandwidth, $\widehat{h}_{CV}$. Details on cross-validation estimation of h for the fixed discrete beta kernel estimator are given in Mazza and Punzo (2011).

5 Simulation Results

As previously mentioned, exposure may vary enormously across the age range, directly influencing the variability of the crude rates. Thus, we expect that a model that makes explicit allowance for exposure in the definition of the local bandwidth (3) should perform better.

In order to evaluate the gain in performance of the local-bandwidth discrete beta kernel estimator in (4) with respect to the fixed-bandwidth one in (1), we have performed a simulation study based on real data. Data we consider, composed of the number of exposed to risk e_x and the crude mortality rates $\mathring{q}_x$, with ω set to 85, are referred to the male Italian population for the year 2008.[1]

The scheme of the simulations can be summarized as follows:

1. First of all, we have graduated the $\mathring{q}_x$ via the well-known parametric model of Heligman and Pollard (1980). The graduated rates q_x will be hereafter referred to as the "true" mortality rates;
2. For each replication performed and for each age x, the simulated rates are obtained by dividing the d_x generated from a Bin (e_x, q_x) by e_x;
3. For each replication and for each age x, once fixed a grid of 51 equally-space d values for s ranging from 0 to 1, the global bandwidth h of the adaptive discrete beta kernel estimator $\widehat{q}_x$ in (4) is obtained by minimizing the cross-validation statistic $CV(h|s)$ in (5).
4. For each replication and for each value of s, the comparison between the smoothed and the "true" mortality rates is dealt via the sum of the squares of the proportional difference

$$S^2 = \sum_{x=0}^{85} \left(\frac{q_x}{\widehat{q}_x} - 1 \right)^2,$$

[1]Istat: data available from http://demo.istat.it/

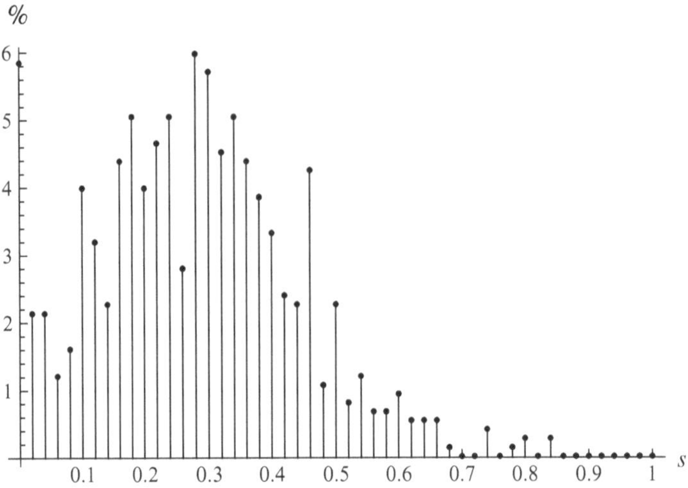

Fig. 1 Bar plot of the % of times in which the corresponding value of s on the x-axis has originated the minimum S^2 value

that is a commonly used divergence measure in the graduation literature, because since the high differences in mortality rates among ages, we want the mean relative square error to be low (see Heligman and Pollard, 1980).

Simulation results are summarized in Fig. 1, which displays a bar plot with the percentage of times (computed over 1,000 replications) in which the corresponding value of s on the x-axis has originated the minimum S^2 value, with respect to the 51 values of s.

The plot shows that the fixed-bandwidth estimator ($s = 0$) obtains the minimum S^2 in the 5.83 % of the times, while $s = 0.28$ is the value that the most of the times (5.96 %) works better. Although this difference may seem tiny, note that $s = 0.28$ gets a lower S^2 than the fixed-bandwidth estimator 82 % of the times and that for any $s < 0.76$ the local-bandwidth estimator beats the fixed-bandwidth one.

6 An Application to Italian Mortality Data

In this section, mortality data for the Sicily Region (Italy), for the year 2008, are graduated via the adaptive discrete beta kernel estimator. Data, always downloadable from http://demo.istat.it/, consist of values for $\mathring{q}_x$ and average e_x and are classified by age (ranging from 0 to 100 or older) and sex. The attention is here focused only on the male population. As before, we have chosen to take a range of ages between 0 and $\omega = 85$; this allows to make the graphical inspection of the next plots easier.

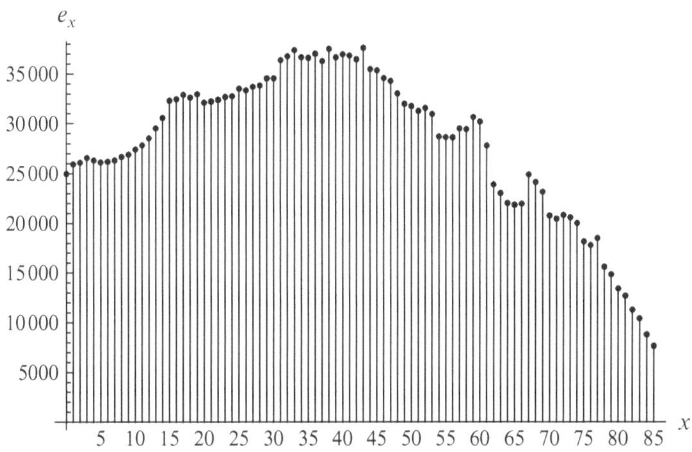

Fig. 2 Bar plot of the average male exposure for the year 2008 in the Sicily Region

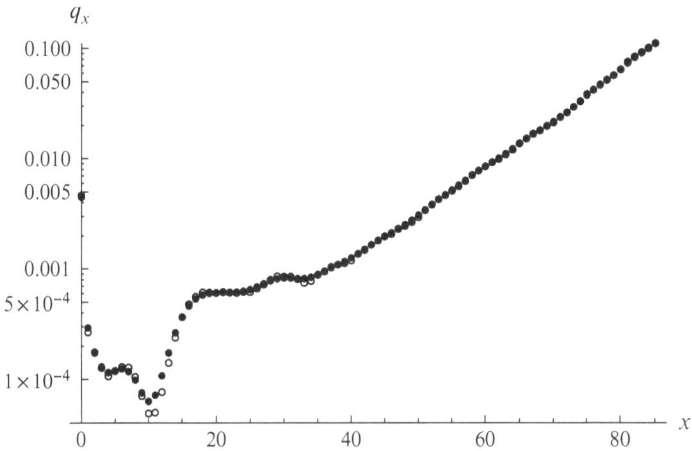

Fig. 3 Observed (○) and graduated (●) mortality rates in logarithmic scale for the year 2008 in the Sicily Region

In the period and age range under study, the distribution of the male population at risk is displayed by the bar plot in Fig. 2. The great variation in exposure, over the age range, shows the usefulness of the adaptive approach. Note that one offhanded change in exposure is visible in the age ranges 60–62, due to the Second World War.

Figure 3 shows, in logarithmic scale, the crude mortality rates and superimposes them the graduated counterparts obtained via the adaptive discrete beta kernel estimator in (4). While the sensitivity parameter, according to the results in Sect. 5, has been fixed to $s = 0.28$, the bandwidth h has been estimated by minimizing the cross-validation statistic in (5); the estimated result is: $\widehat{h}_{CV} = 0.00206$. It is easy to note that the graduated points have a more regular behavior than the observed ones, above all for the age range between 0 and 15. Moreover, a small but prominent hump, peaking around 18 years of age, is also visible; this "excess mortality rate", known in literature as accidental hump, is typically observed especially in males and it is probably due to an increase in a variety of risky activities, the most notable being to obtain a driver's license.

7 Concluding Remarks

In this paper an adaptive version of the discrete beta kernel estimator introduced in Mazza and Punzo (2011) has been proposed for the graduation of mortality rates. This proposal allows for the estimated rates of mortality to include explicitly the extra information provided by the changing amounts of exposure, in addition to the information from the crude rates themselves. A further sensitivity parameter s has been added to allow the user to control the degree of emphasis placed on the relative changing in exposure. The usual bandwidth h is used to control the absolute level of smoothness. Simulations have confirmed the gain in performance of this new approach with respect to the fixed-bandwidth one. Finally, it is important to note that the resulting adaptive discrete beta kernel graduation is conceptually simple and so is its implementation.

References

Copas, J. B., & Haberman, S. (1983). Non-parametric graduation using kernel methods. *Journal of the Institute of Actuaries, 110*, 135–156.

Debòn, A., Montes, F., & Sala, R. (2006). A comparison of nonparametric methods in the graduation of mortality: Application to data from the Valencia region (Spain). *International Statistical Review, 74*(2), 215–233.

Gavin, J., Haberman, S., & Verrall, R. (1995). Graduation by kernel and adaptive kernel methods with a boundary correction. *Transactions of the Society of Actuaries, 47*, 173–209.

Härdle, W. (1992). *Applied nonparametric regression, volume 19 of econometric society monographs*. Cambridge: Cambridge University Press.

Heligman, L., & Pollard, J. (1980). The age pattern of mortality. *Journal of the Institute of Actuaries, 107*, 49–80.

Mazza, A., & Punzo, A. (2011). Discrete beta kernel graduation of age-specific demographic indicators. In S. Ingrassia, R. Rocci, & M. Vichi (Eds.), *New perspectives in statistical modeling and data analysis, studies in classification, data analysis and knowledge organization* (pp. 127–134). Berlin/Heidelberg: Springer.

Punzo, A. (2010). Discrete beta-type models. In H. Locarek-Junge, & C. Weihs (Eds.), *Classification as a tool for research, studies in classification, data analysis and knowledge organization* (pp. 253–261). Berlin/Heidelberg. Springer.

Punzo, A., Zini, A. (2012). Discrete approximations of continuous and mixed measures on a compact interval (to appear). *Statistical Papers, 53*(3), 563–575.

Modelling Spatial Variations of Fertility Rate in Italy

Massimo Mucciardi and Pietro Bertuccelli

Abstract Standard regression model parameters are assumed to apply globally over the entire territory where measured data have been taken, under the assumption of spatial stationarity in the relationship between the variables under study. In most cases this assumption is invalid. Instead, geographically weighted regression (GWR) explicitly deals with the spatial non-stationarity of empirical relationships. Considering a georeferenced dataset on provincial total fertility rate (TFR) in Italy, GWR technique shows a significant improvement in model performance over ordinary least squares (OLS). We also discuss about the test for spatial non-stationarity.

1 Introduction

Classical regression techniques are empirical approaches that have been commonly applied in the field of demography. Model parameters are assumed to apply globally over the entire territory where measured data have been taken, under the assumption of spatial stationarity in the relationship between the variables under study. Sadly, in most cases, this assumption is invalid. If collected data are georeferenced, a better understanding of underlying spatial relationships can be achieved through geographically weighted regression (GWR), which explicitly deals with the spatial variability of empirical relationships (Fotheringham et al., 2002). The technique provides a weighting of information that is locally associated and allows regression model parameters to vary in space. This can help to reveal spatial variations in the empirical relationships between variables that would otherwise be ignored in the

M. Mucciardi (✉) · P. Bertuccelli
Department of Economics, Statistics, Mathematics e Sociology, University of Messina, Via T. Cannizzaro n°278-98122 Messina, Italy
e-mail: mucciard@unime.it; pbertuccelli@unime.it

A. Giusti et al. (eds.), *Classification and Data Mining*, Studies in Classification, Data Analysis, and Knowledge Organization, DOI 10.1007/978-3-642-28894-4_30,

overall analysis. Although GWR approach is mainly applied in studies which are related to ecology and earth sciences, where the territorial influence has a big impact on the interaction between variables (see the works of Foody (2003) and Wang et al. (2005)), in this paper we attempt to apply GWR in the field of demography. We should remember that the growing number of applications in spatial demography addresses space in several ways, ranging from visualization of one or more variables in a map, to sophisticated spatial statistical models that seek to explain why a particular spatial pattern is observed (Mucciardi and Bertuccelli, 2011). The paper is organized as follows: in Sect. 2 we describe the GWR technique; in Sect. 3, considering a real dataset on provincial total fertility rate (TFR) in Italy, we estimate a GWR model showing significant improvement in model performance over OLS.

2 The GWR Model

GWR extends the traditional regression model by allowing the estimation of local parameters, so that the model can be written as:

$$y_i = \beta_0(u_i) + \sum_k \beta_k(u_i)x_{ik} + \epsilon \quad for\ i = 1, \ldots, n \tag{1}$$

where (u_i) denotes the i-th point in the space and $\beta_k(u_i)$ is a realization of a continuous function $\beta_k(u)$ at point i. In other words, the continuous function $\beta_k(u)$ is a surface of parameters in which we take measurements at certain points to evidence the spatial variability of the surface. As can be seen, if the parameters are spatially invariant, the form of Eq. (1) is equivalent to that of the standard OLS:

$$y_i = \beta_0 + \sum_k \beta_k x_{ik} + \epsilon \tag{2}$$

As matter of fact, GWR provides a way to recognize and to measure spatial relationships between observations. Although there could be problems in calibrating equation (1), because the form of function $\beta_k(u)$ is unknown, in statistical literature models of this kind are quite common, because GWR belongs to the class of varying coefficient models introduced by Hastie and Tibshirani (1993). A way to overcome the estimation problem due to the unknown functional form of $\beta_k(u)$ is to calibrate the model through non parametric techniques. Specifically, Fotheringham et al. (2002) suggested to calibrate n local models (one for each location point) introducing a kernel weighting function. The estimation of the models is performed using a WLS (Weighted Least Square) approach, applying a different weight matrix for each of the n reference points. The formula for the estimation of the parameter at a specific spatial location is given by:

$$\hat{\beta}_k(u_i) = (X'W(u_i)X)^{-1}X'W(u_i)y \tag{3}$$

where $\hat{\beta}(u_i)$ represents an estimate of $\beta(u_i)$ and $W(u_i)$ is an n by n matrix whose diagonal elements are the weights of each of the n observed data for regression point i and all the other off-diagonal elements are zero. The $\hat{\beta}(u_i)$ value can be interpreted as point-wise estimate of $\beta(u_i)$ in the measuring point u_i. Actually, the adoption of a weighting scheme that keeps in count the distance between observations has a very specific meaning: we assume that values taken by parameter β in the neighbourhood of u_i are more similar to $\beta(u_i)$ than values related to points away from u_i. Under this assumption, the weights are chosen in order to assign greater importance to nearest observations by using a kernel function. Two of the most used weighting functions in the GWR technique are the so called Gaussian or near-Gaussian and bi-square function. The near-Gaussian weighting function is given by:

$$w_{ij} = e^{-\frac{1}{2}\left(\frac{d_{ij}}{b}\right)^2} \tag{4}$$

where d_{ij} is the distance between location point u_i and location point u_j and b is a bandwidth parameter. If i and j coincide than the weight associated at that point will be 1. When the distance d_{ij} increases, the weighting of the data will decrease according to a Gaussian curve. The bi-square kernel is instead defined as:

$$\begin{cases} w_{ij} = [1 - (d_{ij}/b)^2]^2 \text{ if } d_{ij} < b \\ w_{ij} = 0 \text{ otherwise} \end{cases} \tag{5}$$

Equation (5) provides a near-gaussian weighting function for all the points whose distance from i is lower than b and sets to 0 all the other weights. As can be noted from the (3), if all the weights on the diagonal of $W(u_i)$ are 1, then the GWR estimator is equivalent to the OLS estimator. From this point of view, OLS can be seen as special case of the GWR estimator. Although an unbiased estimate of the local parameters is not possible, through a calibration process we can obtain estimates with a small amount of bias. If coefficients continuously vary across space, a WLS regression will hardly provide an unbiased estimate of parameter beta at given point u_i. This happens because for each spatial location there will be a different value of $\beta(u_i)$, but WLS regression produces a single value $\beta(\hat{u}_i)$ that is the same for all the points. A quite small value of bandwidth parameter b can assure a small bias of $\beta(u_i)$ because the points falling in the circle with radius b centered at u are few respect to the entire sample and, allegedly, their theoretical betas will be similar to $\beta(u_i)$ value. With a small bandwidth standard error will be high due to the low number of observations that are included in WLS regression (bisquare kernel) or have a high weight associated. In fact as can be seen from gaussian kernel formula, the weights quickly go to zero when the distance between observations is more than b. On the other hand, a wide distance bandwidth will produce a reverse effect: bias will increase but standard error will be lower. Therefore we must find an equilibrium between bias and variance (bias-variance trade off), in order to balance the effects of excessive variability of estimates or of a severe distortion of the parameters. A solution to this problem is to choose the bandwidth through a cross validation approach suggested for local regression by Cleveland (1979) and for kernel estimation by Bowman (1984). The score of this function:

$$CV(b) = \sum_{i=1}^{n}[y_i - \hat{y}_{\neq i}(b)]^2 \tag{6}$$

is used in the calibration process to find the optimal bandwidth, where $\hat{y}_{\neq i}$ is the fitted value of y_i with the observation at point i omitted from the computation. The choice of the optimal bandwidth will be done minimizing (6) with respect to b. The minimization is carried out through optimisation techniques such as Golden Section Search (Greig, 1980). As can be seen from (6), the cross-validation formula is essentially the sum of the estimated predicted square errors. It can be thought as a measure of the overall performance of a particular bias/variance combination: b represents the bandwidth value that offers the best compromise between bias and variance for a given data sample. Another method suggested by Fotheringham and Brunsdon to choose b concerns the implementation of an "adaptive bandwidth kernel". In fact, if the data points are not homogeneously distributed in space (e.g. the points are concentrated only in certain areas while in other areas are scarce), a classic kernel method, which uses a fixed bandwidth distance, cannot be able to adequately capture the possible spatial effects on parameters. In this case, the weights can be determined by adopting an algorithm of this kind:

$$\begin{cases} w_{ij} = [1 - \left(d_{ij}/b\right)^2]^2 & \textit{if } j \textit{ is one of Nth nearest neighbours of } i \\ w_{ij} - 0 & \text{otherwise} \end{cases} \tag{7}$$

This is a k-nn method which determines the optimal number of neighbours, where N represents the number of points to be included within the local calibration of the model. The kernel used, as reported in Eq. (7), is a bi-square function. In this case b is the distance from nearest Nth neighbour to point i and changes along the chosen reference point for the local model. There are other approaches for the bandwidth selection of the GWR, such as the generalized cross-validation criterion, which is described in Loader (1999), or the method based on the minimisation of the AICc (Akaike Information Criterion, Hurvich et al., 1998).[1]

3 Performance Comparison on Socio-demographic Data Set

The data employed in the analysis come from a georeferenced provincial data set (see for more information Mucciardi and Bertuccelli, 2011). To show a performance comparison between OLS and GWR methods, we decided to carry out

[1]For GWR is defined as: $AIC_c = 2n\ ln(\hat{\sigma} + n\ ln(2\pi) + n\left\{\frac{n+tr(S)}{n-2tr(S)}\right\}$, where n is the sample size, $\hat{\sigma}$ is the estimated standard deviation of he error term and tr(S) is the trace of the hat matrix (Fotheringham et al., 2002).

an analysis of the relationships between the TFR of the 103 Italian provinces for the year 2005 (Fig. 1a) and some well known social and demographic determinants. The independent variables at the provincial level, which have low values of multicollinearity, include: share of foreign women (SFW); mean age at childbearing for mother (MACM); indirect expense for childbearing and maternity per capita (IECM); internal migration rate (IMR); marriage rate for 1,000 inhabitants (MAR) and civil marriage rate (CMR). First we model the provincial TFR with OLS regression. The analysis of the global model indicates that the TFR exhibits a statistically significant negative relationship with MACM and IMR and statistically positive one with SFW, IECM and MAR. The global model yields a value of AIC (−227.28) and a reasonable global fit of 0.61. In this step the variable CMR is not significant (Table 1). Instead in the distribution of the residuals we found clear evidence of spatial instability with a value of Moran test (I_{OLSres}) of $0.21 (p < 0.01)$.

In the second step, the variables chosen in the first step were used for the construction of the local model. For the GWR model we have considered: (a) the centroids of the provinces for the distance calculation between spatial units; (b) the adaptive kernel technique for bandwidth selection (Fotheringham et al., 2002); (c) the AICc and the RSS (Residual Sum of Squares) for models evaluation; (d) the Leung test to reveal the spatial non-stationarity. This test is used to verify if local parameters can be considered variable across the space. It uses estimated parameters sample variance to assess spatial variation by defining two approximated χ^2 distributions, both for the parameter variance and the residuals variance, in order to build an F test (Leung et al., 2000). Among the many estimated models for the explanation of provincial TFR, we chose a GWR model with an adaptive kernel (Fotheringham et al., 2002), the number of neighbours estimated by algorithm was 22. Finally, the full results are shown in Table 1. The analysis and the interpretation of the GWR estimates is done by keeping in count the global model (OLS), the tests for spatial non-stationarity and the mapping of the coefficient (choropleth maps). Now, in the GWR model we analyse how these relations change from one province to another and find out possible differences that remain hidden in the global model (Table 1).

The results obtained indicate that local model significantly improved the OLS results with AIC value dropping from to −227.28 to −245.50 and R^2 rising from 0.61 to 0.77 ($p < 0.01$). The spatial autocorrelation of residuals disappears in the GWR model (see Fig. 2d). In fact the Moran test is 0.09 (value not significant). It can also be seen from the BFC99 test (Brunsdon, 1999), an F test which measures the difference between GWR and OLS residuals, that GWR model has a significant improvement with respect to OLS. At this point, local relationship between TFR and individual determinants can be done. Although there exists, at the national level, a negative highly significant relationship between the MACM and TFR, the GWR results make it clear that this relationship is highly variable in space. It may be observed from Fig. 1b that the inverse relationship between MACM and TFR is stronger in the south of Italy and less elsewhere. So in the southern provinces a decrease of one age year of MACM can cause an increase of TFR up to 0.136.

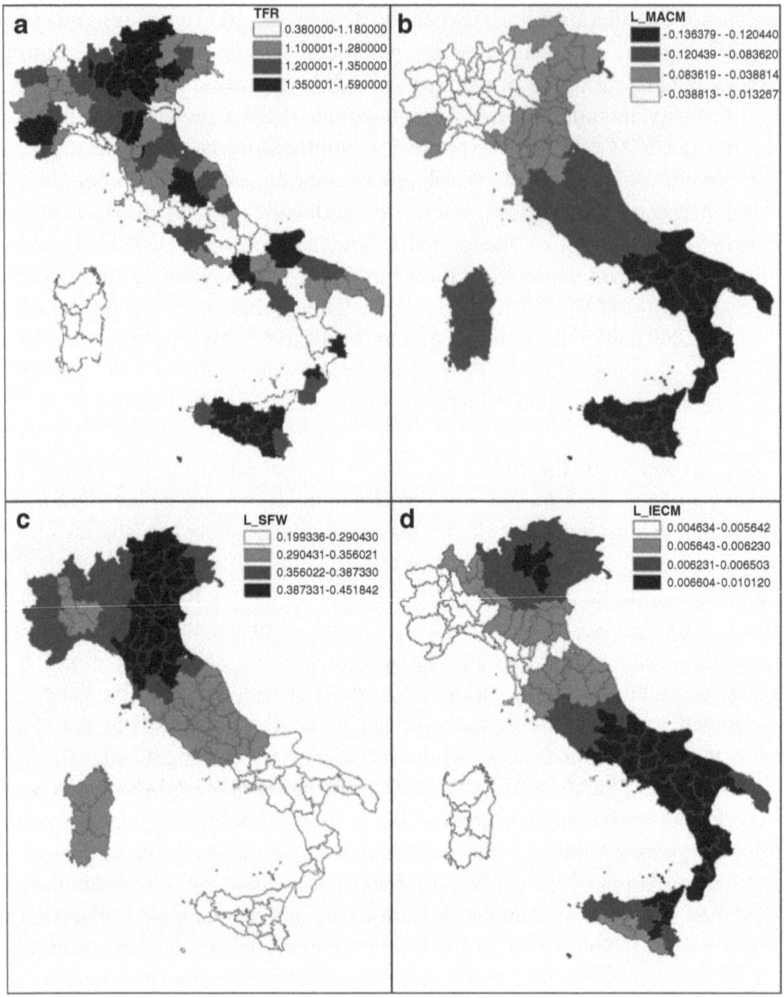

Fig. 1 Distribution of TFR (**a**) and MACM, SFW, IECM parameters by quartile ranges(**b**–**d**)

Despite this fact, even for SFW we can notice a strong difference in the local parameter values (they range from 0.199 to 0.452, see Table 1 and Fig. 1c). The IECM variable, employed here as provincial administration capacity to finance, through conventions, delegations and contracts, services directed to childbearing and maternity, reveals the existence of positive significant relationship in the global model with fertility, even if the local model doesn't provide any significant spatial non-stationarity (see the small range of the parameter and the Leung test on Table 1 and Fig. 1d). This result confirms the positive effect of the indirect expense on TFR for all the Italian territory. The situation for IMR is more complex. Despite the OLS model implies a negative effect on TFR, the GWR shows that this effect is due to

Table 1 Mean values of OLS and GWR parameters

Variable	Global model (OLS)	Min (GWR)	1stQu. (GWR)	Median (GWR)	3rdQu. (GWR)	Max (GWR)	Spatial non-stationarity Leung test (sig. level)
Intercept	4.491***	1.764	2.392	3.632	4.573	4.971	***
MACM	−0.114***	−0.136	−0.120	−0.084	−0.039	−0.013	***
SFW	0.316***	0.199	0.294	0.356	0.387	0.452	***
IECM	0.006***	0.005	0.006	0.006	0.007	0.010	n.s.
IMR	−0.006*	−0.014	−0.008	−0.004	−0.002	0.000	**
MAR	0.041*	−0.033	−0.010	0.007	0.061	0.089	***
CMR	−0.064 *n.s.*	−0.248	−0.145	−0.058	−0.017	0.035	***
$R^2_{OLS} = 0.61$		$R^2_{GWR} = 0.77$		$AIC_{OLS} = -227.28$		$AIC_{cGWR} = -245.50$	
$I_{OLSres} = 0.21^{**}$		$I_{GWRres} = 0.09\ n.s.$		$(RSS_{OLS} - RSS_{GWR}) = 0.20^{***}$ (BFC99 test)			
$*** = p < 0.001$		$** = p < 0.01$		$* = p < 0.05$		*n.s. = not significant*	

Software used for model calibration: R package spgwr. Software used for Moran test: S-Joint (Mucciardi and Bertuccelli 2011). Software used for choropleth maps: ArcGis 9.1

the central Italian provinces (Fig. 2a). However, more demographic analyses should be done to explain this phenomenon. The variable MAR is globally correlated with TFR but this correlation is stronger in the central-southern provinces than in the north of Italy. In fact the parameter in the local model is higher in magnitude and significant in the centre and south of Italy, confirming the hypothesis of a strong cultural and traditional substratum of these territories (Fig. 2b). The last covariate we considered in the model, CMR, often used by scholars as an indicator of secularization, exhibits an interesting behaviour. Although the global model does not evidence any significant effect on the TFR, the GWR model shows a strong spatial correlation of this variable. As we can see (Fig. 2c), the negative impact of CMR on TFR is very strong only in the northern provinces. In this case we account for the highest local variation of the parameter between the covariates.

4 Conclusion

In this study we applied GWR technique to model provincial TFR in Italy. First we estimated an OLS model and a GWR model and then we compared the outputs of the models. The results confirm that the advantage of GWR over OLS is mainly due to the consideration of the true spatial variation of the relationship between fertility and socio-demographic determinants. So the GWR model performed better and provided significant improvement over the global regression models. Global statistical methods like OLS sometimes may ignore the local information and even present a false relationship. As reported above, the OLS model obtained a non significant association between CMR and TFR, while GWR model revealed the existence of a negative relationship in the northern Italian provinces. Our next

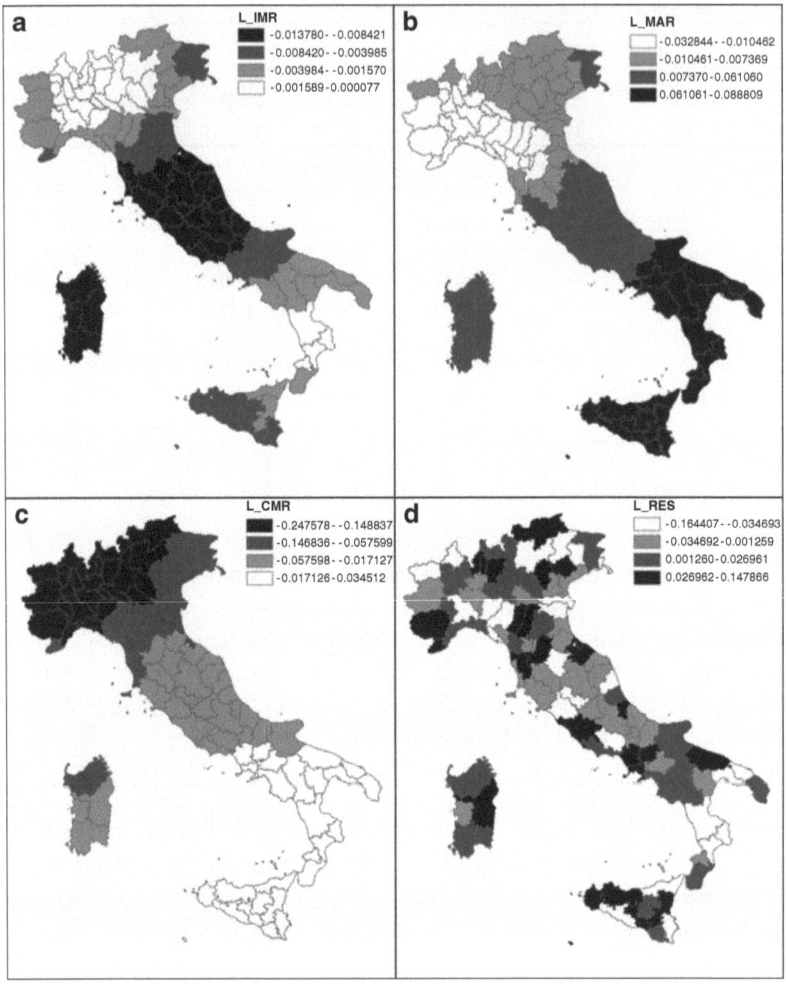

Fig. 2 Distribution of IMR, MAR, CMR parameters (**a–c**) and GWR residuals by quartile ranges (**d**)

objective will be to consider ways in which the technique could be improved. One option could be to consider a mixed GWR in which some coefficients vary locally, while others are the same everywhere. Another option could be to use a multivariate kernel, which seems to be appropriate for GWR model, because spatial units generally have from two to three spatial dimensions. A research on models of this kind is currently in progress by authors.

Acknowledgements The authors would like to thank the anonymous reviewers who provided detailed feedback on the earlier version of this article.

References

Bowman, A. W. (1984). An alternative method of cross valifdation for the smoothing of density estimates. *Biometrika, 71*, 353–360.

Brunsdon, C., Fotheringham, A. S., & Charlton, W. (1999). Some notes on parametric signficance tests for geographically weighted regression. *Journal of Regional Science, 39*(3), 497–524.

Cleveland, W. S. (1979). Robust locally weighted regression and smoothing scatterplots. *Journal of the American Statistical Association, 74*, 829–836.

Foody, G. M. (2003). Geographical weighting as a further refinement to regression modeling: An example focused on the NDVI – rainfall relationship. *Remote sensing of Environment, 88*, 283–293.

Fotheringham, A. S., Charlton, M., & Brundson, C. (2002). *Geographically weighted regression: The analysis of spatially varying relationships*. Chichester: Wiley.

Greig, D. M. (1984). *Optimisation*. London: Longman.

Hastie, T., & Tibshirani, R. (1993). Varying-coefficient models. *Journal of the Royal Statistical Society, Series B, 55*, 757–796.

Hurvich, C. M., Simonoff, J. S., & Tsai, C.-.L. (1998). Smoothing parameter selection in nonparametric regression using an improved Akaike information criterion. *Journal of the Royal Statistical Society, Series B, 60*, 271–293.

Leung, Y., Mei, C. -L., & Zhang, W. -X. (2000). Statistical tests for spatial non-stationarity based on the geographically weighted regression model. *Environment and Planning A, 32*, 9–32.

Loader, C. (1999). *Local regression and likelihood*. New York: Springer.

Mucciardi, M., & Bertuccelli, P. (2011). A GWR model for local analysis of demographic relationships. *China-USA Business Review, 10*(3), 236–244.

Wang, Q., Ni, J., & Tehnhunen, J. (2005). Application of a geographically-weighted regression analysis to estimate net primary production of Chinese forest ecosystems. *Global Ecology and Biogeography, 14*, 379–393.

Visualisation of Cluster Analysis Results

Hans-Joachim Mucha, Hans-Georg Bartel, and Carlos Morales-Merino

Abstract We present some methods for (multivariate) visualisation of cluster analysis results and cluster validation results. Visualisation is essential for a better understanding of results because it operates at the interface between statisticians and researchers. Without loss of generality, we focus on visualisation of clustering based on pairwise distances. Here, usually one can start with "dimensionless" heatmaps (fingerprints) of proximity matrices. The Excel "Big Grid" spreadsheet is both a distinguished depository for data/proximities and a plotting board for multivariate graphics such as dendrograms, plot-dendrograms, informative dendrograms and discriminant projection plots. Informative dendrograms are ordered binary trees that show additional information such as stability values of the clusters. In this way, graphics can be a very useful and much simpler aid for the reader.

1 Introduction

First, we introduce the problem of finding clusters in a set of objects and the problem of cluster validation. Using special randomized weights of objects one can easily perform built-in validations of cluster analysis results via bootstrapping techniques (Mucha, 2007). The stability of cluster analysis results (e.g., hierarchies,

H.-J. Mucha (✉)
Weierstrass Institute for Applied Analysis and Stochastics (WIAS), 10117 Berlin, Germany
e-mail: mucha@wias-berlin.de

H.-G. Bartel
Department of Chemistry, Humboldt University Berlin, Brook-Taylor-Straße 2, 12489 Berlin, Germany
e-mail: hg.bartel@yahoo.de

C. Morales-Merino
Curt-Engelhorn-Zentrum Archäometrie, D6, 3, 68159 Mannheim, Germany
e-mail: carlos.morales-merino@cez-archaeometrie.de

A. Giusti et al. (eds.), *Classification and Data Mining*, Studies in Classification, Data Analysis, and Knowledge Organization, DOI 10.1007/978-3-642-28894-4_31,

partitions, individual clusters and degree of cluster membership) can be assessed based on measures of correspondence between partitions and/or between clusters.

Secondly, we focus on visualisation of cluster analysis and of cluster validation results. In general, clusters can be visualised in "dimensionless" heatmaps of distance matrices (because they are independent on the number of variables) or/and in fingerprints of data matrices. Here, an appropriate ordering of the objects is essential (Mucha et al., 2005). Other well-known graphics are (informative) dendrograms, density plots, principal components analysis plots and discriminant projection plots (Mucha, 2009; Mucha et al., 2002). Informative dendrograms are ordered binary trees (Mucha et al., 2005) that show additional information such as stability values or other descriptive statistics. The programming language is Visual Basic for Application (VBA). Excel 2010 offers new kinds of built-in visualisations such as sparklines (small cell-sized graphics). Sparklines can be used, for example, to visualise statistics of clusters inside informative dendrograms.

2 Cluster Analysis and Graphical Representation

Clustering is a method of unsupervised learning where only unlabeled observations are given. Generally, cluster analysis methods map a data set on a collection of its subsets. For example, this mapping can be remarkably visualised in dendrograms, see Fig. 1.

Our starting point for data clustering is a $I \times J$ data matrix $\mathbf{X} = (x_{ij})$ with I observations and J variables. The aim of clustering techniques is to form groups of objects so that similar objects are grouped in the same cluster and dissimilar ones come in different clusters. Without loss of generality the focus here is mainly on (visualisation of) clustering of observations. Two families of methods will be considered: hierarchical cluster analysis and partitional clustering. Both can often be formulated as pairwise data clustering where instead of $\mathbf{X}$ a distance matrix $\mathbf{D} = (d_{il})$ is used (or more generally, pairwise proximities are given). The heatmap visualisation of such a distance matrix often reveals structure in a high dimensional data set (see Fig. 2, for further details/examples see also Mucha (2009) and Bartel (2009)).

For simplicity (also in view of bootstrapping in Sect. 3 below), we would like to focus on Gaussian model-based cluster analysis in its simplest setting. This results in the sum of squares (SS) or the logarithmic SS criterion that has to be minimised by both hierarchical and partitional methods. The starting point is a distance matrix $\mathbf{D} = (d_{il})$ with pairwise (weigthed) squared euclidean distances as elements. The SS criterion based on $\mathbf{D}$ has to be minimised with respect to a fixed number of clusters K:

$$V_K = \sum_{k=1}^{K} \frac{1}{M_k} \sum_{i \in \mathscr{C}_k} m_i \sum_{l \in \mathscr{C}_k, l>i} m_l d_Q(\mathbf{x}_i, \mathbf{x}_l). \tag{1}$$

Here M_k is the mass of the kth cluster $\mathscr{C}_k$, and m_i is the mass of the observation i. This form comes without an explicit specification of expected value of clusters.

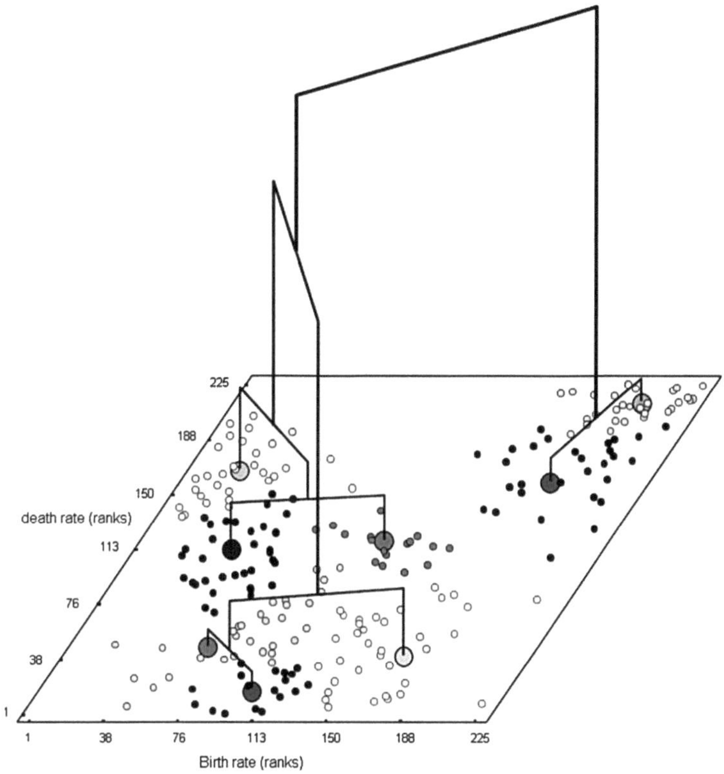

Fig. 1 The plot-dendrogram shows the unique result of hierarchical clustering by the Ward's method. The demographic data behind consists of 227 observations (countries) with the two variables birth rate and death rate in 1999. These variables are part of the population statistics published by CIA World Factbook (1999). Here, instead of the original quantitative variables, only their rank values are used, so that no structure is obviously guaranteed in the univariate case. This real data gives an extraordinary example of getting a clear cluster structure when going to multivariate (bivariate) considerations

Different (soft) bootstap techniques are put into effect simply by playing with the mass m_i of the observation i, see Sect. 3 and Fig. 5.

An equivalent formulation of the logarithmic SS based on pairwise distances can be derived, for details see Mucha et al. (2002). This criterion is more general because here the volumes of the clusters can have different sizes.

Application to archeometry. In the following example, an application of clustering to archeometry is chosen for visualisation purposes (Morales-Merino et al., 2010). More than 300 samples of clay sediments were collected within a radius of about 5 km from the vicinity of the archaeological site of Troia in west Turkey. The clay deposits in the plain of Troia consist predominantly of alluvial sediments of two rivers. These sediments provide the possible sources for the ancient pottery production. The concentrations of the following 26 elements were determined by

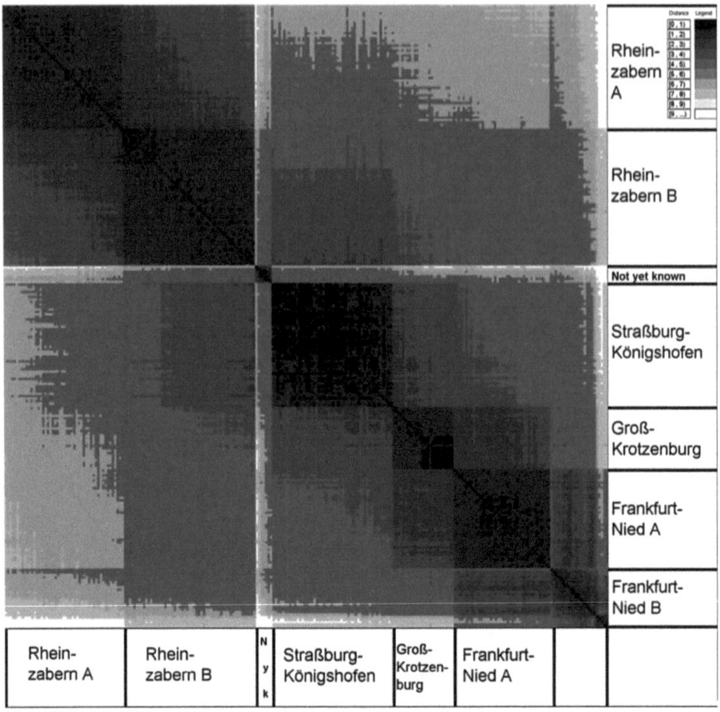

Fig. 2 Heatmap of a distance matrix of Roman bricks and tiles from different findspots in the Rhine area of Germany. The archaeometric data describes about the objects by the following 19 chemical oxides and elements Fe_2O_3, MnO, SiO_2, CaO, TiO_2, MgO, Al_2O_3, Na_2O, K_2O, Cr, Sr, Zr, Zn, Y, Ni, Nb, Rb, V, and Ba. The chemical composition was measured by X-ray fluorescence analysis. These bricks and tiles are coarse ceramics used for buildings. You can find a corresponding fingerprint of the data matrix in Bartel (2009). There one can find many other visualisations concerning this application of clustering to archaeometry

instrumental neutron activation analysis: Na, K, As, Sb, Ba, La, Sm, Yb, Lu, U, Sc, Cr, Fe, Co, Ni, Zn, Rb, Zr, Cs, Ce, Nd, Eu, Tb, Hf, Ta, and Th. Figure 3 shows the result of hierarchical clustering based on criterion (1) (Ward's method). In case of more than two-dimensional data, projection methods such as principal components analysis (PCA) can be used to be able to draw such a plot-dendrogram.

3 Validation of Cluster Analysis by Bootstrapping

Cluster analysis results depend on the given data. In hierarchical cluster analysis, for instance, the question arises: how you chose the number of clusters at which the dendrograms should be cut in Figs. 1 and 3? Bootstrapping can help to answer

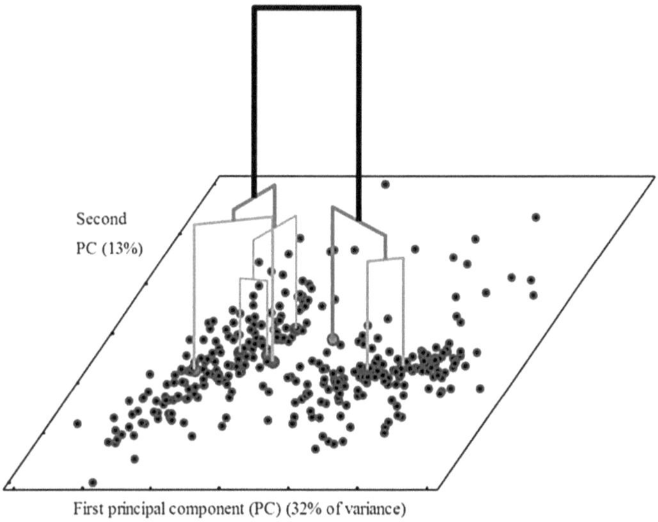

Fig. 3 Plot-dendrogram of 336 observations from the region around Troia. The dendrogram is drawn on the plane of the first two principal components. The quality of this projection by principal components analysis (about 45 % of total variance) is high with respect to 26 variables, see also the nonparametric density plot in Fig. 10 below. For further details on this application see Morales-Merino et al. (2010)

this question. It is a very common validation technique in applications of cluster analysis to life sciences. Here we do not consider special properties like isolation and compactness (Jain and Dubes, 1988). Finding the appropriate number of clusters is the main task apart from individual cluster validation. One gets many bootstrap results instead of an usual unique result of hierarchical cluster analysis. Figure 4 shows the criterion values of merging clusters by the hierarchical Ward's method for 250 bootstrap samples.

Stability strongly depends on how homogeneous and how well separated the clusters are. In the same clustering, however, the stability of individual clusters may be extremely different. Therefore, our proposed built-in validation technique evaluates additionally the stability of each cluster and the degree of membership of each observation to its cluster.

There are several measures of similarity between two clusterings (Hubert and Arabie, 1985) and between sets (Hennig, 2007). In any case, a confusion matrix $\mathbf{N}$ is the common basis: it crosses two partitions g and f (in matrix notation based on the corresponding Boolean assignment matrices $\mathbf{G}$ and $\mathbf{F}$: $\mathbf{N} = \mathbf{F}^T\mathbf{G}$), see Fig. 5.

Well-known measures of correspondence between two partitions such as the adjusted Rand index can be expressed by a confusion matrix instead of the original notation based on counting the four possible combinations of pairs of observations.

Some measures are based on the comparison of pairs of observations concerning their class membership. Examples are the Rand index and the adjusted Rand index R, see Hubert and Arabie (1985). Alternatively, these measures can be expressed by

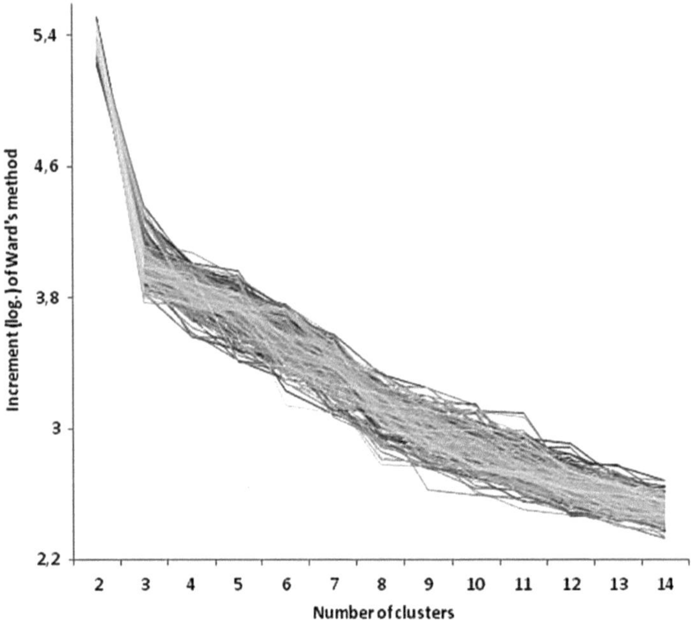

Fig. 4 Levels of merging clusters (i.e., the increment of within-cluster variance) versus the number of clusters. These are the values at which the specified number of clusters is reached in the dendrogram. The criterion values (ordinate) are in logarithmic scale. Obviously, the two cluster solution has a small variation in comparison to the range of values for a number of clusters greater than two

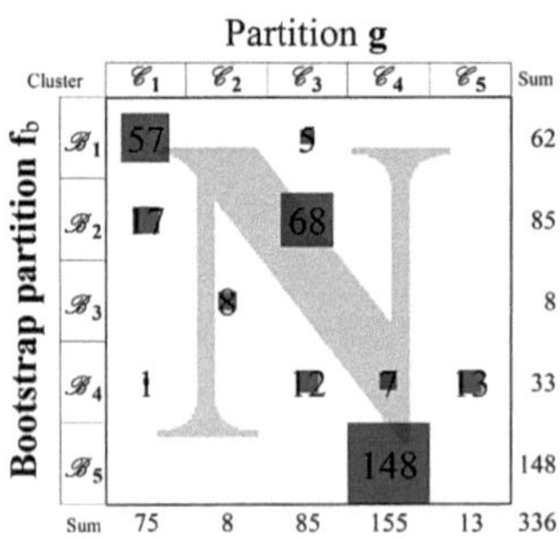

Fig. 5 Example of a confusion matrix that results from crossing two partitions of the same data set. The partition **g** is the original unique result of Ward's method (see Fig. 8). Here, the other one is the result of Ward's method applied to a (soft) bootstrap sample. The size of a square is proportional to the count in the corresponding cell of the confusion matrix

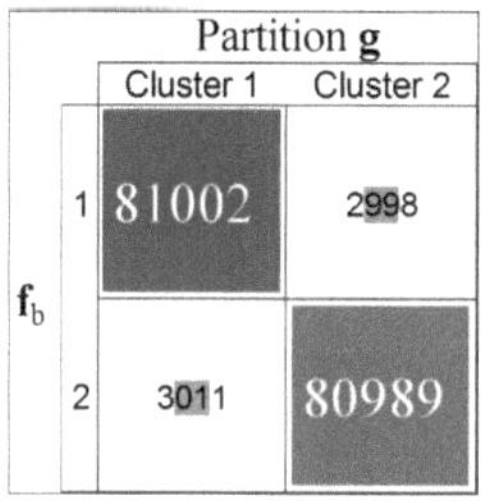

Fig. 6 Cumulative confusion matrix. Here the case of two clusters of hierarchical Ward's cluster analysis is investigated. The size of the squares are proportional to the corresponding numerical values. Without any doubt, the two clusters are very stable because they can be reproduced to a high degree

using a confusion matrix. Other well-known indexes measure the similarity between two sets (clusters) $\mathcal{E}$ and a cluster $\mathcal{F}$ such as

$$\gamma(\mathcal{E}, \mathcal{F}) = \frac{|\mathcal{E} \cap \mathcal{F}|}{|\mathcal{E} \cup \mathcal{F}|} \tag{2}$$

$$\eta(\mathcal{E}, \mathcal{F}) = \frac{|\mathcal{E} \cap \mathcal{F}|}{|\mathcal{E}|} \tag{3}$$

$$\tau(\mathcal{E}, \mathcal{F}) = \frac{2\,|\mathcal{E} \cap \mathcal{F}|}{|\mathcal{E}| + |\mathcal{F}|} \tag{4}$$

($\mathcal{E}$ and $\mathcal{F}$ are nonempty subsets of some finite set.) Hennig (2007) suggests the Jaccard coefficient γ (2). The latter and the measure of Dice τ (4) are symmetric and they attain their minimum 0 only for disjoint sets and their maximum 1 only for equal ones. Obviously, we have $\gamma \leq \eta$. The asymmetric measure η assesses the rate of recovery of subset $\mathcal{E}$ by the subset $\mathcal{F}$. It attains its minimum 0 only for disjoint sets and its maximum 1 only if $\mathcal{E} \subseteq \mathcal{F}$ holds.

The latter three measures γ, η, and τ evaluate the stability of individual clusters. By repeating resampling techniques, one gets many values of similarity. Figure 6 shows the cumulative confusion matrix after 250 bootstrap runs of the Troia data set. Here the two-cluster solution is investigated. Figure 7 shows statistics of the adjusted Rand index R. Figure 8 shows the main results of the bootstrap validation as a so-called informative dendrogram. It is difficult to fix an appropriate threshold to consider a cluster as stable. The numerical values at bottom of the Figure can be used to decide about the number of clusters. They are obtained by averaging the corresponding cluster-wise stability values of individual clusters of a partition. For instance, the total Jaccard measure is obtained by averaging the Jaccard-stability values of individual clusters that are contained in the dendrogram above. Both measures the adjusted Rand index R and the total Jaccard index give a strong recommendation to choose the two cluster solution as outstanding stable.

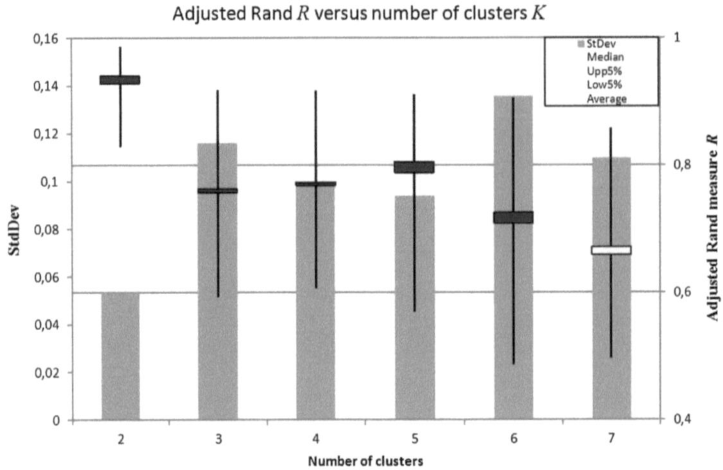

Fig. 7 Statistics of the adjusted Rand index R versus the number of clusters. For interpretation and further details see, for example, Mucha (2004, 2007)

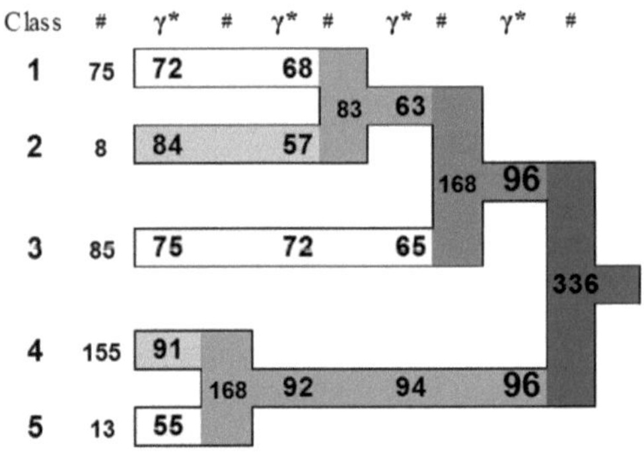

K	**5**	**4**	**3**	**2**
Total γ^*	**81**	**81**	**80**	**96**
τ^*	88	88	87	98
η^*	90	91	93	98
Adjusted R	**79**	**77**	**76**	**93**
R	91	89	88	96
F&Mallow	86	85	85	96

Fig. 8 Informative dendrogram of hierarchical cluster analysis of 336 observations (the cardinality of clusters is given below the symbol #). Here the Jaccard measure γ (values are in %) evaluates the stability of every cluster. The numerical values of similarity at the bottom (also in %), especially the adjusted Rand index R and the total Jaccard index, can be used to decide about the number of clusters

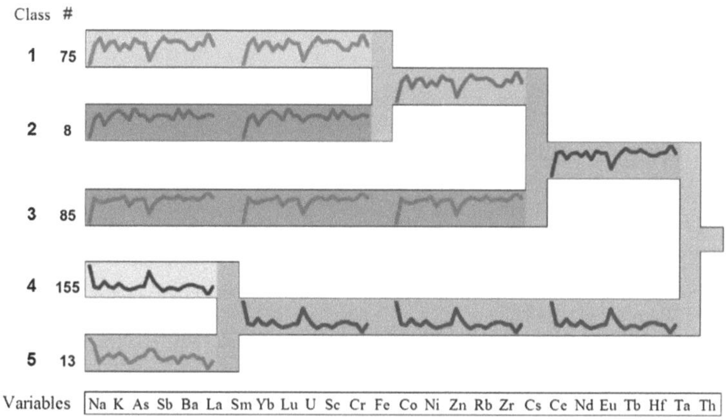

Fig. 9 Informative dendrogram with sparklines that present the centroids of the clusters. Informative dendrograms are ordered binary trees that show additional information, see also Fig. 8. Sparklines can be used for visualisation of descriptive statistics of clusters such as mean values

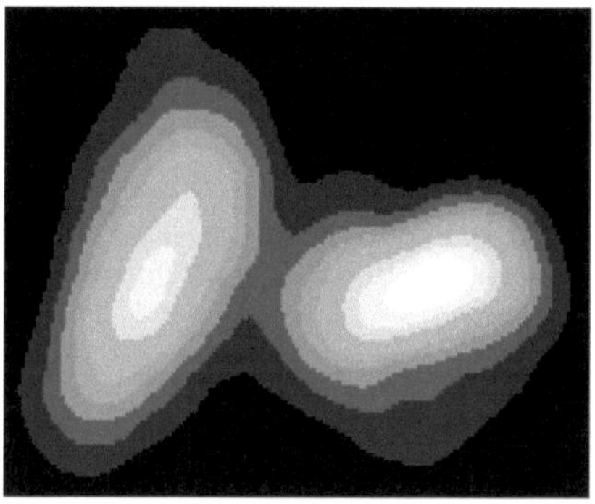

Fig. 10 Bivariate density estimation of principal components projection of archaeometric data from Troia. The same data as in Fig. 3 is used here

4 The "Big Grid" Spreadsheet Plotting Board of Excel

Excel 2007 and later now supports 1,048,576 rows and 16,384 columns. Moreover, there are new or improved built-in graphics tools like sparklines (small cell-sized graphics) for better understanding the data and the clustering results, see Fig. 9. These "Big Grid" spreadsheets are a distinguished depository for data/proximities and a convenient plotting board for (multivariate) graphics based on Visual Basics for Application code (VBA) (Mucha, 2009). For example, Fig. 10 shows a density plot that also suggests two clusters. Practically, Excel allows boundless possibilities of visualisation of cluster analysis results. So, the cell-based visualisation of huge distance matrices is possible.

References

Bartel, H. -G. (2009). Archäometrische Daten römischer Ziegel aus *Germania Superior*. In H.-J. Mucha, & G. Ritter (Eds.), *Classification and clustering: Models, software and applications* (Rep. No. 26), WIAS, Berlin, pp. 50–72.

CIA World Factbook. (1999). Population by country. http://www.geographic.org.

Hennig, C. (2007). Cluster-wise assessment of cluster stability. *Computational Statistics and Data Analysis, 52*, 258–271.

Hubert, L. J., & Arabie, P. (1985). Comparing partitions. *Journal of Classification, 2*, 193–218.

Jain, A. K., & Dubes, R. C. (1988). *Algorithms for clustering data*. Englewood Cliffs. Upper Saddle River, NJ: Prentice-Hall, Inc.

Morales-Merino, C., Mucha, H. -J., Bartel, H. -G., & Pernicka, E. (2010). Clay sediments analysis in the troad and its segmentation. In O. Hahn, A. Hauptmann, D. Modarressi-Tehrani, & M. Prange (Eds.), *Archäometrie und Denkmalpflege* (Metalla, Sonderheft 3, pp. 122–124). Bochum: Dt. Bergbau-Museum Bochum.

Mucha, H. -J. (2004). Automatic validation of hierarchical clustering. In J. Antoch (Ed.), *Proceedings in Computational Statistics, COMPSTAT 2004, 16th Symposium* (pp. 1535–1542). Heidelberg: Physica-Verlag.

Mucha, H. -J. (2007). On validation of hierarchical clustering. In R. Decker, H. -J. Lenz (Eds.), *Advances in data analysis* (pp. 115–122). Berlin: Springer.

Mucha, H. -J. (2009). Cluscorr98 for Excel 2007: Clustering, multivariate visualization, and validation. In H. -J. Mucha, G. Ritter (Eds.), *Classification and clustering: Models, software and applications* (Rep. No. 26, pp. 14–40). Berlin: WIAS.

Mucha, H. -J., Simon, U., & Brüggemann, R. (2002). *Model-based cluster analysis applied to flow cytometry data of phytoplankton* (Tech. Rep. No. 5). Berlin: WIAS.

Mucha, H. -J., Bartel, H. -G., & Dolata, J. (2005). Techniques of rearrangements in binary trees (Dendrograms) and applications. *Match, 54*(3), 561–582.

The Application of M-Function Analysis to the Geographical Distribution of Earthquake Sequence

Eugenia Nissi, Annalina Sarra, Sergio Palermi, and Gaetano De Luca

Abstract Seismicity is a complex phenomenon and its statistical investigation is mainly concerned with the developing of computational models of earthquake processes. However, a substantial number of studies have been performed on the distribution of earthquakes in space and time in order to better understand the earthquake generation process and improve its prediction. The objective of the present paper, is to explore the effectiveness of a variant of Ripley's K-function, the M-function, as a new means of quantifying the clustering of earthquakes. In particular we test how the positions of epicentres are clustered in space with respect to their attributes values, i.e. the magnitude of the earthquakes. The strength of interaction between events is discussed and results for L'Aquila earthquake sequence are analysed.

1 Introduction

Statistical seismology originated when many statistical methods and stochastic models were applied to the seismic investigation. In particular, spatial point statistical theory has proved to be a crucial statistical tool to summarise seismic data. Some of early applications of point processes in this context can be found in the pioneeristic work of Vere-Jones (1970). Many authors also focused on the

E. Nissi (✉) · A.L. Sarra
Department of Economics, "Gabriele d'Annunzio" University of Chieti-Pescara, Pescara, Italy
e-mail: nissi@unich.it; a.sarra@dmqte.unich.it

S. Palermi
ARTA (Agenzia Regionale per la Tutela dell'Ambiente dell'Abruzzo), Pescara, Italy
e-mail: s.palermi@artaabruzzo.it

G. De Luca
Istituto Nazionale di Geofisica e Vulcanologia-Centro Nazionale Terremoti, Rome, Italy
e-mail: gaetano.deluca@ingv.it

A. Giusti et al. (eds.), *Classification and Data Mining*, Studies in Classification, Data Analysis, and Knowledge Organization, DOI 10.1007/978-3-642-28894-4_32,

spatial distribution of seismicity using a diverse range of techniques. Anyway, all of them recognise that an earthquake sequence can be interpreted as a realization of a stochastic point process in a multidimensional space since each earthquake may be identified by a point in space (epicentral coordinates), in time and in the magnitude domain. Statistical studies show that distribution of earthquakes in a region is usually dominated by significant clusters in both space and time: earthquakes influence the timing and location of subsequent events. The lack of spatial independence in seismic data has been traditionally perceived as a problem obscuring the ability to separate the background seismicity from clustering patterns (Ogata et al., 2002), as required in one of the most used seismological point process model: the Epidemic-Type-After-Shock-Sequence (ETAS) model (Ogata, 1998). Accordingly, a relevant role in the study and the comprehension of seismic process and its realisation is played by the second-order properties of point processes, as the description of seismic events requires the relaxation of any assumption about statistical independence of earthquakes. The well-known Ripley K-function (Ripley, 1976) may give valuable information about interdependence among events and is a powerful tool to assess whether or not a point pattern satisfies the Poisson model, where events are assumed to be statistically independent. Many authors have considered the use of second-order statistics in space-time processes as a useful means for the comprehension of point patterns properties. Among others, see for example, Schoenberg (2004), Adelfio and Schoenberg (2009) and Adelfio and Chiodi (2009). In some of these works applications to earthquake data are provided. In particular, Schoenberg (2004) used the second order statistics for testing the relation between magnitude and space-time location of a spatial temporal marked point process whereas a weighted version of the second order statistics, defined for not necessarily stationary Poisson models, has been proposed by Adelfio and Schoenberg (2009). Hence, over the last decades, methods based on Ripley's K-function have been undergone a rapid growth. To account for first-order effects (spatial heterogeneity) inhomogeneous and space time K- function have been developed (see Baddeley et al., 2000; Gabriel and Diggle, 2009, respectively). However, these methods have resulted not easily tractable as they require the estimate of local density. A more simpler solution is given by Marcon and Puech (2003a,b) who introduce the so-called M-function which represents a generalisation of Ripley's K-function. As far as we know the empirical studies of this function have been mainly confined to the field of geographical epidemiology and spatial economics to assess the concentration of childhood leukaemia in the North England and the geographic concentration of industries in a non-homogeneous spatial framework respectively. This study puts forward that the M-function is also an appropriate tool for examining the clustering features of seismic data. The reviewed technique is employed to explore whether the foreshock and aftershock sequences of L'Aquila strong earthquake (April 6th 2009 Mw 6.3) exhibit spatial clustering. This paper is outlined as follows. In Section 2, the definitions of the ordinary K-function and its reformulation as M-function are provided. The M-function is then employed in Section 3 to study the sequences of foreshock and aftershock with predefined magnitude of L'Aquila earthquakes. In that section we also discuss the merits and limitations of the method in comparison with the techniques currently available. Section 4 provides some concluding remarks and directions for future studies.

2 The M-Function: a Variant of Ripley's K-Function

In this section, we restrict our attention to the second-order properties of a point process, based on the distribution of distances between pairs of points. According to Diggle's definitions, we call *event* a point of the process and *point* an arbitrary spatial location. Moreover, we refer to completely mapped spatial point process data: i.e. the locations of all events in a defined study area are considered. Informally, the interpoint interaction can be investigated by the second moment function, better known as Ripley's K-function. It is also called second-order analysis to indicate that the focus is on variance, or second moment, or inter-event distance. The K-function is typically defined as the expected number of further points in a circle centred at an arbitrary point (which is not counted) divided by the overall rate (the intensity λ or mean number of events for unit area) of the pattern, as formalised in (1):

$$K(r) = \lambda^{-1} E[\text{ extra events within distance r of randomly chosen events}] \quad (1)$$

Essentially it describes the characteristics of point processes at many distance scales, taking into consideration the density of points, the borders and the sample size. For many point processes the expectation in the numerator of $K(r)$ can be analytically evaluated. Tractable second order properties exist for a Homogenous Poisson Process which generates pattern consistent with complete spatial randomness (CSR). For homogenous and isotropic point patterns, the second-order characteristic depends only on distance r, but not on the direction or the location of points and it can be expressed as:

$$K(r) = \pi * r^2 \quad (2)$$

The classical exploratory approach for univariate point pattern is to compare a given point pattern to null hypothesis of CSR to test if the point process under consideration exhibits clustering or inhibitory behaviour. A solution to characterise non-homogeneous point processes is given by Marcon and Puech (2003a,b), who introduce the so called M-function. In what follows we summarise the theoretical framework of this new function. According to Marcon and Puech proposal, the definition of the M-function, allowing the analysis of non-homogeneous point patterns, involves the following steps. First of all, it is essential to define a probabilistic estimator of the K-function, hence its reformulation in a heterogeneous space is necessary, finally point weights can be attributed to each point by using proper probability laws. In defining the M-function, we start considering the following edge correction of Ripley's K-function estimator:

$$\hat{K}(r) = \frac{A}{N * (N-1)} \sum_{i=1}^{N} \frac{\pi * r^2}{A_{ir}} \sum_{j=1 i \neq j}^{N} I(i, j, r) \tag{3}$$

Let us denote with N the total number of points. The studied domain area is A, whereas the mean number of events for unit area, say λ, is estimated by $\frac{N}{A}$. In establishing the Ripley's K-function estimator we know that some circles can be partially outside the research area and some adjustments are required for those circles. We denote with A_{ir} the inside area of circle; $I(i, j, r)$ stands for a dummy: its value is 1 if the point j is located inside the circle and 0 otherwise. Following the edge-effect correction method proposed by Getis (1984), the number of neighbours is multiplied by a correction factor defined as the ratio of the area of circle $\pi * r^2$ and the part of circle inside the area A_{ir}. The formulation of Ripley's K-function estimator in Eq. (3) can be normalised by the area of the circle of radius r:

$$\frac{\hat{K}(r)}{\pi * r^2} = \frac{\frac{\sum_{i=1}^{N} \frac{\sum_{j=1 i \neq j}^{N} I(i,j,r)}{A_{ir}}}{N}{N-1}}{A} \tag{4}$$

In that way an adimensional, neutral and more intuitive expression of the K-function is obtained. It is easy to argue that the numerator is the average local density of neighbours whereas the denominator is the density of neighbours on the whole domain, used as a benchmark. The Eq. (4) can be viewed as the ratio of two probability laws, arising from two Bernoulli proofs. The first Bernoulli proof consists in searching a neighbour around a point i in an elementary area in the circle of radius r whilst the second one is defined by searching a neighbour around the point i but this time on the whole domain. The aim is to find a particular neighbour. In our context that particular neighbour is the epicenter with a predefined magnitude. We call cases those epicentres with a predefined magnitude and denote them with mk. In such a way we shall focus on space heterogeneity and the Ripley's K-function can be easily adapted for that purpose, as follows:

$$M'_{mk}(r) = \frac{\frac{\sum_{j=1}^{N_{mk}} \frac{\sum_{j=1, i \neq j}^{N_{mk}} I_{mk}(i,j,r)}{\sum_{j=1, i \neq j}^{N} I_{mk}(i,j,r)}}{N_{mk}}}{\frac{N_{mk}-1}{N-1}} \tag{5}$$

where I_{mk} is the indicator function whose value is equal to 1 if both points i and j are cases and the distance between them is at most r; I_{mk} is equal to 0 otherwise. It is worth noting that the numerator in (5) is the average value (on all cases) of the ratio of neighbour cases to neighbour points (controls plus cases); the denominator is estimated in the same way but on the entire territory. In constructing the M-function we can associate a weight for each dummy obtaining a function written as:

$$M_{mk}(r) = \sum_{i=1}^{N_{mk}} \frac{\frac{\sum_{j=1,i\neq j}^{N_{mk}} w_{mk} I(i,j,r)}{\sum_{j=1,i\neq j}^{N} w_{mk} I(i,j,r)}}{\sum_{j=1}^{N_{mk}} \frac{W_{mk}-w_i}{W-w_i}} \tag{6}$$

where $w_{mk}(i, j, r)$ denotes a weight, W_{mk} is the summed weight of points belonging to m_k and W is the total weight of all points. Numerator and denominator in (6) retain the same meaning of Eq. (5) but expressed in terms of weights. In this paper, as a preliminary step, we choose the same weight (equal to 1) for each case. Through the M-function we change the way of evaluating the possible locations: we are able to compare the number of certain type of events to the total number of events. In the seismic setting, the Marcon and Puech solution implies to draw a circle with radius r around each epicentre with a predefined magnitude. In each of those circles the number of epicentres with a certain magnitude is counted as well as the number of all epicentres. First we compute the quotient of those two numbers in a circle; next the quotients are employed to compute the average over all those circles. Finally, that average is divided by the total number of events on the whole research area. We exploit the main features of the M-function to test if the positions of epicentres are clustered in space with respect to their attribute values, that is the magnitude of the earthquakes.

3 Details of Application

3.1 Data Set

The earthquake sequence investigated belongs to the area of Central Italy (L'Aquila) bounded in longitude by 13.034° and 13.749° East and in latitude by 42.113 and 42.634 North. The library SpatialEpi in the statistical package R are used to convert the standard WGS84 epicentral coordinates into the Euclidean planar coordinates. The distance unit in the Euclidean plane is km. The earthquake catalogue provided by the National Institute of Geophysics and Vulcanology (INGV: http://bollettinosismico.rm.ingv.it) was the data source of our analysis. A total number of 17,928 occurrences is taken into account. The events pertain to a time period spanning from 1st October 2008 to 30th October 2009.

3.2 Results and Discussion

We analyse a posteriori the seismicity before and after the L'Aquila earthquake, with the aim to detect the spatial scale at which foreshocks and aftershocks sequences with different magnitudes can interact. The results of this study were obtained by considering a target area of radius of 60 km around the epicentre of the mainshock of 6 April 2009. The mainshock splits the database in two parts: the foreshock and aftershock sequences. Obviously the phenomenon appears clustered: the foreshock and aftershock sequences are characterised by spatial aggregation as revealed by the conventional Ripley's K-function which exhibit spatial clustering under 30 km in both the earthquake sequences analysed. However the Ripley's K-function assumes that epicentres can be located anywhere in the study region. In the analysis of seismicity of L'Aquila area events are more likely to occur on known faults, supporting the hypothesis of spatially inhomogeneous patterns. For that reason, the M-function analysis, described in the previous section, can be a more appropriate tool to study the density of particular events. In particular, the focus here is on quantifying the scale at which clustering of epicentres with a certain magnitude takes place. For both foreshock and aftershock sequences we consider earthquakes with magnitude 2.5 or larger and earthquakes with magnitude 3.0 or larger. The choice of the thresholds is essentially driven by the need of ensuring a more accurate completeness of earthquake catalogue: lower intensity earthquakes seem to be characterized by a greater uncertainty in the source parameters (latitude, longitude, depth, time). On the other hand, the increase of 1 unit in magnitude leads to a drastic reduction in the number of events. In a word, in the paper, we choose those earthquake intensities as they represent a compromise between the two aforesaid considerations and guarantee a better reliability of statistical results. In the weaker foreshock sequence, considering events with magnitude 2.5 or larger, the spatial concentration, measured by the M-function, is observed at a radius less than 500 m. When earthquakes with magnitude 3.0 or larger are investigated the M-function analysis reveals significant concentration of this kind of events up to 3 km. The findings for the foreshock sequences are not displayed. Summarising the characteristics of aftershocks sequences by means of M-function, we observe a valuable concentration in a radius of about 30 km for both the aftershocks sequences analysed. Figures 1 and 2 shows the results for the aftershock sequences and they can be interpreted as follows. The confidence interval is determined using Monte Carlo simulations. The simulations are set up by preserving the epicentres locations and randomly assigning the magnitude to each location. Basically, we generate a large number of simulations of random data corresponding to the null hypothesis of homogeneous spatial distributions. Here 100 simulations are generated and their results sorted at each distance. The 95 % confidence interval of M-function for each value of r is delimited by the outer 5 % of randomly generated values (Mmin and Mmax in the legend). In such a way it is possible to ascertain whether the results differs significantly from the null hypothesis of homogeneous distribution. The aforementioned results are valuable to explore the occurrence rules of the L'Aquila strong earthquake. In particular, the empirical findings allow us to highlight that for the foreshock activities we obtain a dense concentration of epicentres in a very narrow area and this would indicate well enough the epicentral area of the forthcoming mainshock. By contrast, we find that the two investigated aftershock sequences are spatially distributed within the entire seismogenic area.

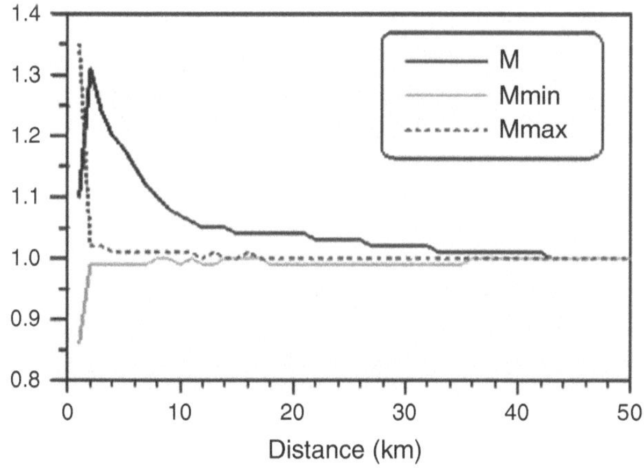

Fig. 1 Estimated M-function of L'Aquila aftershocks sequence of events with magnitude 2.5 or greater with the envelopes (MMin-MMax). A set of Monte Carlo simulations is performed. The distance scales r are chosen from 1 to 50 km with step of 1 km

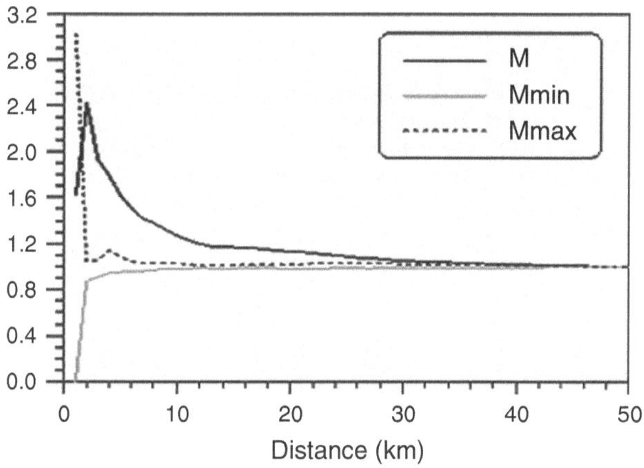

Fig. 2 Estimated M-function of L'Aquila aftershocks sequence of events with magnitude 3 or greater with the envelopes (MMin-MMax). A set of Monte Carlo simulations is performed. The distance scales r are chosen from 1 to 50 km with step of 1 km

4 Concluding Remarks

In this paper we employed the M-function as a new method for the spatial analysis of earthquakes distribution. The reviewed approach is a generalisation of Ripley's K-function and is based on a multiscale study of neighbourhood relationship. It allows to describe the density of neighbours in comparison to the total events occurred in the study region and takes into account the heterogeneous distribution of occurrences. In our setting the neighbours are the epicentres with a predefined magnitude. The use of M-function for seismic investigation is expected to contribute to a better description of characteristics of earthquake events, i.e. foreshock and aftershock sequences and to explore in further detail the possible determinants of spatial interaction. Insightful differences among earthquake sequences are detected thanks to the proposed procedure. In particular, the M-function analysis of the foreshock sequences of L'Aquila strong earthquake reveals a dense concentration of events in a very narrow area instead this quantitative measure of aftershock activities seems to indicate an aggregation of events in the entire seismogenic area. It is worth noting that the method provides comparability of concentration measurements across earthquake sequences and remains unbiased concerning different scales. Further it can be modified depending on desiderated significance level. Anyway, this paper just performs a preliminary study of L'Aquila earthquake sequence. A topic for future research regards the possibility to take into account point weights, giving for example, a different importance to events characterised by some uncertainty in the measurements values.

References

Adelfio, G., & Chiodi, M. (2009). Second-order diagnostics for space-time point processes with application to seismic events. *Environmetric, 20*, 895–911.

Adelfio, G., & Schoenberg, F. P. (2009). Point process diagnostics based on weighted second-order statistics and their asymptotic properties. *Annals of the Institute of Statistical Mathematics, 61*, 929–948.

Baddeley, A., Møller, J., & Waagepetersen, R. (2000). Non and semi-parametric estimation of interaction in inhomogeneous point patterns. *Statistica Neerlandica, 54*(3), 329–350.

Gabriel, E., & Diggle, P. J. (2009). Second-order analysis of inhomogeneous spatio-temporal point process data. *Statistica Neerlandica, 63*(1), 43–51.

Getis, A. (1984). Interaction modeling using second-order analysis. *Environmental and Planning A, 16*, 173–183.

Marcon, E., & Puech, F. (2003a). Evaluating the geographic concentration of industries using distance-based methods. *Journal of Economic Geography, 3*(4), 409–428.

Marcon, E., & Puech, F. (2003b). *Measures of the Geographic Concentration of Industries: Improving Distance-based Methods,* Working paper. Universitè Paris I, Cahiers de la MSE.

Ogata, Y. (1998). Space-time point-process models for earthquake occurrences. *Annals of the Institute of Statistical Mathematics, 50*(2), 9–27.

Ogata, Y., Zhuang, J., & Vere-Jones, D. (2002). Stochastic declustering of space-time earthquake occurrences. *Journal of the American Statistical Association, 97*(458), 369–380.

Ripley, B. D. (1976). The second-order analysis of stationary point processes. *Journal of Applied Probability, 13*(2), 255–266.

Schoenberg, F. P. (2004). Testing separability in multi-dimensional point processes. *Biometrics, 60*, 471–481.

Vere-Jones, D. (1970). Stochastic models for earthquake occurrence. *Journal of the Royal Statistical Society, Series B, 32*, 1–62.

Energy Consumption – Gross Domestic Product Causal Relationship in the Italian Regions

Antonio Angelo Romano and Giuseppe Scandurra

Abstract Despite of increasing amount of literature available on the relationship between energy consumption and Gross Domestic Product on a multi-countries framework, empirical analysis about this relationship in a spatial disaggregate level remains scarce. In this paper we investigate the relationship between energy consumption and real income by means of panel dataset of Italian regions using annual data covering the period 1980–2007. The panel co-integration and panel vector error correction models are employed to infer the dynamic directions of the causality between the two variables. Based on the panel co-integration test by Westerlund (Oxf Bull Econ Stat 69:709–748, 2007) we individuate a long-run equilibrium relationship between *Gross Domestic Product* (GDP) and *Electricity Consumption* (CE). Furthermore, the results of a panel Vector Error Correction Model suggest the presence of a bi-directional causality between variables both in the long-run and in short-run.

1 Introduction and Background

Recently, a new debate has sparked Italy on the need to invest in new energy sources to increase domestic production and to better respond to the growing demand comes from sectors of economic activity and from Italian families. Italy, as is known, is heavily dependent on foreign resources in the supply of energy, especially oil and gas, but also electricity. The launch of new investments and increasing domestic production of electricity would reduce imports by decreasing, therefore, the energetic dependence. The new investments will also provide new input to increase consumptions and would allow greater economic growth.

A.A. Romano (✉) · G. Scandurra
Department of Statistics and Mathematics for Economic Research, University of Naples "Parthenope", 40, Medina, I – 80133, Naples, Italy
e-mail: antonio.romano@uniparthenope.it; giuseppe.scandurra@uniparthenope.it

A. Giusti et al. (eds.), *Classification and Data Mining*, Studies in Classification, Data Analysis, and Knowledge Organization, DOI 10.1007/978-3-642-28894-4_33,

The debate around the relationship between energy consumption and economic growth has therefore turned on and has become increasingly intense in recent years. Many studies examine the relationship between economic growth and energy consumption in order to test if the energy consumption stimulates growth or, conversely, if economic growth increases the demand for energy.

The search for the causal link between energy consumption and economic growth leads to the formulation of four hypotheses trying to explain the effect and cause. The growth hypothesis postulates that energy consumption can directly impact economic growth. The presence of unidirectional causality from energy consumption to economic growth confirms this hypothesis. Second, the conservation hypothesis is supported if there is unidirectional causality from economic growth to energy consumption. Third, if a bi-directional causality relationship exists between economic growth and energy consumption we confirm a feedback hypothesis. Fourth, the two variables are not interrelated. We do not find a causality relationship between energy consumption and economic growth. This is the case of the neutrality hypothesis.

The literature focuses mainly on analysis by a panel of countries. Al-Iriani (2006), for example, investigates the causality relationship between Gross Domestic Product (GDP) and Energy Consumption in six countries of the *Gulf Cooperation Council* (GCC). He finds a unidirectional causality running from GDP to energy consumption. Mahadevan and Asafu-Adjaye (2007) study this relationship in energy exporting developed and developing countries. They find that in developed countries there is both short-run and long-run bi-directional causality between economic growth and energy consumption while for the energy exporting developing countries, energy consumption causes economic growth only in the short-run. In various papers, Apergis and Payne (2009a,b, 2010) analyze the relationship between energy consumption and economic growth. They reveal a causal relationship between energy and GDP in a different direction depending on the panel of countries that they investigate. These results suggest that energy conservation may harm economic growth in the short-run and/or in the long-run. A wide survey of the empirical literature on the causal relationship between energy consumption and economic growth could be found in Payne (2010).

While investigating the relationship between energy consumption and economic growth in a specific country, or in group of countries, is an ongoing research area, we have not found empirical research about the causal relationship in the various realities that make up a heterogeneous country. It was therefore decided to study the causal link between the variables considered in the panel of the Italian regions.

These regions are quite different with respect to the GDP. In this paper we investigate the direction of the nexus between economic growth and energy consumption in Italy in a simple bivariate framework. In particular we test if the two variables have a long-run equilibrium relationship using a panel dataset of the 20 Italian regions. To the best of our knowledge, there is a lack of empirical works at disaggregate level. Given the economic heterogeneity of the Italian regions, in which exist a strong economic gap, it would be reasonable to expect the relationship between energy consumption and economic growth that in some cases support the hypothesis of neutrality (Yu and Choi, 1985) that the energy consumption should not significantly affect the economic growth of one region and other cases in which energy consumption plays, however, a major role. From the statistical point of view is whether there is a unidirectional causality of energy consumption on income, or if that link takes on the characteristic bidirectional as a sign of complementarity between the two variables. In order to investigate the direction of panel causation we estimate a *panel dynamic Vector Error Correction Model* using the *Generalized Methods of Moments* (GMM) estimator proposed by Arellano and Bond (1991). The organization of the paper is as follow. Section 2 describes data and the empirical approach. In Sect. 3 we analyze the main results. Section 4 closes with some concluding remarks.

2 Data and Methods

In this paper we investigate the causal relationship between changes in electricity consumption in Italy and the variation of its GDP. It is well knows, in fact, that there are different assumptions about the nature of economic growth in relation to energy consumption in general, and the electricity consumption in particular. The use of electricity consumption instead of energy consumption is due to the short time series available for energy. However, electricity consumption is an important percentage of total energy consumption thanks to the network infrastructure significantly extended in a country like Italy and, of course, the fact that its measure is characterized by extreme accuracy, timeliness and spatial granularity.

The data are the annual time series from 1980 to 2007 of *Gross Domestic Product* (GDP) in constant 2,000 € price and *Electricity Consumption* (CE) in the 20 Italian regions. All variables are expressed through natural logarithms. A panel dataset of Italian regions is used in order to limit the effect of the small time span of the aggregated data. There are three main issues that we can solve using a panel dataset. In fact, a panel dataset allows to: (1) solve some problems of non-standard distributions of test statistics used for the identification of unit roots in the regression equations, (2) have more informative data and (3) to reduce collinearity between the variables.

First, we have computed heterogeneity (i.e. variation of the intercept over regions and time) test by using the Breush-Pagan Lagrange Multiplier test. The presence of heterogeneity suggests the use of Im, Pesaran e Shin (Im et al., 2003) panel unit root test to determine the stationarity properties of the variables before testing for co-integration. The null hypothesis is that each series in the panel contains a unit root while the alternative hypothesis is that at least one of the individual series in the panel is stationary. Table 1 reports the Im, Pesaran e Shin (IPS) panel unit root test statistic for dataset. The panel unit root test reveals that each variable is integrated of order 1 (*I(1)*).

In order to test for co-integration between electricity consumption and gross domestic product we use the test procedure proposed by Westerlund (2007) because

Table 1 Im, Pesaran and Shin (Im et al., 2003) panel unit root test

Variables	Level	First difference	Decision
GDP	2.7048	−15.2512[a]	*I(1)*
CE	1.8897	−14.7230[a]	*I(1)*

[a] Significant at 1%

Table 2 Westerlund tests for the null of no co-integration between GDP and electricity consumption in the Italian regions

Test	Test statistics	P-value	Decision
G_t	−2.571	0.000	*Cointegrated*
G_a	−11.836	0.000	*Cointegrated*
P_t	−8.749	0.011	*Cointegrated*
P_a	−8.423	0.000	*Cointegrated*

Table 3 Fully modified ordinary least squares (FMOLS) long-run estimates

	Dependent variable	
Independent variable	GDP	CE
GDP	–	1.11[a]
CE	0.58[a]	–
Constant	4.77[a]	−2.44[a]

[a] Significant at 1%

it is considered more robust than the Pedroni's co-integration test (Pedroni 1999, 2004) in the case of small samples. Westerlund proposes four tests of the null hypothesis of no co-integration that does not impose any common factor restriction on the data and that uses the available information more efficiently than residual based tests. The tests proposed by Westerlund are panel extensions of those proposed in the time series context by Banerjee et al. (1998). As such, they are designed to test the null hypothesis of no co-integration by inferring whether the error correction term is a conditional error correction model (ECM) is equal to zero. If the null hypothesis of no error correction is rejected, than the null hypothesis of no co-integration is also rejected. Table 2 reports the Westerlund's test statistics.

The four tests fail to accept the null hypothesis of no co-integration. Thus, the evidence suggests that in panel dataset there is a long-run equilibrium relationship between GDP and electricity consumption. Having established a co-integrating relationship, we estimate the long-run equilibrium relationship by using the *Fully Modified Ordinary Least Squares* (FMOLS) proposed by Pedroni (2000).

In Table 3 we report the estimated coefficients of the long-run regression. The FMOLS estimator is applied to as many single equation as the number of the variables included in the VECM that are *I(1)* and co-integrated.

All the coefficients are positive and significant at 1% level. The coefficients can be interpreted as elasticity estimates. The 1% increase in electricity consumption increases real GDP by 0.58% while a 1% increase in real GDP increases electricity consumption by 1.11%. To infer the causal relationship between the variables, a panel Vector Error Correction Model is estimated. The Error Correction Term

(ECT) is represented by the residuals of the first stage long-run model that was previously estimated by FMOLS, following the two-step procedure proposed by Engle and Granger (1997). The following dynamic error correction model is estimated:

$$\Delta GDP_{i,t} = \alpha_{1,j} + \beta_{1,i} ECT_{i,t-1} + \sum_{k=1}^{q} \gamma_{1,i,k} \Delta GDP_{i,t-k} + \sum_{k=1}^{q} \theta_{1,i,k} \Delta CE_{i,t-k} + \epsilon_{1,i,t} \tag{1}$$

$$\Delta CE_{i,t} = \alpha_{2,j} + \beta_{2,i} ECT_{i,t-1} + \sum_{k=1}^{q} \gamma_{2,i,k} \Delta GDP_{i,t-k} + \sum_{k=1}^{q} \theta_{2,i,k} \Delta CE_{i,t-k} + \epsilon_{2,i,t} \tag{2}$$

where $i = 1, \ldots, N$ for each regions in the panel and $t = 1, \ldots, T$ refers to the time period. The parameter $\alpha_{i,j}$ allows for the possibility of regions-specific fixed effects, Δ is the first-difference operator; k is the lag length; $\gamma_{i,k}$ and $\theta_{i,k}$ are the short-run adjustment coefficients; *ECT* are the lagged residuals derived from the long-run co-integrating relationship and $\epsilon_{i,t}$ are disturbance terms assumed to be uncorrelated with mean zero.

In order to estimate Eqs. (1) and (2) a widely used estimator is based on the panel generalized method of moments (GMM) proposed by Arellano and Bond (1991). Therefore, this panel data model is estimated using instrumental variable to deal with the correlation between the error term and the lagged dependent variables.

3 Main Results and Discussion

The existence of a co-integrating relationship between GDP and *Electricity Consumption* suggests that there must be Granger causality in at least one direction, but it does not indicate the direction of temporal causality between the variables. It is possible, therefore, highlight the relationships between the variables considered. Table 4 reports the estimates of the panel VECM.

The error correction terms are significant in both equations. The adjustment coefficients have the expected negative sign, which implies that they indeed reflect an error correction mechanism that tends to bring the system closer to its long-run equilibrium. In the short-run, the GDP depends directly by the present values of energy consumption. The electricity consumption has a direct relationship, as expected, with present value of the GDP. Therefore, in the short-run, consumption of electricity is influenced by the wealth produced in the Italian regions.

The directions of panel causation can be identified by testing for the significance of the coefficient of each of the dependent variables in Eqs. (1) and (2). In Eq. (1), short-run causality from energy usage to real GDP is tested based on $H_0 : \gamma = 0$. In Eq. (2), short-run causality from real GDP to energy consumption is tested based on $H_0 : \theta = 0$. Masih and Masih (1996) and Asufu-Adjaye (2000) interpreted the weak Granger causality as a short-run causality in the sense that the dependent

Table 4 VECM estimates

Independent variable	Dependent variable Δ GDP	ΔCE
ΔGDP	–	0.366[a]
ΔGDP_{-1}	0.100[b]	0.002
ΔGDP_{-2}	0.006	−0.005
ΔCE	0.292[a]	–
ΔCE_{-1}	0.029	−0.341[a]
ΔCE_{-2}	–	−0.170[a]
ECT	−0.249[a]	−0.432[a]
Constant	0.006[a]	0.035[a]

[a] Significant at 1%
[b] Significant at 5%

Table 5 Strong causality test

Dependent variable		Direction
ΔGDP	ΔCE	
$H_0 : ECT - \Delta\text{CE} = 0$[a]	$H_0 : ECT - \Delta\text{GDP} = 0$[a]	$\longleftrightarrow$

[a] Significant at 1%

variable responds only to short term shocks to the stochastic environment. Next, for long-run effect, we look the significance of the speed of adjustment β which are the coefficients of the error correction terms in Eqs. (1) and (2). Finally, it is also desirable to check whether the two sources of causation are jointly significant: we use the joint test to check for strong causality (Oh and Lee, 2004) (*ECT* and ΔGDP; *ECT* and ΔCE) where the variables bear the burden of a short-run adjustment to re-establish a long-run equilibrium, following a shock to the system. If there is no causality in either direction, the neutrality hypothesis holds, otherwise, univocal or bi-directional causality exists. Since all the variables are entered into the model in stationary form, a Wald test with a chi-squared statistic distribution can be used to test the null hypothesis of no causality (or weak exogeneity of the dependent variable) (Table 5).

The results show that Italian regions grow through increasing consumption of electricity. They do not seem to pursue energy savings policies to support growth rates. Unless structural changes of the current pattern of consumption, is expected a growing energy demand needed to sustain growth levels and international competitiveness.

4 Conclusion and Further Researches

This paper presents some preliminary results of a new empirical insights into the analysis of causal relationship between economic growth and energy consumption when considering the Italian regions, in a historical period marked by strong willingness to initiate a series of new investments in alternative production energy sources (like nuclear energy) to ensure a national demand less dependent on foreign sources. The panel cointegration test indicates there is a long-run equilibrium relationship between GDP and *Electricity Consumption*. The long-run elasticities estimated are positive and statistically significants. Furthermore, the estimation of panel vector error correction model reveals a bi-directional causality relationship both in the long-run and in the short-run. Thus, results lend to support feedback hypothesis. We can conclude that the interdependence between GDP and *Electricity Consumption* suggests a country where growth requires more energy available and it does not invest in policies aimed at reducing energy consumption. In fact, during the years 1980–2007, we do not observe a significant increase in capital productivity or Total Factor Productivity (TFP) in Italy. As well known, Italy is among the last countries in the OECD for investment in *research and development* aimed at increasing productivity. The new investments promised by policy makers in new energy production sources can increase the wealth and reduce the dependence on imports but they should have a low environmental impact in order to reduce the greenhouse gas emission and to respect the Kyoto protocol. Also should be conducive private investments in energy saving capital goods.

There are still some interesting questions to pursue in future research. It is useful to investigate: (1) the effect that inclusion of new variables could have in this analysis because of the interaction, for example, of labour and capital on the economic growth; (2) the environmental impact of new investments and the relationship with region's growth prospects; (3) the causal relationship between energy consumption and GDP considering the different productive structure of Italian regions. These aspects, not secondary, are objective of further research.

References

Al-Iriani, M. A. (2006). Energy–GDP relationship revisited: An example from GCC countries using panel causality. *Energy Policy, 34*, 3342–3350.

Apergis, N., & Payne, J. E. (2009a). Energy consumption and economic growth in Central America: Evidence from a panel cointegration and error correction model. *Energy Economics, 31*, 211–216.

Apergis, N., & Payne, J. E. (2009b). Energy consumption and economic growth: Evidence from the commonwealth of independent states. *Energy Economics, 31*, 641–647.

Apergis, N., & Payne, J. E. (2010). A panel study of nuclear energy consumption and economic growth. *Energy Economics, 32*, 545–549.

Arellano, M., & Bond, S. R. (1991). Some tests of specification for panel data: Monte Carlo evidence and an application to employment equations. *Review of Economic Studies, 58*, 277–297.

Asufu-Adjaye, J. (2000). The relationship between energy consumption, energy prices and economic growth: Time series evidence from Asian developing countries. *Energy Economics, 22*, 615–625.

Banerjee, A., Dolado, J. J., & Mestre, R. (1998). Error correction mechanism tests for cointegration in a single equation framework. *Journal of Time series analysis, 19*(3), 267–283.

Engle, R. F., & Granger, C. W.(1987). Cointegration and error correction: Representation, estimation and testing. *Econometrica, 55*, 251–276.
Im, K. S., Pesaran, M. H., & Shin, Y. (2003). Testing for unit roots in heterogeneous panels. *Journal of Econometrics, 115*, 53–74.
Mahadevan, R., & Asafu-Adjaye, J. (2007). Energy consumption, economic growth and prices: A reassessment using panel VECM for developed and developing countries. *Energy Policy, 35*, 2481–2490.
Masih, A. M. M., & Masih, R. (1996). Energy consumption, real income and temporal causality: Results from a multi-country study based on cointegration and error correction modelling techniques. *Energy Economics, 18*, 165–183.
Oh, W., & Lee, K. (2004). Causal relationship between energy consumption and GDP revisited: The case of Korea 1970–1999. *Energy Economics, 26*, 51–59.
Payne, J. E. (2010). Survey of the international evidence on the causal relationship between energy consumption and growth. *Journal of Economic Studies, 37*(1), 53–95.
Pedroni, P. (1999). Critical values for cointegration tests in heterogeneous panels with multiple regressors. *Oxford bullettin of Economics and Statistics, 61*, 653–670.
Pedroni, P. (2000). Fully modified OLS for the heterogeneous cointegrated panels. *Advances in Econometrics, 15*, 93–130.
Pedroni, P. (2004). Panel cointegration: Asymptotic and finite sample properties of pooled time series tests with an application to the PPP hypothesis. *Econometric Theory, 4*, 579–625.
Westerlund, J. (2007). Testing for error correction in panel data. *Oxford Bulletin of Economics and Statistics, 69*(6), 709–748.
Yu, E. S. H., & Choi, J. Y. (1985). The causal relationship between energy and GNP: An international comparison. *Journal of Energy and Development, 10*, 249–272.

MIX
Papier aus verantwortungsvollen Quellen
Paper from responsible sources
FSC® C105338

If you have any concerns about our products,
you can contact us on
ProductSafety@springernature.com

In case Publisher is established outside the EU,
the EU authorized representative is:
Springer Nature Customer Service Center GmbH
Europaplatz 3, 69115 Heidelberg, Germany

Printed by Libri Plureos GmbH
in Hamburg, Germany